Physician's Guide to
ARTHROPODS
OF
MEDICAL IMPORTANCE
FIFTH EDITION

Physician's Guide to
ARTHROPODS
OF
MEDICAL IMPORTANCE
FIFTH EDITION

Jerome Goddard, Ph.D.

Medical Entomologist
Bureau of Environmental Health
Mississippi Department of Health
and
Assistant Professor of Medicine
School of Medicine
The University of Mississippi Medical Center
Jackson, Mississippi

 **CRC Press**
Taylor & Francis Group
Boca Raton London New York

CRC Press is an imprint of the
Taylor & Francis Group, an informa business

EMAIL: Jerome Goddard @ msdh.state.ms.us

CRC Press
Taylor & Francis Group
6000 Broken Sound Parkway NW, Suite 300
Boca Raton, FL 33487-2742

© 2007 by Taylor & Francis Group, LLC
CRC Press is an imprint of Taylor & Francis Group, an Informa business

---
Library of Congress Cataloging-in-Publication Data

Goddard, Jerome.
   Physician's guide to arthropods of medical importance / Jerome Goddard. -- 5th ed.
     p. ; cm.
   "A CRC title."
   Includes bibliographical references and index.
   ISBN-13: 978-0-8493-8539-1 (hardcover : alk. paper)
   ISBN-10: 0-8493-8539-3 (hardcover : alk. paper)
   1. Arthropod vectors. I. Title. II. Title: Arthropods of medical importance.
   [DNLM: 1. Arthropod Vectors. 2. Parasitic Diseases. 3. Arthropods--pathogenicity. 4. Bites and Stings--therapy. WC 695 G578p 2007]

RA641.A7G63 2007
614.4'32--dc22   2006102922

Visit the Taylor & Francis Web site at
http://www.taylorandfrancis.com

and the CRC Press Web site at
http://www.crcpress.com

DEDICATIONS

Dr. Paul K. Lago
Thank you for your love of entomology, maintenance of high academic standards, and acts of patience and kindness. You are truly an entomologist *par excellence*.

Millard and Betty Lothenore
Thank you for an example of integrity and unconditional love.

E.J. and Clarene
Thank you each for a parent's love and more; for faith, hope, strength, phone calls ... Words cannot express my appreciation to you both.

Rosella, my wife
Thank you for being a saint; for unswerving support through thick and thin ... for loving me.

FOREWORD

Is the bug an insect or an arachnid?

As I wrote in a previous foreword, those who write forewords for books always say they are honored to be asked to do so. I really mean it when I say I am. Over the years, I have gotten to know Dr. Jerome Goddard very well as we have collaborated on a number of research projects of mutual interest. This fifth edition of the "bug book" is a product of his exceptional knowledge and experience which I have noted firsthand. In 2004, the fourth edition won the "highly commended" designation (second place) in the Medicine Category of the British Medical Association's Medical Book of the Year competition. This edition should be a contender for first place.

Dr. Goddard has had a unique career in what may be best termed "applied medical entomology." For a number of years after completing his graduate education in entomology, he served in the School of Aerospace Medicine, Brooks Air Force Base, San Antonio, Texas. During that time, he taught and consulted and, in doing so, developed a global perspective on the role of arthropods in human diseases. That perspective has been preserved in each edition of this book, a text that has been used by military, public health, and other medical providers, both at home and on assignment, over the years. That explains why so many copies of the previous editions were purchased for use outside of the United States.

As an allergist–immunologist, I see many patients who seek evaluation for conditions thought to be related to insect stings or bites. This book is an especially useful compendium of the typical and atypical responses to these insects and other arthropods, and the entomology is a great help for those who are distant from college zoology courses. Patients are greatly relieved to know their health care provider is informed in this area and are more willing to accept recommendations offered in that context.

The fifth edition continues the previous tradition, in that the information provided is practical and presented in a format optimal for use on an as-needed basis. The addition of several new color photographs and 35 new black-and-white photographs is a major plus. Two more clinical case histories enliven the text. General guidelines to treatment of conditions resulting from exposure to arthropods and appropriate references to detailed material are also provided. Another novel addition, the "Bug Coach" CD-ROM, provides over 100 line drawings, figures, and photos in three sections: (1) arthropods, (2) arthropod-induced diseases and conditions, and (3) an identification guide.

With globalization of commerce, same-day air travel between continents, and military activity in the most remote parts of the world, there is increasing contact with arthropod species otherwise not commonly encountered. This text has come of age at a time when it is needed most. It will especially be valued by health care providers who find themselves in unknown territory, both geographically and medically.

Now anybody can tell an insect from an arachnid.

Richard D. deShazo, M.D.
Billy S. Guyton Distinguished Professor
Professor of Medicine and Pediatrics
Chairman, Department of Medicine
University of Mississippi Medical Center

Among the scientific disciplines, the relative importance of medical entomology continues to increase. Malaria affects hundreds of millions of people annually, and development of the much-needed malaria vaccine seems as distant as ever. Also, dengue fever inflicts pain and suffering on millions of persons each year and now is threatening the southern United States. There are even "new" or emerging vector-borne diseases being recognized. Lyme disease was unknown 30 years ago. Now it is the most commonly reported vector-borne disease in the United States, with about 20,000 cases reported each year. Human cases of tick-borne ehrlichiosis were first described in the United States in 1986; now at least three different ehrlichial agents have been found affecting humans.

Age-old vector-borne diseases such as epidemic typhus, plague, yellow fever, and relapsing fever are still around. These agents remain endemic in many parts of the world, and under the right conditions (e.g., war or disaster) can quickly erupt into epidemics. Arthropod adaptability, combined with ecological and environmental change and frequent air travel, ensure that vector-borne diseases will continue to be a problem for humankind.

For most people living in the industrialized nations, the threat from insects, spiders, or mites lies primarily with stings and bites of various species and the reactions, both allergic and nonallergic, to them. For example, due to the ubiquity of honey bees or fire ants (in many areas), almost every person is occasionally stung. Allergic reactions can be severe, resulting in death. In addition, for unknown reasons, fire ant invasions into nursing homes — resulting in attacks on patients — seem to be increasing in the southern United States.

This book was written to provide physicians, other health care providers, and public health officials with a reference of these insects, mites, scorpions, and spiders of public health importance as well as topics related to these organisms. Voluminous works could be developed on many of these topics. However, a deliberate effort has been made to keep extraneous information to a minimum. Also, as in all areas of science, entomology includes controversies over certain points and "facts." In many cases these facts are constantly changing and being revised. Accordingly, I have chosen to streamline the references in this book and to provide views that represent a general consensus of the current status of each subject.

The primary focus of this arrangement is to provide easy, almost instant access to essential information concerning these topics. It is not the intent of this reference to make entomologists out of the readership. Specialists should be consulted whenever

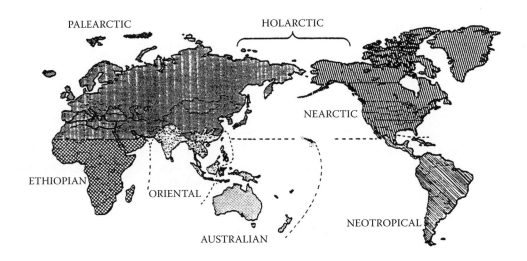

Figure 1
Zoogeographic regions of the world. (From U.S. Department of Health Education and Welfare, Public Health Service.)

possible for definitive identification of an arthropod. Extensive technical jargon has been avoided as much as possible in the "General Description" sections. However, a glossary is included to aid the reader in defining and locating descriptive terms and characters.

The volume begins with several chapters on the pathologic conditions caused by arthropods, and the principles of treating those conditions. These are provided because a physician may first have to identify the nature of an arthropod-caused problem (sting, bite, blistering, etc.). Chapter 1, Principles of Treatment, includes the rationale behind the various treatment regimes. This should be helpful because, although specific recommendations may change through time, the underlying principles of controlling the immune response will not. When arthropods are mentioned in Part I, there will be a parenthetical reference indicating where in Part III the reader can go for more detailed information. Part II consists of a chapter on identification of arthropods and a chapter on common signs and symptoms of vector-borne disease.

The third major part of the book is an alphabetical arrangement of the arthropods of medical importance with clearly marked subheadings for easy information access. To find a topic or insect section, the reader should look for that name or topic alphabetically. Keep in mind that all flies are grouped together, as are all lice, all mites, etc. A person wishing to find the topic "screwworm fly" would consult the flies chapters. Also, it is important to remember that common names vary with locality. A "blue-tailed darner" may mean one thing to the author and something totally different to someone else. Often the geographic distributions of the arthropods are given in relation to zoogeographic region. Figure 1 is provided to reacquaint the reader with these regions. The index includes the various pathologic conditions and as many of the common names as possible to aid the reader in finding a particular topic or insect. Also, it is important to remember that if a patient brings in to the clinic an insect, mite, or spider associated with a particular health problem, it is prudent to deal with the problem (with this reference, hopefully) at hand but also to submit the specimen to a university or health department entomologist for definitive identification. This might be important for later follow-up, consultation, or legal matters.

In *Physician's Guide to Arthropods of Medical Importance, Fifth Edition*, the chapters have been updated with much of the latest information and current references. The mosquito chapter was revised with the help of Dr. Bruce Harrison at the North Carolina Department of Environment and Natural Resources. Lastly, many of the older photographs have been replaced with new, improved ones (or line drawings), and some color photographs were replaced with better ones. More importantly, a CD-ROM was developed to accompany the new edition. The CD-ROM contains helpful identification aids, additional reading materials, and more photos. I am very excited about this new edition and wish to extend appreciation to my colleagues for continued interest in this book.

Finally, every effort has been made to ensure that the treatment recommendations herein are current and widely recognized as appropriate. However, it must be emphasized that treatment recommendations may change over time and should not be construed to be the sole specific treatment guidelines for any one case. Physicians should consult appropriate medical literature (Conn's Current Therapy, *for example) and/or drug package inserts for the most up-to-date treatment recommendations.*

Jerome Goddard, Ph.D.

ACKNOWLEDGMENTS

FIRST EDITION

This book would not have been possible without the help and advice of numerous individuals. My special thanks are due to two medical entomologists. Chad McHugh, a uniquely insightful civilian United States Air Force (USAF) entomologist (Brooks Air Force Base, Texas), most generously read every chapter (sometimes more than once) and offered invaluable advice and comments. LTC Harold Harlan (an outstanding Army entomologist at the Uniformed Services University of the Health Sciences, Bethesda, Maryland) also read the entire book, giving helpful advice and additional information. Both of these individuals were more than willing to take time out of their busy schedules to work through a quite voluminous manuscript.

A few physicians with whom I work directly or indirectly reviewed portions of the manuscript and/or offered much-needed comments: Drs. Mary Currier and Tom Brooks (Mississippi Department of Health), and Drs. David Conwill and John Moffitt (University of Mississippi Medical Center).

During the formative stages of the manuscript the following individuals reviewed specific chapters or subject areas: Dr. Hans Klompen (Institute of Arthropodology and Parasitology, Georgia Southern University), Dr. Paul Lago (Biology Department, University of Mississippi), Dr. Robert Lewis (Department of Biology/Zoology, Iowa State University), Maj. Tom Lillie (a U.S. Air Force entomologist), Tim Lockley (Imported Fire Ant Lab, Animal and Plant Health Inspection Service [APHIS], United States Department of Agriculture [USDA]), and Dr. Hal Reed (Biology Department, Oral Roberts University).

Information on specific arthropods and/or photographs were provided by Dr. Virginia Allen (Geisinger Medical Center, Pennsylvania); Steve Bloemmer (Tennessee Valley Authority, Land Between the Lakes); Dr. Tom Brooks (University of Mississippi Medical Center and Mississippi Department of Health); Drs. Richard Brown, Clarence Collison, and Bob Combs (Entomology Department, Mississippi State University); Ian Dick (Environmental Health Department, The London Borough of Islington), Sandra Evans (U.S. Army Environmental Hygiene Agency, Aberdeen Proving Ground), Harry Fulton (Bureau of Plant Industry, Mississippi Department of Agriculture), LTC Harold Harlan (a U.S. Army entomologist), Dr. James Jarratt (Entomology Department, Mississippi State University), Dr. Hans Klompen (Institute of Arthropodology and Parasitology, Georgia Southern University), John Kucharski (Agricultural Research Service,

USDA), Dr. Paul Lago (Biology Department, University of Mississippi), Maj. Tom Lillie (a U.S. Air Force entomologist), Tim Lockley (Fire Ant Lab, APHIS, USDA), Dr. Chad McHugh (Civilian U.S. Air Force entomologist), Dr. Hal Reed (Biology Department, Oral Roberts University), Dr. Richard Robbins (Armed Forces Pest Management Board, Defense Pest Management Information Analysis Center), Dr. John Schneider (Entomology Department, Mississippi State University), and Sue Zuhlke (Mississippi Gulf Coast Mosquito Control Commission).

The USAF medical entomology facts sheets (from the Epidemiology Division of the USAF School of Aerospace Medicine) were instrumental in writing some chapters, as were some written sections and illustrations from the Mississippi Department of Health publication *The Mosquito Book* by Ed Bowles. Les Fortenberry (Mississippi Department of Health) did most of the original art work. Much of Chapter 28 (Ticks) was taken from a previous military manual written by the author entitled *Ticks and Tick-Borne Diseases Affecting Military Personnel*. Art work in that publication was originally done by Ray Blancarte (USAF School of Aerospace Medicine), and some photography was done by Bobby G. Burnes (also of the USAF School of Aerospace Medicine). The Centers for Disease Control "Key to Arthropods of Medical Importance," which is revised and included as a figure in Chapter 8, was originally written by H. D. Pratt, C. J. Stojanovich, and K. S. Littig.

My wife, Rosella M. Goddard, did much of the typing and encouraged me during the three years of manuscript preparation. I owe a great deal of gratitude to her.

SECOND EDITION

Chad McHugh (USAF civilian entomologist), Dr. David Conwill (University of Mississippi Medical Center), and Dr. Mary Currier (Mississippi Department of Health) provided helpful comments.

The following persons provided photographs or permission to use their material: Dr. Mary Armstrong, Ralph Turnbo, and Tom Kilpatrick (all at the Mississippi Department of Health), Dr. Alan Causey (University of Mississippi Medical Center), Mike and Kathy Khayat (Pascagoula, Mississippi), and Dr. Gary Groff (Pascagoula).

As always, my wife, Rosella, and my sons, Jeremy and Joseph, helped me immensely. Many of the sting or bite lesions were photographed by my boys as we spent time in the field collecting specimens.

THIRD EDITION

As scientific knowledge continues to expand at an unprecedented rate, it is obvious that no one person can hope to keep up. Accordingly, I continue to utilize several scientists/physicians as resource persons. Their help is critical; I could not keep this book up-to-date without their help. They are Dr. Chad McHugh (USAF civilian entomologist), Dr. Hans Klompen (currently at Ohio State University), Drs. David Conwill and William Lushbaugh (University of Mississippi Medical Center), Dr. Fernando de Castro (Dermatology Associates, Lexington, Kentucky), and Drs. Mary Currier and Risa Webb (Mississippi Department of Health). Phyllis Givens (Jackson, Mississippi) and George Allen (Jackson) provided photographs or specimens.

Again, my wife, Rosella, and my sons, Jeremy and Joseph, helped me immensely. Fourteen of the pictures in this book are of Jeremy or Joseph, either to illustrate lesions or to demonstrate a particular activity.

FOURTH EDITION

Dr. Chad McHugh (USAF civilian entomologist), Dr. Chris Paddock (CDC), Dr. Mary Currier (Mississippi Department of Health), and Drs. John Moffitt and Richard deShazo (University of Mississippi Medical Center) provided helpful comments. Dr. deShazo was invaluable in helping me update the allergy sections and graciously allowed me to use a brief portion of his writing in Chapter 1 under "Mechanisms of Allergic Reactions."

The following persons provided photographs, specimens, or permission to use their material: Dr. Mike Brooks (Laurel, Mississippi), Dr. Barry Engber (North Carolina Department of Health), Dr. James Jarratt (Mississippi State University), Dr. Richard Russell and Stephen Doggett (Westmead Hospital, Westmead, Australia), Sheryl Hand and Dr. Sally Slavinski (Mississippi Department of Health). I am especially indebted to James Jarratt, a long-time friend who has helped me through the years photograph specimens and allowed me to use his (much better) photos.

FIFTH EDITION

Dr. Bruce Harrison (North Carolina Department of Environment and Natural Resources) graciously helped me revise the mosquito chapter and Dr. Chris Paddock (CDC Pathology Activity) provided comments on insect bite pathology. David Notton and Nigel Wyatt (Department of Entomology, The Natural History Museum, London) allowed me to examine/study several specimens of African diptera for this book. Several of the best photographs included in this edition could not have been possible without the help of my assistant, Wendy Varnado (Mississippi Department of Health). My son, Jerome Goddard, II, single-handedly developed the CD-ROM for inclusion in this book. I never cease to be amazed at his computer programming abilities. Many of the pictures on the CD are from the CDC, Armed Forces Pest Management Board, Armed Forces Institute of Pathology, or the USDA.

The following persons provided photographs, specimens, or permission to use their material: Kailen Austin (student, Mississippi College), Dr. Michael Brooks (Laurel [Mississippi] Ear, Nose, and Throat Surgical Clinic), Mallory Carter (student, Mississippi College), Rachel Freeman (student, Mississippi College), Dr. Blake Layton (Mississippi State University), Margaret Morton (Mississippi Department of Health), Joe MacGown (Mississippi Entomological Museum, Mississippi State University), Wendy Varnado (Mississippi Department of Health), and Gretchen Waggy (Grand Bay National Wildlife Refuge).

AUTHOR

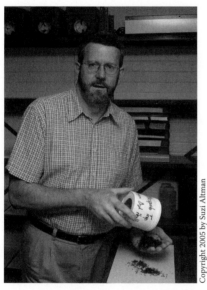

Jerome Goddard is a medical entomologist in the Bureau of Environmental Health, Mississippi Department of Health, and also holds two appointments in the School of Medicine, The University of Mississippi Medical Center: clinical assistant professor of preventive medicine and assistant professor of medicine.

He received his bachelor's and master's degrees in biological science from the University of Mississippi in 1979 and 1981, and his Ph.D. degree in medical entomology from Mississippi State University in 1984. In December of 1985 he was commissioned as an officer in the U.S. Air Force and served as a medical entomologist in the Epidemiology Division of the USAF School of Aerospace Medicine, Brooks AFB, Texas, for three and a half years. In 1988 he was named Best Academic Instructor in the Residents in Aerospace Medicine Course and Company Grade Officer of the Year. Since March of 1989 Dr. Goddard has been serving in the capacity of State Medical Entomologist at the Mississippi Department of Health, Jackson, Mississippi, where he designs, implements, and supervises all vector control programs relating to public health throughout the state of Mississippi.

Dr. Goddard has authored or coauthored over 150 scientific publications in the field of medical entomology and is the author of *Ticks and Tick-Borne Diseases Affecting Military Personnel,* published by the USAF, as well as *Infectious Diseases and Arthropods,* published by Humana Press. He was featured in *Reader's Digest* and on a series entitled "Living with Bugs" on the Learning Channel. In 2001, Dr. Goddard published a novel about a mosquito-borne disease outbreak entitled, *The Well of Destiny,* and in 2006, published another novel, this time about virus behavior, entitled, *Vital Forces.* Dr. Goddard frequently presents seminars and guest lectures nationally and internationally on "Arthropods and Medicine" and related topics. He is a member of the American Association for the Advancement of Science, the Mississippi Mosquito and Vector Control Association, and the Mississippi Entomological Association. His main research interests are the ecology and epidemiology of tick-borne diseases, but he also publishes on a wide range of medically important arthropods. Physicians are encouraged to submit comments or entomological questions to Dr. Goddard via his website: www.jeromegoddard.com.

CONTENTS

PART III
Arthropods of Medical Importance

PART IV
Personal Protection Measures Against Arthropods

PART I

PATHOLOGICAL CONDITIONS CAUSED BY ARTHROPODS AND PRINCIPLES OF THEIR TREATMENT

PRINCIPLES OF TREATMENT FOR ARTHROPOD BITES, STINGS, AND OTHER EXPOSURE

TABLE OF CONTENTS

I. INTRODUCTION

Arthropods adversely affect humans in a number of ways. There are direct, nonallergic effects such as tissue damage due to stings and bites, as well as vesicating fluid exposure and tissue infestation by the bugs themselves (e.g., myiasis). Additionally, some venoms produce necrosis in human tissues, and others produce neurological effects. Indirect effects on human health include disease transmission, and allergic reactions to bites and stings as well as to arthropod skins or emanations.

Because there are different underlying mechanisms that produce the pathological reactions associated with arthropods, it is imperative that attending physicians properly categorize a reaction in order to counteract those ill effects. This chapter is designed to present a brief overview of the underlying principles of treating arthropod exposure, focusing on the different types of pathological conditions produced by arthropods. No effort is made to explain in detail the immunological and physiological bases underlying these types of pathology; instead, a general overview of the mechanisms involved and ways to counteract or control them is given. No mention is made of the immunopathological consequences of arthropod-borne parasitic infections.

II. DIRECT EFFECTS OF ARTHROPOD EXPOSURE

Certainly, bees or wasps can sting and produce toxic effects in humans by their venom alone, regardless of hypersensitivity. Venoms in many social wasps and bees contain substances that produce pain and release histamine (directly, not IgE mediated; see Chapter 3 for a more detailed discussion). Stings or bites cause some tissue damage and inflammation. Inflammation is a result of at least three events: (1) an increase in blood supply to the affected area; (2) an increase in capillary permeability, allowing larger molecules to cross through the endothelium; and (3) leukocytes, mostly neutrophils and macrophages, migrating into the affected tissues.

It has often been estimated that between 500 and 800 honey bee stings could cause death in humans owing to toxic effects of the venom alone. One author calculated that 1500 honey bee stings would constitute the median lethal dose for a 75-kg person based on extrapolation from the LD_{50} of bee venom for mice.[1] These direct toxic effects (from honey bees or other social Hymenoptera) would include release of histamine, contraction of smooth muscle, increase in capillary permeability, vasodilation with a resulting drop in blood pressure, destruction of normal tissue barriers, destruction of red blood cells, and pain. Severe cases would probably result in renal failure.[1] Treatment strategies would include symptomatic treatment until the venom effects were diminished. As histamine is a component of bee, wasp, and hornet venoms, and as melittin (found in honey bee venom) causes histamine to be released from cells, administration of antihistamines would be indicated. In addition, therapeutic agents to counteract the ill effects of histamine release, e.g., bronchodilators, would also be helpful.

Biting insects produce direct effects on humans as well. Mosquitoes are a nuisance because of their biting behavior, and they may produce tiny punctate hemorrhages (with or without a halo) or persistent papular lesions.[2] Sometimes, large wheals with gross surrounding edema are produced owing to sensitization. Significant blood loss may also occur from mosquito bites. Snowpool mosquitoes in northern Canada emerge by the trillions each spring and landing counts on the human forearm have been reported as high as 300 per min.[3] This rate of biting could reduce the total blood volume in a human body by half in 90 min.[3] Black flies, attacking by the thousands, may cause severe annoyance and small itchy papules and swelling.[2] Ceratopogonid midges also bite in vast numbers, causing irritation and numerous minute papular lesions that may persist for several days.[2] Other biting insects that affect human health directly include bed bugs, kissing bugs, horse and deer flies, stable flies, fleas, and lice. Treatment principles for the direct effects of biting insects generally involve palliative antipruritic lotions or creams and a brief course of systemic corticosteroids if necessary.[2]

Some caterpillars possess poison-filled spines that break off in human skin upon handling or other contact (see Chapter 5 and Chapter 14). These spines release venomlike substances into the skin upon contact, and pathology similar to a sting may develop. Except in systemic reactions, treatment generally involves topical application of palliatives. In addition, the embedded broken-off spines themselves may need to be removed. This may be done with clear adhesive tape in a repeated "stripping" action.

Myiasis, the invasion of human tissue by living fly maggots, is also a direct effect of arthropods on human health (see Chapter 6 and Chapter 21). Although inflammation and secondary infection may be involved, the primary treatment is to remove the maggots. Most pathology associated with myiasis resolves fairly readily after removal.

Blister beetles contain the vesicating agent, cantharidin, which produces water-filled blisters on human skin a few hours after exposure. Blisters resulting from exposure are generally not serious but may require efforts to prevent secondary infection.

III.   HYPERSENSITIVITY REACTIONS TO ARTHROPOD VENOM OR SALIVA

Sometimes the human immune system produces undesirable results in trying to protect the body. In a hypersensitive or "allergic" person, a relatively innocuous antigen elicits an out-of-proportion immune reaction. Thus, the tissue damage resulting from hypersensitivity is worse than the damage produced by the salivary secretion, venom, or other antigen itself.

A.   Hypersensitivity Reactions

Hypersensitivity reactions fall into two principal categories, reflecting the two major subdivisions of the immune system. The first category includes reactions initiated by antibody (for instance, immediate hypersensitivity reactions), in which symptoms are manifest almost immediately after exposure to antigen by a sensitized person. The second category includes reactions initiated by T lymphocytes (delayed hypersensitivity), and symptoms are usually not obvious for a number of hours or days.

Some authors break hypersensitivity down into four types: I, II, III, and IV. Types I to III involve antibody-mediated reactions. Type I reactions are IgE-mediated immediate hypersensitivity reactions. A systemic reaction to a honey bee sting is a good example of type I hypersensitivity. Because the allergen is directly introduced into the blood or tissue, a severe reaction can occur, such as anaphylactic shock. Type II reactions are antibody-mediated "cytotic" reactions similar to those occurring with some hemolytic reactions. Type III reactions are mediated by circulating antibody–antigen complexes and cause clinical syndromes such as serum sickness. Type IV reactions are mediated by T lymphocytes and macrophages and occur independently of antibody.

Allergens and specific antibodies produced in response to allergens do not by themselves cause the pathological symptoms associated with immediate hypersensitivity. Instead, the chemical substances (called *mediators*) released or activated in the host's tissues, resulting from the antigen–antibody binding in solution or on the cell membranes, cause the characteristic tissue damage associated with hypersensitivity.

In atopic persons, the initial exposure to an allergen stimulates an immunoglobulin E (IgE) response. IgE is a minor component of normal blood serum, having a concentration of approximately 1 μg/ml. IgE levels are generally higher in atopic persons than in normal individuals of the same age; however, a normal IgE level does not exclude atopy. IgE levels are also elevated in persons with parasitic worm infections, which indicates its beneficial role in humans. IgE-sensitized mast cells in the gut mucosa provide a good defense against the worms attempting to traverse the gut wall. The IgE produced in atopic individuals in response to allergens sensitizes mast cells, which degranulate upon future exposure to allergen.

Mast cells are similar structurally and functionally to basophils. They are found in association with mucosal epithelial cells as well as in connective tissue. Mast cells characteristically contain approximately 1000 granules, which upon degranulation release pharmacological mediators causing the allergic symptoms.

Mechanisms of allergic reactions.   Having "allergies" reflects an autosomal dominant pattern of inheritance with incomplete penetrance. This pattern of inheritance shows up as a propensity to respond to allergen exposure by producing high levels of allergen-specific IgE. Excess production of IgE appears to be controlled by various immune response genes located in the major histocompatibility complex (MHC) on chromosome 6.

IgE response is dependent on prior sensitization to allergen. The allergen must first be internalized by antigen-presenting cells, including macrophages, dendritic cells, activated T lymphocytes, and B lymphocytes. After allergen processing, peptide fragments of the allergen are presented with class II (MHC) molecules of host antigen-presenting cells to CD4+T lymphocytes. These lymphocytes have receptors for the particular MHC–peptide complex. This interaction results in release of cytokines by the CD4+ cell. T-helper lymphocytes (CD4+) are apparently of two classes: T_H1 and T_H2. If the CD4+ cells that recognize the allergen are of the T_H2 class, a specific group of mediators is released, including interleukin-4 (IL-4), IL-5, and IL-9. Other cytokines such as IL-2, IL-3, IL-10, IL-13, and granulocyte-macrophage stimulating factor (GM-CSF), are also released in the process of antigen recognition but are not specific to the T_H2 class. Cytokines such as IL-4, IL-5, and IL-6 are involved in B-cell proliferation and differentiation. Activated B lymphocytes (with bound allergen) are stimulated by these cytokines to multiply and secrete IgM antibody. IL-4, IL-6, IL-10, and IL-13 from T_H2 cells promote B-cell isotype switching to IgE production. Thus, atopy appears to be a result of predisposition toward T_H2-type responses, resulting in production of large quantities of allergen-specific IgE.

IgE antibodies specific for a certain allergen bind to mast cells and basophils. When these "sensitized" cells are re-exposed to the offending allergen, IgE molecules attached to the surface of mast cells and basophils become cross-linked by allergen, leading to a distortion of the IgE molecules and a subsequent series of enzymatic reactions and cell degranulation that release mediators into the bloodstream and local tissues. The most important preformed mediator is histamine, which causes vasodilation, increased vascular permeability (leading to edema), and mucous secretion (respiratory tract). Other mediators are formed during degranulation such as Prostaglandin D_2 (PGD_2), the sulfidopeptide leukotrienes LTC_4, LTD_4, and LTE_4 (slow-reacting substance of anaphylaxis), platelet-activating factor (PAF), and bradykinin. PAF is a potent chemotactic factor, and the sulfidopeptide leukotrienes and bradykinin are vasoactive compounds. Cross-linking of IgE on mast cells also activates phospholipase A_2 and releases arachidonic acid from the A_2 position of cell membrane phospholipids. Mast cells then metabolize arachidonic acid through the cyclooxygenase pathway to form prostaglandin and thromboxane mediators or through the lipoxygenase pathway to form leukotrienes.

Once the allergic reaction begins, mast cells amplify it by releasing vasoactive agents and cytokines such as GM-CSF, tumor necrosis factor α (TNF-α), transforming growth factor β, IL-1 to IL-6, and IL-13. These cytokines lead to further IgE production, mast cell growth, and eosinophil growth, chemotaxis, and survival. For instance, IL-5, TNF-α, and IL-1 promote eosinophil movement by increasing their expression of adhesion receptors on endothelium. Then, arriving eosinophils secrete IL-1, which favors T_H2 cell proliferation and mast cell growth factor IL-3. Eosinophils release oxygen radicals and proteins that are toxic to affected tissues.

B. Local Hypersensitivity Reactions

Local allergic reactions involve the nose, lung and, occasionally, the skin. These areas where allergen makes contact with sensitized (IgE-"loaded") tissues are usually the only ones affected in these reactions. Allergic and perennial rhinitis, as well as asthma, may be due to arthropods or their emanations. A good example of this is house-dust mite allergy (see Chapter 24).

C. Systemic Hypersensitivity Reactions

Systemic allergic reactions are more likely to occur when the allergen reaches the blood or lymph circulations and involve several organ systems. *Anaphylaxis* is the term often used to describe the rapid, sometimes lethal sequence of events occurring in certain cases upon subsequent exposure to a particular allergen. Initial signs of anaphylaxis are often cutaneous, such as generalized pruritus, urticaria, and angioedema. If the reaction continues, excessive vasodilation and increased vascular permeability caused by histamine and the other mediators may lead to irreversible shock. When angioedema affects the larynx, oropharynx, or tongue, the upper airway can become occluded. Pulmonary edema and bronchial constriction may lead to respiratory failure.

D. Late Hypersensitivity Reactions

A cutaneous late-phase IgE-mediated response in allergic individuals may appear 2 to 48 h after challenge and is characterized by a second wave of inflammatory mediators and dramatic influx of immune and inflammatory cells to the site of antigen exposure. These reactions, also called *large local reactions*, are pruritic, painful, erythematous, and edematous, and often peak within 12 h after stings. The edema from large local reactions can, in extreme cases, be severe enough to cause compression of nerves or blood vessels to an extremity.[4] Late-phase asthma and anaphylaxis occur via similar mechanisms.

E. Delayed Hypersensitivity Reactions

An allergic dermatitis, characterized by eczema-like eruptions on the skin, may develop in response to insect or mite body parts, saliva, or feces secondary to the immediate reaction. Delayed-type hypersensitivity reactions typically appear over a period of several days, perhaps not maximal until 48 or 72 h after antigen exposure. This is cell-mediated immunity wherein CD4-positive T lymphocytes react with antigen and release lymphokines into tissues. These lymphokines may serve as attractants for monocytes.

F. Treatment Principles for Hypersensitivity Reactions

Antihistamines block most, if not all, of the effects of histamine release. This is accomplished by competing for histamine at its receptor sites, thus preventing histamine from attaching to these receptor sites and producing an effect on body tissues. Oral administration of antihistamines is often recommended for local reactions. In treating generalized systemic or anaphylactic reactions, epinephrine remains the most important treatment and can be life-saving. Antihistamines such as diphenhydramine hydrochloride are given parenterally.

Localized wheal and flare reactions to mosquito bites may even be prevented by use of oral antihistamines. One study demonstrated that persons who had previously had dramatic cutaneous reactions to mosquito bites, when taking cetirizine (Zyrtec®), had a 40% decrease in the size of the wheal response at 15 min and the size of the bite papule at 24 h.[5]

Corticosteroids have an anti-inflammatory effect. They act by various mechanisms including vasoconstriction, decreasing membrane permeability, decreasing mitotic activity of epidermal cells, and lysosomal membrane stabilization within leukocytes and monocytes. In antigen-dependent T cell activation reactions (delayed hypersensitivity), steroids inhibit antigen-specific lymphocyte activation and proliferation. Also, inhibition of the influx of inflammatory cells by glucocorticoids leads to inhibition of the appearance of inflammatory mediators during the late phase. Applied topically to the skin, steroids deplete Langerhans cells of CD1 and HLA-DR molecules, blocking their antigen-presenting function.

In certain arthropod-related allergies (including asthma) such as dust mite or cockroach allergies, inhaled steroids, leukotriene antagonists, or cromolyn sodium may sometimes be used. Cromolyn stabilizes mast cells against degranulation, thus preventing release of histamine, leukotrienes, and other pharmacological mediators. The use of epinephrine in severe or systemic hypersensitivity reactions acts to suppress (stabilize) mediator release from mast cells and basophils and reverses many of the end-organ responses to the pharmacological mediators of anaphylaxis. Thus, there is bronchodilation and relaxation of smooth muscle. The prompt use of epinephrine can often lead to complete resolution of the clinical manifestations of anaphylaxis within minutes.[6]

Other specific interventions may be needed to manage anaphylaxis (see Chapter 2 for more detail). These include actions such as supplemental inspired oxygen, endotracheal intubation, cricothyrotomy, adrenergic stimulants (such as isoproterenol, dopamine, norepinephrine, nebulized β_2 agonists), glucagon (for β-blocked patients), H_1 and H_2 antihistamines, and glucocorticoids.[6,7] Careful monitoring of each individual case, with particular attention to the intensity and relative progression of the anaphylaxis, should enable the attending physician to decide which of these additional measures are indicated.

IV.   NEUROTOXIC VENOMS

A.   Mechanisms of Toxicity

Widow spiders and some scorpions produce ill effects in humans by neurotoxic venoms. Widow spider (*Latrodectus* spp.) venom is a neuromuscular-damaging protein that affects ion transport. It produces sweating, piloerection, muscular spasm, weakness, tremor, and sometimes paralysis, stupor, and convulsions. This type of venom may not produce obvious skin lesions but will primarily produce the systemic reactions.

Scorpion venom is also neurotoxic. It contains multiple low-molecular-weight basic proteins (the neurotoxins), mucus (5 to 10%), salts, and various organic compounds such as oligopeptides, nucleotides, and amino acids. Unlike most spider and snake venoms, scorpion venom contains little or no enzymes. The low molecular weight proteins increase permeability through the neuronal sodium channels. These toxins directly affect the neuronal portion of the neuromuscular junction, causing depolarization of the nerve and myocyte. They may also increase permeability of neuronal sodium channels in the autonomic nervous system. Systemic symptoms of scorpion envenomation include blurred vision, sweating, spreading partial paralysis, muscle twitching, abnormal eye movements, excessive salivation, hypertension and, sometimes, convulsions. Death (if it occurs) is usually a result of respiratory paralysis, peripheral vascular failure, and myocarditis.

B.   Treatment Principles for Neurotoxic Venoms

Strategies for treating an arthropod bite or sting that is neurotoxic in nature involve counteracting the effects of the venom and supportive treatment. Antivenins are commercially available for many of the widow spider venoms and the venoms of some scorpion species. Muscle relaxants, calcium gluconate, and/or antivenin are often used for widow spider bites (see Chapter 29). Antivenin is sometimes used in treating scorpion stings along with anticonvulsants, vasodilators, assisted ventilation, and other supportive measures as needed (see Chapter 28).

V.   NECROTIC VENOMS

A.   Mechanisms of Toxicity

In contrast to the widow spiders, violin spiders (brown recluse is one of the most notable) have venom that is necrotic in activity coupled with hyaluronidase that acts as a spreading factor. Brown recluse spider venom contains a lipase enzyme, sphingomyelinase D, which is significantly different from phospholipase A in bee and wasp venoms. This specific lipase is the primary necrotic agent involved in the formation of the typical lesions (see Chapter 29). It is possible that neutrophil chemotaxis is induced by sphingomyelinase D.[8] The subsequent influx of neutrophils into the area is critical in the formation of the necrotic lesion.

B.   Treatment Principles for Necrotic Venoms

Treatment of a necrotic arthropod bite (e.g., brown recluse) is controversial because controlled studies are lacking and the severity of the bite is variable.[9] Currently, it may involve antibiotic therapy, antivenin (if available), and dapsone, if the patient is not glucose-6-phosphate dehydrogenase deficient[8–12] (see Chapter 29). King[13] said that application of ice packs may be very important in limiting necrosis because activity of the necrotic enzyme in brown recluse venom is related to temperature.

REFERENCES

1. Camazine, S., Hymenopteran stings: reactions, mechanisms, and medical treatment, *Bull. Entomol. Soc. Am.*, Spring Issue, 17, 1988.

2. Alexander, J.O., *Arthropods and Human Skin*, Springer-Verlag, Berlin, 1984, chap. 9.

3. Foster, W.A. and Walker, E.D., Mosquitoes, in *Medical and Veterinary Entomology*, Mullen, G.R. and Durden, L.A., Eds., Academic Press, New York, 2002, p. 203.

4. Moffitt, J.E. and deShazo, R.D., Allergic and other reactions to insects, in *Rich's Clinical Immunology: Principles and Practice*, 2nd ed., Rich, R.R., Fleisher, W.T., Kotzin, B.L., and Schroeser, H.W., Eds., Mosby, New York, 2001, p. 47.3.

5. Reunala, T., Brummer-Korvenkotio, H., Karppinen, A., Coulie, P., and Palosuo, T., Treatment of mosquito bites with cetirizine, *Clin. Exp. Allerg.*, 23, 72, 1993.

6. Sullivan, T.J., Treatment of reactions to insect stings and bites, in *Monograph on Insect Allergy*, 2nd ed., Levine, M.I. and Lockey, R.F., Eds., American Academy of Allergy and Immunology, Milwaukee, WI, 1986, chap 7.

7. Wasserman, S.I., Anaphylaxis, in *Rich's Clinical Immunology: Principles and Practice*, 2nd ed., Rich, R.R., Fleisher, W.T., Kotzin, B.L., and Schroeser, H.W., Eds., Mosby, New York, 2001, p. 46.7.

8. King, L.E., Spider bites, *Arch. Dermatol.*, 123, 41, 1987.

9. Buescher, L.S., Spider bites and scorpion stings, in *Conn's Current Therapy*, Rakel, R.E. and Bope, E.T., Eds., Elsevier Saunders Publishing, Philadelphia, PA, 2005, p. 1302.

10. Delozier, J.B., Reaves, L., King, L.E., and Rees, R.S., Brown recluse spider bites of the upper extremity, *South. Med. J.*, 81, 181, 1988.

11. Masters, E., Sams, H., and King, L.E., Loxoscelism, *New Engl. J. Med.*, 339, 1944, 1998.

12. Rees, R.S., Campbell, D., Reiger, E., and King, L.E., The diagnosis and treatment of brown recluse spider bites, *Ann. Emergency Med.*, 16, 945, 1987.

13. King, L.E., Brown recluse bites: stay cool, *JAMA*, 254, 2895, 1986.

ALLERGY TO ARTHROPODS AND THEIR VENOMS

TABLE OF CONTENTS

CASE HISTORY

ALLERGIC REACTION TO FIRE ANT STING?

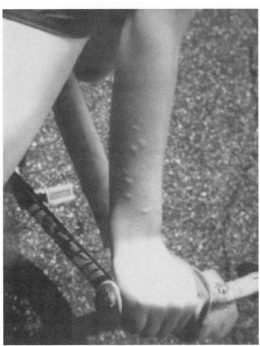

Figure 1
Typical fire ant lesions approximately 30 min after sting. (Reprinted from Lab. Med. 25, 366, 1994. Copyright 1994 by the American Society of Clinical Pathologists. With permission.)

A woman called saying she was having an allergic reaction to an ant sting. From her description of the event and the mound, the specimen was likely a fire ant. She quickly described her lesion — a small pustule — and how she had felt since the sting. The sting had happened the day before. What to do? Go to the hospital? Before I heard anything more about the case, I told her that if she thought she was having an allergic reaction to a sting she should go to the doctor immediately. She persisted in telling the story. It seemed obvious that she was not having an allergic reaction. It had happened the day before, and there was neither swelling nor any systemic effect. Wheal and flare are common initial signs of fire ant stings; pustular lesions are normal 24 h later (Figure 1 and Figure 2).

I. ALLERGY TO STINGS OR BITES

A. Introduction and Medical Significance

People encounter insects almost everywhere. Inevitably, thousands of persons are stung or bitten daily. For most people, local pain, swelling, and itching are the only effects, and they gradually abate. For others, life-threatening allergic reactions occur. More people die each year in the U.S. from bee and wasp stings than from snake bites.[1] Why? Probably because more people are exposed to stinging insects than to poisonous snakes; therefore, some individuals become hypersensitive to such stings. Consider fire ants. They are so numerous and widespread in the southern U.S. that all persons in that area are at high risk of being stung. In a 1988 survey, 2,022 physicians in 13 southern states reported treating 20,755 patients for reactions to fire ant stings. These included 13,139 (63%) patients who had local reactions, 395 (2%) who were treated for cellulitis or sepsis, and 413 (2%) who suffered anaphylactic shock.[2] Freeman[3] reported that fire ants were responsible for 42% of visits to an allergy clinic in San Antonio, TX.

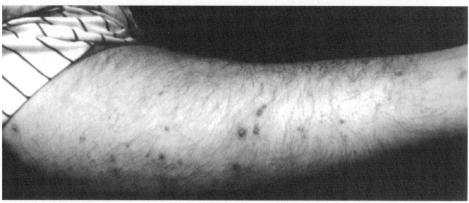

Figure 2
Pustules resulting from fire ant stings, 24 h after sting. (Photo courtesy of Ralph Turnbo.)

Comment: Fire ants are responsible for thousands of human stings in the southern U.S. each year. Whenever their mound is disturbed, they boil out aggressively looking for the intruder. There are generally three types of reactions to stings: normal, large local, and systemic. Large local reactions can occur for several days after a sting, and are characterized by extensive swelling over a large area. For example, if a person is stung on the hand, he or she may swell past the elbow. A systemic reaction — generalized urticaria, angioedema, anaphylaxis — usually begins 10 to 20 min after the sting. However, very rarely, symptoms may not start for several hours. In the case (under discussion), the woman may have been confused about terminology. Sometimes people trying to describe a bite or sting site use words that have totally different meanings to a health care provider.

Source: Adapted from *Lab. Med.* 25, 366, 1994. Copyright 1994 by the American Society of Clinical Pathologists. With permission.

Stinging insects in the order Hymenoptera such as bees, wasps, and ants can kill people in two ways: by the sheer numbers of stings producing toxic effects, and by the allergic reactions in susceptible individuals. It generally takes 500 or more bee stings to kill an individual by the toxic effects of the venom alone (see Chapter 1 for discussion of direct effects), but just one sting may prove fatal for the person with bee sting allergy.

Numerous arthropods can cause allergic reactions in persons by their stings, including various wasps, bees, ants, scorpions, and even caterpillars. However, the ones most commonly involved are paper wasps, yellowjackets, honey bees, and fire ants (see Chapter 31, Chapter 11, and Chapter 10, respectively, for discussions of each of these groups).

In addition to stings, bites from some arthropods may produce allergic reactions, including anaphylaxis and other systemic effects (Figure 1). However, systemic hypersensitivity reactions to arthropod bites are much less common (almost rare) than those resulting from stings. The groups most often involved in producing systemic effects by their bites are the kissing bugs (genus *Triatoma*), black flies, horse flies, and deer flies.[4,5] Mosquitoes, to a lesser extent, are involved, with several reports in the literature of

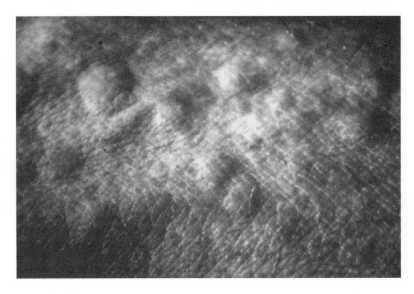

Figure 1
Hypersensitivity reaction to numerous mosquito bites. (Photo courtesy of Dr. Elton Hansens.)

large local reactions, urticaria, angioedema, headache, dizziness, lethargy, and even asthma.[6] Tick bites may sometimes cause extensive swelling and rash. Ticks reported to do so are the hard ticks, *Ixodes holocyclus* and *Amblyomma triguttatum*, and the soft tick, *Ornithodoros gurneyi*. Arthropod saliva from biting insects contains anticoagulants, enzymes, agglutinins, and mucopolysaccharides. Presumably, these components of saliva serve as sensitizing allergens.

Normal reaction to stings or bites.   A normal reaction to one or a few stings involves only the immediate area of the sting and appears within 2 to 3 min. Usually, it consists of redness, itching, swelling, pain, and formation of a wheal at the site. The reaction usually abates within 2 h or so. If a person is stung by numerous hymenopterans, the acute toxic reaction (nonallergic) resulting from large amounts of venom can be severe. Murray[7] describes a man who was stung over 2000 times by bees and exhibited signs of histamine overdosage — severe headache, vomiting, diarrhea, and shock.

Severe local (or large local) reaction to stings or bites.   Large local reactions are characterized by painful, pruritic swelling of at least 5-cm diameter (but still contiguous with the sting site), and may involve an entire extremity. Large local reactions usually peak within 48 h and last as long as 10 d. Most patients with large local reactions have detectable venom-specific IgE antibodies. Large local reactions have not been shown to significantly increase the risk for anaphylaxis upon subsequent stings. In fact, the risk of a systemic reaction in patients who experience large local reactions is no more than 5 to 10%.[8] Venom immunotherapy has been shown to be effective in preventing large local reactions to some hymenopterans, but is rarely required.

Systemic reaction to stings or bites.   Systemic allergic reactions produce symptoms in areas other than the sting site. Thus, the allergic person may have both the local pain, wheal, and itching from the sting, as well as generalized pruritus, urticaria, angioedema, respiratory difficulty, syncope, stridor, gastrointestinal distress, and hypotension (Table 1). Systemic reactions usually begin with widespread cutaneous symptoms such as angioedema or urticaria. These skin manifestations may be the extent of the systemic reaction, or there may be progression to a generalized pruritus, widespread

Table 1
Signs and Symptoms of Anaphylaxis

General	Apprehension, uneasiness, weakness
Cutaneous	Erythema, pruritus, urticaria, angioedema
Gastrointestinal	Abdominal cramps, vomiting, diarrhea
Genitourinary	Urinary or fecal incontinence, uterine cramps
Respiratory	Chest tightness, cough, dyspnea, stridor, wheezing
Cardiovascular	Dizziness, lightheadedness, syncope, hypotension

Source: From Monograph on Insect Allergy, 2nd ed., Levine, M.I. and Lockey, R.F., Eds., American Academy of Allergy and Immunology, 1986. With permission

edema, and upper respiratory distress. In severe reactions, shock begins to develop with a rapid pulse and low blood pressure. The victim may feel a constriction in his throat and chest, and breathing continues to become difficult. Sometimes, a severe allergic reaction results in anaphylactic shock and death within 10 or 15 min, although 20 to 30 min is more common. In a study of 641 deaths from Hymenoptera stings in the U.S., respiratory conditions accounted for 53% of the deaths.[9] Autopsies revealed characteristic laryngeal, epiglottal, and pulmonary edema, along with both serous and mucoid secretions.

Cross-reactivity among venoms. There seems to be a consensus that considerable cross-reactivity occurs among the vespid venoms (yellowjacket, paper wasp, and hornet), meaning that a person sensitized to one vespid venom could have a serious reaction to a sting from other members of the group.[10] Limited studies have indicated cross-reactivity between honey bee and bumble bee venoms. Generally, however, honey bee-allergic individuals do not exhibit cross-reactivity to vespid venoms. (Note: There are exceptions — some individual patients have shown cross-reactivity between honey bee and yellowjacket venoms.[10]) Immunologic cross-reactivity among fire ant venoms and some vespid venoms has been demonstrated, but there is no evidence of cross-reactivity in the clinical setting.

B. *Management and Treatment*

Normal reaction. An extremely useful algorithm for the management of insect sting reactions is provided by Moffitt et al.[8] Treatment for a normal or mild local reaction involves the use of ice packs or pain relievers to minimize pain and washing the wound to lessen the chances of secondary infection. Oral antihistamines may help counteract the effects of histamine (IgE-mediated or not) in the affected tissues resulting from certain venom components. Topical antihistamines and corticosteroids may also be helpful.[11] Alexander[12] recommended calamine products (Caladryl® is often used in the U.S.)

The use of meat tenderizer containing the enzyme papain is of no therapeutic value.[13] The theory behind the use of papain is valid because *in vitro* incubation of papain and venom leads to destruction of the venom activity. However, in a laboratory experiment with mice, there was no marked inhibition of lesion development in mice receiving papain or Adolph's® meat tenderizer by intradermal injection or topical application.[14]

Severe local (or large local) reaction. In the case of a large local reaction characterized by considerable swelling and tenderness around the sting site, rest and elevation of the affected limb may be needed. The patient should avoid exercise because it may exacerbate the swelling. If the sting site is on or near the throat, nose, or eye area, or if there is widespread swelling, patients should definitely seek medical care. Treatment involves analgesics, topical high-potency steroids, oral antihistamines to relieve itching, and systemic steroids (such as prednisone) if swelling is severe.[15] Superimposed infections such as cellulitis or septicemia, unusual with hymenoptera envenomation, require aggressive treatment that may include hospitalization, incision, and systemic antibiotics.[16] If the offending arthropod is a biting fly such as a mosquito, cutaneous reactions may even be prevented by use of high-potency topical or oral antihistamines. One study demonstrated that persons who had previously had dramatic cutaneous reactions to mosquito bites, when taking cetirizine (Zyrtec®), had a 40% decrease in the size of the wheal response at 15 min and the size of the bite papule at 24 h.[17]

Systemic reaction. Persons who experience a generalized allergic reaction (even mild) may be at risk of a severe reaction and possible death upon the next sting (days, weeks, or months later). In the event of a systemic reaction, the most important aspect of care is for that person to get to an emergency facility for immediate treatment. If the individual has an epinephrine kit, it should be used. An ice pack on the sting site may delay absorption of venom, and removal of a honey bee stinger may also reduce venom absorption. However, people should be reminded that these measures should not delay seeking emergency treatment in any way.

Physicians often do several things to treat a severe allergic reaction. There may be some minor differences in procedures used (depending on the reference consulted), but the immediate goal is the same — maintain an adequate airway and support the blood pressure. The American Academy of Allergy and Immunology has published steps for the management of anaphylaxis (Table 2).[18] The following is a modification (see Wasserman[19]) of suggestions made by Stafford et al.[20] for treatment of severe reactions to fire ant stings, which is fairly typical for the management of similar reactions to all Hymenoptera stings:

> An immediate subcutaneous injection of 0.3 to 0.5 ml of a 1:1000 solution of epinephrine (preferably intramuscularly in the lateral thigh) should be administered, and repeated, with blood pressure monitoring, at 10-min intervals if necessary. Intravenous epinephrine may be administered at a rate of 2 µg/min for treatment of severe shock or cardiac arrest, but bolus administration should be avoided. The airway must be established and maintained by using endotracheal intubation or cricothyrotomy, if necessary. Intravenous fluids should be given to replenish depleted intravascular volume in the treatment of anaphylactic shock. Norepinephrine, H_1 and H_2 blocking agents may be required. Systemic corticosteroids and both types of antihistamines may prevent recurrent or biphasic anaphylaxis. Glucagon is appropriate for patients on beta-blockers.

Administration of oxygen (see Table 2) may be needed to minimize development of hypoxia, which by itself may contribute to vascular collapse and cerebral edema. Also, wheezing that is refractory to repeated doses of epinephrine can be treated with continuously nebulized beta agonists such as albuterol.[21] However, the administration of epinephrine is the most important element of treatment. It acts to suppress mediator release from mast cells and basophils and reverses many of the end-organ responses to mediators of anaphylaxis. Complete resolution of the clinical manifestations of anaphylaxis often occurs within minutes. The critical and immediate use of epinephrine is the reason why some people who are allergic to bee or wasp stings carry sting kits

Table 2
Management of Anaphylaxis

General therapeutic measures
    Assessment
    Epinephrine (intramuscularly in the lateral thigh)
    Glucagon if on beta blockers

Specific interventions
    Airway obstruction
        Upper airway obstruction
            Supplemental inspired oxygen
            Extension of the neck
            Oropharyngeal airway
            Endotracheal intubation
            Cricothyrotomy
        Lower airway obstruction
            Supplemental inspired oxygen
            Inhaled beta agonists
            Conventional treatment for status asthmaticus
    Hypotension
        Peripheral vascular defects
            Trendelenberg position
            Intravenous isotonic sodium chloride
            Vasopressors if required (dopamine, intravenous
                epinephrine, norepinephrine)
            Diphenhydramine plus cimetidine
        Cardiac dysfunction
            Conventional therapy of dysrhythmias
            Diphenhydramine plus cimetidine

Suppression of persistent or recurrent reactions
    Direct observation for at least 12 h after anaphylaxis
    Systemic glucocorticoids

Formulate plan to minimize future reactions

Educate in insect avoidance techniques

Medic-Alert® tag

Self-injectable epinephrine

Venom immunotherapy

Source: *Adapted from Monograph on Insect Allergy, 2nd ed.,
Levine, M.I. and Lockey, R.F., Eds., American Academy of
Allergy and Immunology, 1986. With permission. Modi-
fied with information in Yates et al.*[15]

containing syringes loaded with the drug. At least two preloaded syringes are available (Ana-Kit®, Miles Labs, Spokane, WA; Epi-Pen®, Center Labs, Port Washington, NY) in both adult and pediatric versions. In case of a sting, the allergic person can give himself an injection that may very well save his life. Alexander[12] recommends that one or two close and reliable relatives of the allergic person should also be carefully instructed in the correct use of the kit, and especially the administration of epinephrine.

It is important to note that just because an individual uses the epinephrine injection in case of a sting does not mean prompt medical treatment is not necessary. It is still vital to get to a hospital or physician as quickly as possible. The sting kits or loaded syringes are meant only to stave off fulminating symptoms long enough for the victim to get to a hospital. This is especially important in light of the fact that sometimes there is a second phase of anaphylaxis 4 to 10 h after the initial reaction.[22]

Insect sting kits and/or autoinjector syringes must be prescribed by a physician. Any person who has suffered even mild symptoms of an allergic reaction should be counseled to obtain a kit and keep it at hand whenever there is a chance of being stung.

It might also be a good idea for all insect-allergic persons to wear a Medic-Alert® (Medic-Alert Foundation, Turlock, CA) tag or card to alert medical personnel of their allergy in case they lose consciousness.

Long-term management of insect sting allergy. Sting-allergic patients and their physicians should also think of long-term management of the problem. There is always the possibility of being stung again. Immunotherapy is a procedure used by allergists to increase the allergic person's tolerance to insect venom. The process works by stimulating serum-venom-specific IgG and decreasing titers of serum-venom-specific IgE. It is accomplished by numerous injections of venom from offending insects (or from extracts from whole bodies in the case of fire ants). Initially, the injections are very weak. The dosages are gradually increased over time until the patient can tolerate approximately the same amount of venom as in a sting. Then, the patient is kept on a maintenance dose to keep up that tolerance. Although 3 to 5 years of treatment was previously recommended,[23] recent studies indicate the need for extended immunotherapy.[11] Graft[24] has shown that immunotherapy is highly effective and safe for prevention of future systemic reactions to Hymenoptera stings. In a study[25] of 65 patients on a maintenance dose of fire ant whole body extract, only 1 (2%) patient of 47 who were subsequently stung by fire ants had an anaphylactic reaction. Physicians deciding whether or not to initiate venom immunotherapy base their decision on clinical history and results of venom skin tests and venom-specific radioallergosorbent tests (RASTs) (see also Table 3). Adults with a history of systemic reaction and a positive venom skin test or RAST should receive immunotherapy.[26] For some persons, it may seem too expensive and inconvenient to undergo immunotherapy, but it is a way for persons with insect allergy to lead a relatively normal life.

C. *Avoidance of Offending Insects*

Here are some ways for both allergic and nonallergic individuals to avoid stinging insects.

1. Each year, have someone eliminate bee, wasp, and fire ant nests around the home — preferably early in the summer before the nests get large. Pest control operators will usually do this for a fee. If the homeowner wishes to accomplish nest elimination, he or she should wait until night or a very cool morning to

Table 3
Selection of Patients for Venom Immunotherapy

Sting Reaction	ST/RAST[a]	Venom Immunotherapy
Systemic, non-life-threatening (child) immediate, generalized, confined to skin (urticaria, angioedema, erythema, pruritus)	+/−	No
Systemic, life-threatening (child) immediate, generalized, may involve cutaneous symptoms, but also has respiratory (laryngeal edema or bronchospasm) or cardiovascular symptoms (hypotension/shock)	+	Yes
Systemic (adult)	+	Yes
Systemic	−	No

Large local
  >2 in. in diameter
  >24 h in duration
	+/−	No

Normal <2 in. in diameter <24 h in duration	+/−	No

[a] *Venom skin test or venom-specific radioallergosorbent test.*

Source: From Monograph on Insect Allergy, 2nd ed., Levine, M.I. and Lockey, R.F., Eds., American Academy of Allergy and Immunology, 1986. With permission.

    minimize the threat of stings (persons who are allergic to insect stings should not attempt this). If nest elimination is accomplished at night, a flashlight should not be used unless a red filter is used. Bees and wasps will zero in on the beam.

2. Wear light- or khaki-colored clothing when outside during warm weather. These colors are less attractive than dark ones. Be careful to avoid bright-colored floral-patterned clothes.
3. While driving a car during the warm weather months, car windows should be closed and the air-conditioning used.
4. Do not go barefoot during the warm weather months. Bees are often found feeding on flowers at ground level, and fire ants have numerous feeding trails (even long distances from their mounds).
5. Wear long pants and long sleeves when working outdoors.
6. Wear gloves when gardening. A lot of people are stung on the hand while picking flowers or vegetables.
7. Avoid the use of scented sprays, perfumes, shampoos, suntan lotions, and soaps when working outdoors.
8. Avoid clover patches, gardens full of blossoms, blossoming trees, fields of goldenrod, and other areas with concentrations of bees, wasps, or ants.
9. Be cautious around rotting fruit, garbage cans, and littered picnic areas, especially in the late summer and early fall. Yellowjackets often feed in those areas.

OFTEN-ASKED QUESTION

MY DOCTOR SAID I WAS ALLERGIC TO KISSING BUGS. HOW CAN THAT BE, AS THEY DO NOT STING?

Most people are aware that you can become sensitized to venoms from many different stinging insects, leading to allergic reactions ranging from mild to severe (including anaphylactic shock). However, there is confusion when it comes to allergic reactions to bites. Arthropod bites may produce allergic reactions as well, though rare, presumably a result of hypersensitivity to salivary components secreted during the biting process. Arthropod saliva contains anticoagulants, enzymes, agglutinins, and mucopolysaccharides, which may serve as sensitizing allergens. Reactions have occurred following bites by many different types of arthropod but most commonly from bites by *Triatoma* (kissing bugs), horse and deer flies, and mosquitoes.

Triatoma allergy. Kissing bugs — so named because of the nasty habit of taking a blood meal from the face — belong to the insect family Reduviidae (hence, the sometimes-used moniker *reduvid bugs*), but specifically, the subfamily Triatominae. Within this subfamily, some (not all) species fall under the genus *Triatoma*; triatomines may also be in other genera. There are at least ten *Triatoma* species found in the U.S., but only about six of these are likely to be encountered.[1,2] Allergic reactions have been reported from bites by five species (*T. protracta, T. gerstaeckeri, T. sanguisuga, T. rubida,* and *T. rubrofasciata.*[3]), although in the U.S., *T. protracta* is the species most often reported in allergic reactions.[4,5] Kissing bug bites may be painless, leaving a small punctum without surrounding erythema, or they may cause delayed local reactions appearing like cellulitis. Anaphylactic reactions include itchy, burning sensations, respiratory difficulty, and other typical symptoms of anaphylaxis.[2]

Triatoma bugs feed on vertebrate hosts such as bats, other small- and medium-sized mammals, birds, and humans. Accordingly, the pests are often found in association with their host nest or habitation: caves, bird nests, rodent burrows, houses, etc. For example, *T. protracta* is found in woodrat nests. Bugs periodically fly away from the nests of their hosts

---

10. Avoid drinking sodas or eating popsicles, ice-cream cones, watermelons, and other sweets outdoors. This may attract bees and yellowjackets.
11. If you see a bee or wasp nest, or encounter a nest before the insects become agitated, retreat slowly. Do not panic. However, once the nest is disturbed, it is best to run immediately even though the hymenopterans are attracted to movement.

II. ALLERGY: IRRITATION CAUSED BY CONSUMING OR INHALING INSECT OR MITE PARTS

A. Introduction and Medical Significance

Several insect or mite species (or their body parts) may cause irritation or allergic reactions when inhaled and, less commonly, when ingested. House dust mites, *Dermatophagoides farinae* (and *D. pteronyssinus*), several species of mayflies and caddisflies, some nonbiting chironomid midges, and cockroach body parts or feces are the major inhalant offenders. As these arthropods die, their decaying cast skins become part of the environmental dust. In addition, insect emanations such as scales, antennae, feces, and (nocturnal cyclical flights) and may be attracted to lights at dwellings, subsequently gain entrance, and try to feed. Some species are able to colonize houses; they seem especially prolific in substandard structures with many cracks and crevices, mud walls, thatch roofs, etc.

Protection from *Triatoma* bites. Personal protection measures from kissing bugs involve avoidance (if possible), such as not sleeping in adobe or thatched-roof huts in endemic areas, and exclusion methods such as erecting bed nets.[6] Domestic or peridomestic kissing bug species (Mexico, Central America, and South America) may be controlled by proper construction of houses, sensible selection of building materials, sealing of cracks and crevices, and precision targeting of insecticides within the home. In the U.S., prevention of bug entry into homes may involve outdoor light management (i.e., lights placed away from the house, shining back toward it, instead of lights on the house), and efforts to find and seal entry points around the home.

REFERENCES

1. Schofield, C.J. and Dolling, W.R., Bed bugs and kissing-bugs, in *Medical Insects and Arachnids,* Lane, R.P. and Crosskey, R.W., Eds., Chapman and Hall, London, 1993, pp. 483–516.
2. Rohr, A.S., Marshall, N.A., and Saxon, A., Successful immunotherapy for *Triatoma protracta* induced anaphylaxis, *J. Allerg. Clin. Immunol.,* 73, 369–375, 1984.
3. Ryckman, R.E., Host reactions to bug bites: a literature review and annotated bibliography, *Calif. Vector Views,* 26, 1–23, 1979.
4. Marshall, N., Liebhaber, M., Dyer, Z., and Saxon, A., The prevalence of allergic sensitization to *Triatoma protracta* in a southern California community, *J. Med. Entomol.,* 23, 117–124, 1986.
5. Marshall, N.A. and Street, D.H., Allergy to *Triatoma protracta* I. Etiology, antigen preparation, diagnosis, and immunotherapy, *J. Med. Entomol.,* 19, 248–252, 1982.
6. Goddard, J., Kissing bugs and Chagas' disease, *Infect. Med.,* 16, 172–175, 1999.

saliva are suspected as being sources of sensitizing antigens. Compounding the problem, the average child spends 95% of his or her time indoors, providing plenty of time for sensitization. As for the digestive route, cockroach vomit, feces, and pieces of body parts or shed skins contaminating food are most often the cause of insect allergy via ingestion.

Reactions via the respiratory route.   Until the mid-1960s physicians simply diagnosed certain people as being allergic to house dust. Subsequently, Dutch researchers made the first link between house dust allergy and house dust mites[27,28] (see also Chapter 24). The mites commonly infest homes throughout much of the world and feed on shed human skin scales, mold, pollen, feathers, and animal dander. They are barely visible to the naked eye and live most commonly in mattresses and other furniture where people spend a lot of time. The mites are not poisonous and do not bite or sting, but they contain powerful allergens in their excreta, exoskeleton, and scales. For the hypersensitive individual living in an infested home, this can mean perennial rhinitis, urticaria, eczema, and asthma, often severe. In fact, Htut and Vickers[29] say that house mites are the major cause of asthma in the U.K. House dust mites can also be triggers for atopic dermatitis.[30]

Recent evidence indicates that early and prolonged exposure to inhaled allergens (such as dust mites and cockroaches) plays an important role in the *development* of both

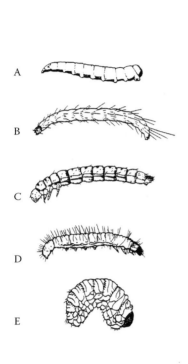

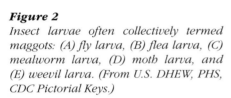

Figure 2

Insect larvae often collectively termed maggots: (A) fly larva, (B) flea larva, (C) mealworm larva, (D) moth larva, and (E) weevil larva. (From U.S. DHEW, PHS, CDC Pictorial Keys.)

Figure 3

Magnified view of hairs from dermestid beetle. Left: hastisetae (175×); right: spicisetae (400×). (From USAF Medical Service Digest. Courtesy of Major Tom Lillie.)

bronchial hyperreactivity and acute attacks of asthma. Accordingly, bronchial provocation with house dust mite or cockroach allergen can increase nonspecific reactivity for days or weeks. So, the root cause of asthma onset is sometimes exposure to house dust mites or cockroaches. Asthma-related health problems are most severe among children in inner-city areas. It has been hypothesized that cockroach-infested housing is at least partly to blame. In one study of 476 asthmatic inner-city children, 50.2% of the children's bedrooms had high levels of cockroach allergen in dust.[31] That study also found that children who were both allergic to cockroach allergen and exposed to high levels of this allergen had 0.37 hospitalizations a year, as compared to 0.11 for other children.[31]

Mayflies and caddisflies are delicate insects that spend most of their lives underwater as immatures. They emerge as adults in the spring and summer in tremendous numbers, are active for a few days, and then die. They do not bite or sting, but body particles from mass emergence of these insects have been well documented as causing allergies.

Nonbiting midges in the family Chironomidae have also been implicated as causes of insect inhalant allergy. A greater prevalence of asthma has been demonstrated in African populations seasonally exposed to the "green nimitti" midge, *Cladotanytarsus lewisi*.[32,33] Kagen et al.[34] implicated *Chironomus plumosus* as a cause of respiratory allergy in Wisconsin.

Figure 4
*Warehouse beetle larvae (Coleoptera: Dermestidae). Left: dorsal view; right: ventral view.
(From USAF Medical Service Digest. Courtesy of Major Tom Lillie.)*

In areas heavily infested with cockroaches, constant exposure to house dust contaminated with cockroach allergens is unavoidable. Accordingly, many people become sensitized and develop cockroach allergy. In a study in Thailand, 53.7% of 458 allergic patients reacted positively to cutaneous tests of cockroach body parts.[35] In a study in New York City the figure was even higher; over 70% of almost 600 allergic patients routinely visiting 7 hospitals reacted positively to cockroach antigen.[36]

Reactions via the digestive tract. Adult beetles and larval flies, moths, or beetles, as well as their cast skins, often contaminate food and may be responsible for irritation and allergic responses through ingestion. The confused flour beetle, *Tribolium confusum*, and rice weevil, *Sitophilus granaries,* have been reported to cause allergic reactions in bakery workers.[37] In addition, physicians are often confronted with parents worried about their children who have inadvertently eaten a maggot in their cereal, candy bar, or other food product. These maggots may be moth, beetle, or fly larvae (Figure 2), and generally cause no problems upon ingestion. However, some beetle larvae (primarily the family Dermestidae) found in stored food products possess minute barbed hairs (Hastisetae) and slender elongate hairs (Spicisetae) that apparently can cause enteric problems[38] (Figure 3 and Figure 4). The symptoms experienced after ingesting dermestid larvae have been attributed to mechanical action of the hastisetae

Figure 5
Cockroach found in bottle of creamer after top was left open.

and spicisetae resulting in tissue damage or irritation in the alimentary tract. Clinical symptoms include diarrhea, abdominal pain, and perianal itch.[39,40]

Cockroaches seem to be most often involved in allergic responses. Allergens are present in cockroach feces, which can be inadvertently ingested in heavily infested areas (Figure 5). Other allergens are present in cockroach saliva and exoskeletons, which can be introduced into foodstuffs.[37]

B. Management and Treatment

House dust mites. Management of house dust mite allergy may be achieved by immunotherapy, as well as encasing mattresses with plastic, keeping mattresses free of dust, using a synthetic (or washable) pillow, keeping airborne dust levels low, use of tile or wood floors instead of carpet, and efficient and frequent housecleaning. For vacuuming, double-thickness filters or high efficiency particulate air (HEPA) filters are needed for maximum results. The housecleaning tasks are best accomplished by a nonallergic person or family member. As the mites require a relative humidity of 60% to flourish, humidity control — by increasing ventilation or using a dehumidifier — will limit mite numbers in a house. Recent studies have demonstrated the effectiveness of benzyl

CASE HISTORY

BUGS IN PEANUT BRITTLE

A woman brought in a peanut brittle bar purchased at a grocery store, and upon eating it, said she observed bugs in it. She claimed that she immediately threw up and was very ill. She asked that I identify the specimens and document the event for her lawyer. The food item was examined and found to contain larvae and adults of the Indian meal moth, a common food pest (Figure 1). I wrote a letter detailing the product brought in (lot number and other package details), and species identification of the insect and its habits. I also made statements to the effect that the insects in question would not cause immediate nausea, and that I could not verify that the food product was infested at time of purchase.

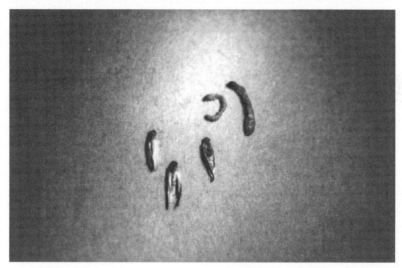

Figure 1
Adults and larval Indian meal moths removed from peanut brittle. (Reprinted from Lab. Med. 25, 366, 1994. Copyright 1994 by the American Society of Clinical Pathologists. With permission.)

Comment: When people bring in opened food products claiming that the product was infested with bugs upon purchase, medical personnel should not say or write things that confirm the allegation. Actual physical illness is usually not the issue — these cases almost always result in lawsuits. Medical personnel generally have no objective information about how long the product had been at the client's home, or how it had been treated, intentionally or otherwise.

Source: Adapted from *Lab. Med.* 25, 371, 1994. Copyright 1994 by the American Society of Clinical Pathologists. With permission.

benzoate or tannic acid for the treatment of mites in carpets. However, results are equivocal, and tannic acid may cause color changes in carpets. Further, even with the use of acaricides, airborne allergen loads may not be reduced below sensitization levels. One thing that does seem to show promise is the combined use of heat and steam treatment of home furnishings. One study showed that a single treatment of home furnishings reduced mite allergen load to below the risk level for sensitization and improved the asthmatic patients' bronchial hyperreactivity (BHR) fourfold.[41]

Cockroaches. Cockroach inhalant allergy may be managed by symptomatic therapy, immunotherapy with cockroach extracts, and intense sanitation and pest control measures to reduce roach populations. A recent study demonstrated the effectiveness of environmental intervention, including extermination visits by pest control personnel. In that study, asthmatic children who participated in the intervention had significantly fewer asthma symptoms compared to those in the control group: an average of 21 fewer days of symptoms in the first year and an average of 16 fewer days during the second.[42] It must be noted, however, that killing cockroaches may not eliminate antigen from the dwelling. The shelf life of cockroach allergen is several years. It is better to live in a dwelling that has never had a cockroach infestation. Treatment of other insect inhalant allergy also includes avoidance and symptomatic therapy. Where avoidance is impossible or impractical because of residential or occupational exposure, antihistamines, inhaled steroids, beta-2 agonists, or cromolyn sodium may be indicated.

Food contamination. To prevent allergic reactions resulting from ingesting insect parts or fluids, sanitation is the answer. Food preparation areas should be thoroughly cleaned and sanitized prior to cooking, even if the areas "look" clean. Cockroaches may have contaminated these surfaces during the night. Leftovers should be properly covered and/or refrigerated to prevent cockroach feeding. Cereal, nuts, candy, flour, and cornmeal should be examined before consumption for evidence of insect infestation such as small beetles or weevils. Staples such as flour, meal, and cereals can be placed in tight-fitting plastic containers to prevent insect infestation.

Both insect inhalant and sting allergy, similar to any other allergy, can usually be effectively managed and treated. By avoiding the offending types or species involved, practicing good sanitation (in the case of inhalant or ingestant allergies), and applying appropriate immunotherapy or antihistamine therapy, allergic persons should be able to conduct their lives in a relatively normal manner.

REFERENCES

1. Parrish, H.M., Analysis of 460 fatalities from venomous animals in the U.S., *Am. J. Med. Sci.*, 245, 129, 1963.

2. Stafford, C.T., Hutto, L.S., Rhoades, R.B., Thompson, W.O., and Impson, L.K., Imported fire ants as a health hazard, *South. Med. J.*, 82, 1515, 1989.

3. Freeman, T., Hymenoptera hypersensitivity in an imported fire ant endemic area, *Ann. Allerg. Asthma Immunol.*, 78, 369, 1997.

4. Moffitt, J.E., Venarske, D., Goddard, J., Yates, A.B., and deShazo, R.D., Allergic reactions to *Triatoma* bites, *Ann. Allerg. Asthma Immunol.*, 91, 122, 2003.

5. Hoffman, D.R., Allergic reactions to biting insects, in *Monograph on Insect Allergy*, 2nd ed., Levine, M.I. and Lockey, R.F., Eds., American Academy of Allergy and Immunology, Milwaukee, WI, 1986, chap. 14.

6. Gluck, J.C., Asthma from mosquito bites: a case report, *Ann. Allerg.*, 56, 492, 1986.

7. Murray, J.A., A case of multiple bee stings, *Cent. Afr. J. Med.*, 10, 249, 1964.

8. Moffitt, J.E., Golden, D.B., Reisman, R.E., Lee, R., Nicklas, R., Freeman, T., deShazo, R., Tracy, J., Bernstein, I.L., Blessing-Moore, J., Khan, D.A., Lang, D.M., Portnoy, J.M., Schuller, D.E., Spector, S.L., and Tilles, S.A., Stinging insect hypersensitivity: a practice parameter update, *J. Allerg. Clin. Immunol.*, 114, 869, 2004.

9. Levine, M.I. and Nall, T.M., Pathologic findings in Hymenoptera deaths, in *Monograph on Insect Allergy*, 2nd ed., Levine, M.I. and Lockey, R.F., Eds., American Academy of Allergy and Immunology, Milwaukee, WI, 1986, chap. 4.

10. Richman, P.G. and Baer, H., Hymenoptera venoms: composition, immunology, standardization, and stability, in *Monograph on Insect Allergy*, 2nd ed., Levine, M.I. and Lockey, R.F., Eds., American Academy of Allergy and Immunology, Milwaukee, WI, 1986, chap. 2.

11. Freeman, T.M., Clinical practice, hypersensitivity to hymenoptera stings, *New Engl. J. Med.*, 351, 1978, 2004.

12. Alexander, J.O., *Arthropods and Human Skin*, Springer-Verlag, Berlin, 1984, chap. 10.

13. Ross, E.V., Badame, A.J., and Dale, S.E., Meat tenderizer in the acute treatment of imported fire ant stings, *J. Am. Acad. Dermatol.*, 16, 1189, 1987.

14. Agostinucci, W., Cardoni, A.A., and Rosenberg, P., Effect of papain on bee venom toxicity, *Toxicon*, 19, 851, 1981.

15. Yates, A.B., Moffitt, J.E., and deShazo, R.D., Anaphylaxis to arthropod bites and stings, *Immunol. Allerg. Clin. North Am.*, 21, 635, 2001.

16. Frazier, C.A., Anaphylactic response to insect stings, *Compr. Ther.*, 2, 67, 1976.

17. Reunala, T., Brummer-Korvenkotio, H., Karppinen, A., Coulie, P., and Palosuo, T., Treatment of mosquito bites with cetirizine, *Clin. Exp. Allerg.*, 23, 72, 1993.

18. Levine, M.I. and Lockey, R.F., Eds., *Monograph on Insect Allergy*, American Academy of Allergy and Immunology, Milwaukee, WI, 1986.

19. Wasserman, S.I., Anaphylaxis, in *Rich's Clinical Immunology: Principles and Practice*, 2nd ed., Rich, R.R., Fleisher, W.T., Kotzin, B.L., and Schroeser, H.W., Eds., Mosby, New York, 2001, p. 46.7.

20. Stafford, C.T., Hoffman, D.R., and Rhoades, R.B., Allergy to imported fire ants, *South. Med. J.*, 82, 1520, 1989.

21. Kemp, S.F. and deShazo, R.D., Prevention and treatment of anaphylaxis, in *Allergens and Allergen Immunotherapy*, 3rd ed., Lockey, R.F., Bukantz, S.C., and Bousquet, J., Eds., Marcel Dekker, New York, 2004, chap. 40.

22. Sullivan, T.J., Treatment of reactions to insect stings and bites, in *Monograph on Insect Allergy*, 2nd ed., Levine, M.I. and Lockey, R.F., Eds., American Academy of Allergy and Immunology, Milwaukee, WI, 1986, chap. 7.

23. Reisman, R.E., Insect stings, *New Engl. J. Med.*, 331, 523, 1994.

24. Graft, D.F., Venom immunotherapy for stinging insect allergy, *Clin. Rev. Allerg.*, 5, 149, 1987.

25. Hylander, R.D., Ortiz, A.A., Freeman, T.M., and Martin, M.E., Imported fire ant immunotherapy: effectiveness of whole body extracts, *J. Allerg. Clin. Immunol.*, 83, 232, 1989.

26. Graft, D.F., Indications for venom immunotherapy, in *Monograph on Insect Allergy*, 2nd ed., Levine, M.I., Lockey, R.F., American Academy of Allergy and Immunology, Milwaukee, WI, 1986, chap. 8.

27. Spieksma, F.T.M., The mite fauna of house dust, with particular reference to the house dust mite, *Acarologia*, 9, 226, 1967.

28. Spieksma, F.T.M., The House Dust Mite, Dermatophagoides Pteronyssinus, Producer of House Dust Allergen, Thesis, University of Leiden, Netherlands, 1967, p. 65.

29. Htut, T. and Vickers, L., The prevention of mite-allergic asthma, *Int. J. Environ. Health Res.*, 5, 47, 1995.

30. Cameron, M.M., Can house dust mite-triggered atopic dermatitis be alleviated using acaricides? *Br. J. Dermatol.*, 137, 1, 1997.

31. Rosenstreich, D.L., Eggleston, P., Kattan, M., Baker, D., Slavin, R.G., Gergen, P., Mitchell, H., McNiff-Mortimer, K., Lynn, H., Ownby, D., and Malveaux, F., The role of cockroach allergy and exposure to cockroach allergen in causing morbidity among inner-city children with asthma, *New Engl. J. Med.*, 336, 1356, 1997.

32. Gad El Rab, M.O. and Kay, A.B., Widespread immunoglobulin E-mediated hypersensitivity in the Sudan to the "green nimitti" midge, *Cladotanytarsus lewisi, J. Allerg. Clin. Immunol.*, 66, 190, 1980.

33. Kay, A.B., MacLean, C.M., Wilkinson, A.H., Gad El Rab, M.O., The prevalence of asthma and rhinitis in a Sudanese community seasonally exposed to a potent airborne allergen, the "green nimitti" midge, *Cladotanytarsus lewisi, J. Allerg. Clin. Immunol.*, 71, 345, 1983.

34. Kagen, S.L., Yunginger, J.W., and Johnson, R., Lake fly allergy: incidence of chironomid sensitivity in an atopic population, *J. Allerg. Clin. Immunol.*, 73, 187, 1984.

35. Choovivathanavanich, P., Insect allergy: antigenicity of the cockroach and its excrement, *J. Med. Assoc. Thailand*, 57, 237, 1974.

36. Cornwell, P.B., *The Cockroach*, Hutchinson and Co., London, 1968, chap. 13.

37. Arlian, L.G., Arthropod allergens and human health, *Annu. Rev. Entomol.*, 47, 395, 2002.

38. Lillie, T.H. and Pratt, G.K., The hazards of ingesting beetle larvae, *USAF Med. Serv. Dig.*, 31, 32, 1980.

39. Jupp, W.W., A carpet beetle larva from the digestive tract of a woman, *J. Parasitol.*, 42, 172, 1956.

40. Okumura, G.T., A report of canthariasis and allergy caused by *Trogoderma, Calif. Vector Views*, 14, 19, 1967.

41. Htut, T., Higenbotta, T.W., Gill, G.W., Darwin, R., Anderson, P.B., and Syed, N., Eradication of house dust mites from the homes of atopic asthmatic subjects: a double blind study, *J. Allerg. Clin. Immunol.*, 107, 55, 2001.

42. Morgan, W.J., Crain, E.F., Gruchalla, R.S., O'Connor, G.T., Kattan, M., Evans, R., III, Stout, J., Malindzak, G., Smartt, E., Plaut, M., Walter, M., Vaughn, B., and Mitchell, H., Results of a home-based environmental intervention among urban children with asthma, *New Engl. J. Med.*, 351, 1068, 2004.

CHAPTER 3

STINGS

TABLE OF CONTENTS

I. INTRODUCTION AND MEDICAL SIGNIFICANCE

As discussed in the first chapter, stings by venomous arthropods can produce direct effects in humans by the toxic action of the venom alone or indirect effects due to allergic reactions (Table 1). Direct toxic effects are very rare but may include cerebral infarction, neuropathies (even optic), and seizures.[1-3] In addition, secondary infection may arise from stings, especially if the lesion is scratched (Figure 1). The direct effects of a sting can be mild such as pain, itching, wheal, flare, etc., or can be serious when numerous stings are received and the large amount of venom injected produces toxic effects. Small children are at a higher risk for developing severe toxicity because of their smaller body weight. One account of a toxic reaction in a child from massive hornet stings described clinical features such as coma, respiratory failure, coagulopathy, renal failure, and liver dysfunction.[4] But for most individuals, the risk of a severe reaction resulting from either a toxic or allergic mechanism is quite low. Lightning claims more lives annually in the U.S. than stinging arthropods,[5] and adverse reactions to penicillin kill seven times as many.[6]

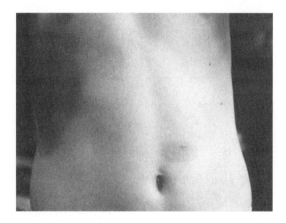

Figure 1
Secondary infection of the fire ant sting lesion due to scratching.

Table 1
Various Reactions to Insect Stings

Reactions	Response	Comments
Normal	Local pain; itching; swelling	Generally subsides in 2 h
Large local	Extensive swelling	Subsides in several days
Anaphylaxis	Life-threatening shock; difficulty in breathing	Immediate medical attention required
Toxic reactions	Headache; vomiting; diarrhea; shock	Nonallergic, caused by direct effect of many stings
Unusual syndromes	Serum sickness Vasculitis Neuritis Reversible renal diseases	May result from multiple stings or sting in or near a nerve

Many arthropods can sting, but several groups are notorious offenders. Parrish[7] analyzed fatalities due to venomous animals in the U.S. from 1950 to 1959, and found that 229 of 460 recorded deaths were due to the stings of yellowjackets, other wasps, ants, and bees. Honey bees, being practically ubiquitous, accounted for most of those numbers. In other countries, combined with these hymenopterans, scorpions cause significant mortality. One reference[8] lists 20,352 deaths from scorpion stings in Mexico alone during the periods 1940–1949 and 1957–1958. In Brazil, there are more than 5000 reported stings and 48 deaths from scorpions each year.[9]

II. PATHOLOGY PRODUCED BY ARTHROPOD STINGS

Alexander[10] described a typical hymenopteran sting (excluding ants) as a central white spot marking the actual sting site surrounded by an erythematous halo. The entire lesion generally is a few square centimeters in area. He also reported an initial rapid dermal edema with neutrophil and lymphocyte infiltration. Plasma cells, eosinophils, and histiocytes appear later. Lesions produced by fire ant stings are characterized by a central wheal with surrounding erythema, followed by the development of a vesicle, and finally a pustule[10] (see Color Figure 3.1 through Color Figure 3.3). According to Caro et al.[11] the pustules are thin-roofed, contain polymorphonuclear cells, and lymphocytes after 24 h; and eosinophils, plasma cells, polymorphonuclear cells, and lymphocytes after 72 h. Of course, the histopathology of arthropod stings varies with the insect and whether or not the victim has preexisting antibody to an insect venom. Most large local reactions reflect the presence of IgE antibody. DeShazo et al.[12] described such reactions in detail in studies of fire ant stings. Whereas the typical wheal and flare reactions followed by a sterile pustule were composed of a nonspecific cellular infiltrate; the erythematous, indurated, and pruritic large local reactions occurring in individuals with venom-specific IgE consisted of an eosinophil-rich mixed cellular infiltrate with densely polymerized fibrin. The fibrin gel structure is manifested by the edematous, indurated quality of these lesions, which take 3 to 5 d to resolve.

III.   STINGING BEHAVIOR

Most stinging wasps and some bees are solitary or subsocial insects, and they use their stings primarily for subduing prey. This *offensive* use of stinging and venom by these species rarely leads to human envenomization, except in a few cases of inadvertent or deliberate handling of the specimens. These venoms generally cause slight and temporary pain to humans.

The social wasps, bees, and ants are a different story. They use the sting primarily as a defensive weapon, and their venom causes intense pain in vertebrates. Workers (sterile, female insects) of all these groups instinctively defend their nest. Figure 2 illustrates how quickly fire ants attack their victims. One study showed that 51% of people were stung by fire ants within 3 weeks of summertime exposure in an infested area.[13] Encountering a single bee, wasp, or ant out foraging for food is generally not dangerous and will not usually result in a sting. However, walking too near a nest will elicit rapid defensive stinging behavior by numerous guard bees, wasps, or ants. In yellowjackets, the numerical response is in proportion to the extent of the disturbance and the defensive flight is brief (1.5 to 5 min) and usually confined to a radius of about 7 m around the nest.[10]

IV.   MORPHOLOGY OF THE STING APPARATUS

In all stinging wasps, bees, and ants the stinger is a modified ovipositor, or egg-laying device, that may no longer function in egg laying. Accordingly, in the highly social Hymenoptera only a queen or other reproductive caste member lays eggs; the workers gather food, conduct other tasks, and sting intruders.

A typical ovipositor (nonstinging device) consists of three pairs of elongate structures, called *valves*, which can insert the eggs into plant tissues, soil, etc. One pair of the valves makes up a sheath and is not a piercing structure, whereas the other two pairs form a hollow shaft that can pierce substrate in order for the eggs to pass down through. Two accessory glands within the body of the female inject secretions through the ovipositor to coat the eggs with a gluelike substance.

Figure 2
How fast fire ants attack a doll placed against a tree: (A) 3 sec, (B) 10 sec, (C) 30 sec.

For the stinging configuration, the ovipositor has several modifications to enable a stinging function (Figure 3 and Figure 4). The genital opening from which the eggs pass is anterior to the sting apparatus, which is flexed up out of the way during egg laying. Also, the accessory glands have been modified. One now functions as a venom gland and the other (the Dufour's gland) is important in production of pheromones. The venom gland leads to a venom reservoir or poison sac, which may contain up to 0.1 ml of venom in some of the larger hymenopterans.

The stinger itself is well adapted for piercing the skin of vertebrates. In the case of yellowjackets there are two lancets and a median stylet that can be extended and thrust

Figure 2 (continued)
How fast fire ants attack a doll placed against a tree: (A) 3 sec, (B) 10 sec, (C) 30 sec.

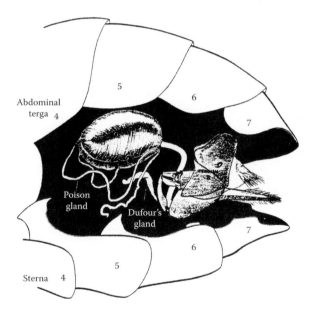

Figure 3
Cutaway view of yellowjacket sting apparatus. (From USDA Agriculture Handbook No. 552.)

into a victim's skin (Figure 5). Penetration is not a matter of a single stroke, but instead by alternate forward strokes of the lancets sliding along the shaft of the stylet. The tips of the lancets are slightly barbed (and actually recurved like a fishhook in the case of honey bees) so that they are essentially sawing their way through the victim's flesh. Contraction of the venom sac muscles injects venom through the channel formed by the lancets and shaft. The greatly barbed tip of the lancets in honey bees prevents the stinger from being withdrawn from vertebrate skin; thus, the entire sting apparatus is torn out as the bee flies away. Other hymenopterans, on the other hand, can sting repeatedly.

Figure 4
Microscopic view of end of wasp abdomen showing sting apparatus.

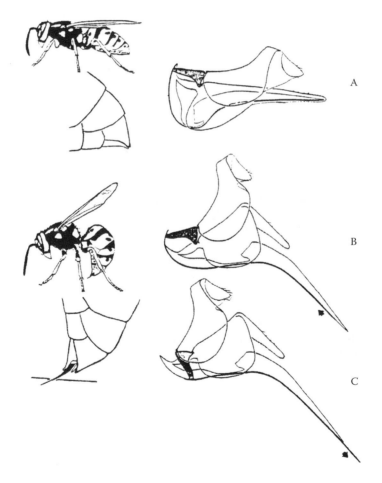

Figure 5
Lateral views of yellowjacket worker sting apparatus, retracted and extended during maximum thrust. (From USDA Agriculture Handbook No. 552.)

Table 2
Protein Composition of Selected Hymenoptera Venoms

	Honey Bees	Vespid Wasps	Fire Ants
Enzymes			
Phospholipase A_1	–	+	+[a]
Phospholipase A_2	10–12%	–	
Phospholipase B	1%	+	+
Hyaluronidase	1–2%	+	–
Acid phosphatase	15%	–	+
Alkaline phosphatase	+	–	
Lipase	–	+	
Esterase	+	+	
Protease	–	+	–
Peptides			
Hemolysins	Melittin 40%	+	
Mastolytic peptides	MCD peptide 2%	Mastoparans	
Neurotoxins	Apamin 3%	+	
Antigen 5	–	+	+
Kinins	–	+	
Group specific allergens	+	+	+

Note: + = present, – = absent, no symbol = not investigated.

a Specificity unknown.

Source: Adapted from Schmidt, J.O., Clin. Exp. Allergy, 24, 511, 1994. With permission.

V. VENOM COMPONENTS AND ACTIVITY

Table 2 lists some of the active constituents of the vespid wasp, honey bee, and fire ant venoms, adapted from Schmidt.[14] Table 3 shows the common names of these allergens based on World Health Organization (WHO) nomenclature. Venoms are highly complex mixtures of pharmacologically and biologically active agents. Because some venoms are similar, there may be cross-reactivity reactions in humans, but not always. Histamine is the most predominant low-molecular-weight component. Serotonin, dopamine, noradrenalin, and acetylcholine are also present. The amount of serotonin seems to be directly related to the painfulness of the sting. Melittin is a protein polypeptide toxin that is a primary constituent of honey bee venom. It is a direct agent of hemolysis. Apamin is one of the smallest polypeptides known (molecular weight: 2038). It is a neurotoxin and its interaction with the spinal cord is well established. MCD (mast cell degranulating) peptide, as a mastocytolytic agent, is very effective in releasing histamine. Interestingly, MCD peptide only comprises approximately 2% of bee venom but can produce the same effects as melittin, which comprises as much as 50% of bee venom. Kinins occupy an intermediary position between biogenic amines and high-molecular-weight compounds. The relative importance of kinins in envenomization is yet to be clarified. Phospholipase A is an enzyme that can attack structural

Physician's Guide to Arthropods of Medical Importance, Fifth Edition

Table 3

World Health Organization (WHO) Nomenclature for Some Common Fire Ant, Vespid Wasp, and Bee Allergens

Arthropod	Allergen	Common Name
Fire ants	Sol i I	Phospholipase
	Sol i II	—
	Sol i III	Antigen 5 group
	Sol i IV	—
Vespid wasps	Dol m I	Phospholipase A$_1$
	Dol m II	Hyaluronidase
	Dol m III	Acid phosphatase
	Dol m V	Antigen 5
Honey bees	Api m I	Phospholipase A$_2$
	Api m II	Hyaluronidase
	Api m III	Melittin
	Api m IV	Acid phosphatase

Source: Adapted in part from Moffitt, J.E. and deShazo, R.D., Allergic and Other Reactions to Insects, in Rich's Clinical Immunology Principles and Practice, 2nd ed., Rich, R.R., Fleisher, W.T., Kotzin, B.L., and Schroeder, H.W., Jr., Eds., Mosby, New York, 2001, p. 47.6.

phospholipids, resulting in damage to biological membranes, mitochondria, and other cellular constituents. There are two types of phospholipase — A$_1$ and A$_2$ (see Table 3 for current names for these allergens). Phospholipase A$_2$ is present in honey bee venom, whereas vespid (wasps, yellowjackets, and hornets) venoms contain phospholipase A$_1$. Hyaluronidase is a spreading factor that opens the way for other venom components to move through host tissues. It works by hydrolyzing hyaluronic acid, which resists the spread of harmful substances through epithelial and connective tissue.

As for which of these venom components serve as allergens, honey bee allergens are phospholipase A$_2$, hyaluronidase, acid phosphatase, and melittin. Allergens contained in vespid venoms are phospholipase A$_1$, hyaluronidase, and a protein called antigen 5.[15]

Imported fire ants (IFA) possess a venom of an alkaloid nature, which is apparently the first venom of animal origin recognized to be of this type. IFA venom exhibits potent necrotoxic activity. About 95% of their venom consists of water insoluble 6-*n*-alkyl, or alkenyl, 2-methyl piperidines, which do not produce allergic reactions in people but are responsible for pain and pustule formation (Figure 6). The other portion of fire ant venom is an aqueous solution of proteins, peptides, and other small molecules that produces the allergic response in hypersensitive individuals. Of the proteins found, four allergens have been isolated (from *Solenopsis invicta*) — Sol i I, Sol i II, Sol i III, and Sol i IV.[16,17] Some patients react to every combination of the four allergens. Sol i II, Sol i III, and Sol i IV are not immunologically related to bee or wasp venom proteins, so there is no cross-reactivity with bee or wasp venoms. However, people sensitized to other Hymenoptera venoms do show *in vitro* sensitivity to Sol i I. The clinical interpretation of this finding is unclear. This may explain how someone living outside an IFA area could have a systemic reaction to their first fire ant sting upon entering a geographic area containing IFAs.

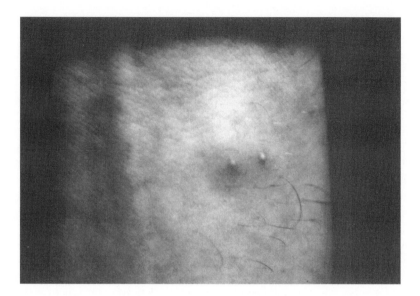

Figure 6
Pustules resulting from fire ant stings.

Harvester ant venom is made up of approximately 70% proteins, including phospholipase A$_2$ and phospholipase B, hyaluronidase, lipase, acid phosphatase, esterases, and others. In the harvester ant species *Pogonomyrmex badius*, high levels of lipase have been found.

Scorpion venom is primarily neurotoxic, containing multiple low-molecular-weight basic proteins (the neurotoxins), mucus (5 to 10%), salts, and various organic compounds, such as oligopeptides, nucleotides, and amino acids.[18] It contains little or no enzymes. Scorpion sting reactions may be either localized (and transitory) or systemic. Localized responses are characterized by pain and swelling, similar to that of a wasp or bee sting. In some localized responses, the sting site might develop into an indurated lesion, even from harmless scorpion species.[19] These local reactions may persist up to 72 h and be followed by development of blood blisters at the sting site. Systemic reactions are highly variable and may not necessarily be life-threatening. Symptoms may range from aching, burning, and numbness in various tissues, to slurred speech, tightness in the chest, and rapid heartbeat. More severe systemic reactions may include profuse sweating, respiratory problems, and convulsions. Death, when it does occur, is often due to cardiac or respiratory failure.[19]

REFERENCES

1. Day, J.M., Death due to cerebral infarction after wasp stings, *Arch. Neurol.*, 7, 184, 1962.
2. Fox, R.W., Lockey, R.F., and Bukantz, S.C., Neurologic sequelae following the imported fire ant sting, *J. Allerg. Clin. Microbiol.*, 70, 120, 1982.
3. Maltzman, J.S., Lee, A.G., and Miller, N.R., Optic neuropathy occurring after bee and wasp sting, *Ophthalmology*, 107, 193, 2000.
4. Watemberg, N., Weizman, Z., Shahak, E., Aviram, M., and Maor, E., Fatal multiple organ failure following massive hornet stings, *Clin. Toxicol.*, 33, 471, 1995.

5. Camazine, S., Hymenopteran stings: reactions, mechanisms, and medical treatment, *Bull. Entomol. Soc. Am.*, Spring 1988 Issue, 17, 1987.

6. Idsoe, O., Guthe, T., Wilcox, R.R., and De Weck, A.L., Nature and extent of penicillin side-reactions, with particular reference to fatalities from anaphylactic shock, *Bull. WHO*, 38, 129, 1968.

7. Parrish, H.M., Analysis of 460 fatalities from venomous animals in the U.S., *Am. J. Med. Sci.*, 245, 129, 1963.

8. Mazzotti, L. and Bravo-Becherelle, M.A., Scorpionism in the Mexican republic, in *Venomous and Poisonous Animals and Noxious Plants in the Pacific Region*, Keegan, H.L. and MacFarlane, W.V., Eds., Pergamon Press, Oxford, 1963, p. 119.

9. Warrell, D.A., Venomous bites and stings in the tropical world, *Med. J. Aust.*, 159, 773, 1993.

10. Alexander, J.O., *Arthropods and Human Skin*, Springer-Verlag, Berlin, 1984, chap. 10.

11. Caro, M.R., Derbes, V.J., and Jung, R.C., Skin responses to the sting of the imported fire ant, *Arch. Dermatol.*, 75, 475, 1957.

12. deShazo, R.D., Griffing, C., Kwan, T.H., Banks, W.A., and Dvorak, H.F., Dermal hypersensitivity reactions to imported fire ants, *J. Allerg. Clin. Immunol.*, 74, 841, 1984.

13. Tracy, J.M., Demain, J.G., Quinn, J.M., Hoffman, D.R., Goetz, D.W., and Freeman, T., The natural history of exposure to the imported fire ant, *J. Allerg. Clin. Immunol.*, 95, 824, 1995.

14. Schmidt, J.O., Let's not forget the crawling hymenoptera, *Clin. Exp. Allerg.*, 24, 511, 1994.

15. Reisman, R.E., Insect stings, *New Engl. J. Med.*, 331, 523, 1994.

16. Hoffman, D.R., Dove, D.E., and Jacobson, R.S., Allergens in Hymenoptera venom XX, isolation of four allergens from imported fire ant venom, *J. Allerg. Clin. Immunol.*, 82, 818, 1988.

17. deShazo, R.D., Butcher, B.T., and Banks, W.A., Reactions to the stings of the imported fire ant, *New Engl. J. Med.*, 323, 462, 1990.

18. Polis, G.A., *The Biology of Scorpions*, Stanford University Press, Stanford, CA, 1990, chap. 10.

19. Mullen, G.R. and Stockwell, S.A., Scorpions, in *Medical and Veterinary Entomology*, Mullen, G.R. and Durden, L.A., Eds., Academic Press, New York, 2002, p. 411.

CHAPTER 4

BITES

TABLE OF CONTENTS

I. INTRODUCTION

Bites by arthropods can be as medically significant as stings, especially for hypersensitive individuals. In this chapter, I have lumped all bites together for discussion. However, the term *bite* probably should be restricted in meaning to purposeful biting by

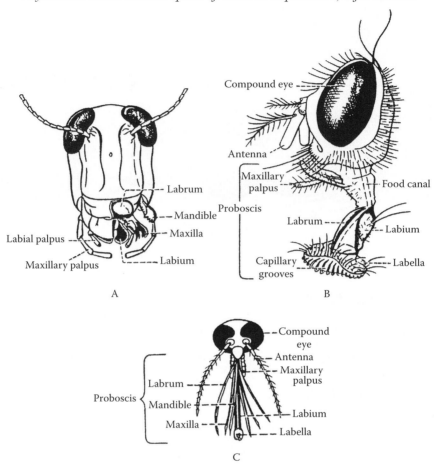

Figure 1
(A) Chewing, (B) sponging, and (C) piercing–sucking mouthparts. (From U.S. DHHS CDC Publ. No. 83:8297.)

species for either catching prey or blood feeding, and not to accidental or inadvertent biting by plant-feeding species. Phytophagous or predaceous insect species may "bite" in self-defense, piercing the skin with their proboscis, but the injury is actually a stab wound and not a true bite.

The method of obtaining blood differs among blood-feeding arthropods. Some species, such as bed bugs, kissing bugs, and fleas, obtain blood directly from venules or small veins — a method termed *solenophagy*. Others, such as ticks, horse flies and deer flies, black flies, and tsetse flies, obtain blood by lacerating blood vessels and feeding from the pool of blood thus formed — a method termed *telmophagy*. The method of blood feeding likely plays a significant role in whether a species is able to acquire and transmit pathogenic microorganisms.

II. MOUTHPART TYPES

Insect mouthparts, at least in the medically important species, can be generally divided into three broad categories: (1) biting and chewing (Figure 1A), (2) sponging (Figure 1B),

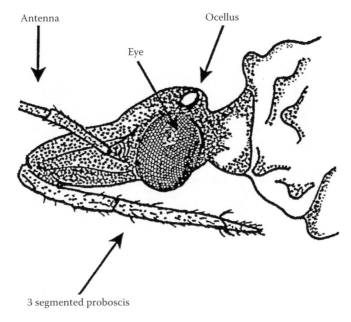

Antenna

Ocellus

Eye

3 segmented proboscis

Figure 2
True bugs (order Hemiptera) often have a three- or four-segmented cylindrical proboscis.

and (3) piercing–sucking (Figure 1C). Within these categories there are numerous adaptations and specializations among the various insect orders. The biting and chewing mouthpart types, such as those in food pest insects, and sponging mouthpart types, found in the filth fly groups, are of little significance regarding human bites; but the piercing–sucking mouthparts, especially the bloodsucking types, are of considerable importance. Insect piercing–sucking mouthparts vary primarily in the number and arrangement of the stylets, which are needlelike blades, and the shape and position of the lower lip of insect mouthparts, the labium. Often, what is termed the *proboscis* of an insect with piercing–sucking mouthparts is an ensheathment of the labrum, stylets, and labium. These mouthparts are arranged in such a way that they form two tubes. One tube is usually narrow, being a hollow pathway along the hypopharynx, and the other is wider, formed from the relative positions of the mandibles or maxillae. Upon biting, saliva enters the wound via the narrow tube, and blood returns through the wider tube by action of the cibarial or pharyngeal pump.

The true bugs (order Hemiptera), such as bed bugs, kissing bugs, and assassin bugs, have the labium formed into a three- or four-segmented cylindrical proboscis (Figure 2). There are four stylets formed from the mandibles and maxillae. In some cases, as in bed bug feeding, the labium folds up above the skin surface, allowing the fascicle (the stylets linked together in a complete unit) to penetrate the skin for feeding.

Sucking lice (order Phthiraptera, suborder Anoplura) do not have an elongated proboscis, but they do have piercing mouthparts that contain several recurved hooks that serve to anchor the mouthparts during feeding. There are no palps. Three stylets make up the fascicle in anoplurans; they are pushed into host tissues by muscular action in the act of biting. Two of the stylets represent the mandible and maxilla, and the other forms the tube functioning as a salivary duct. Salivary secretions are released into the wound, and the pharyngeal pump begins to draw blood into the digestive tract via the food duct.

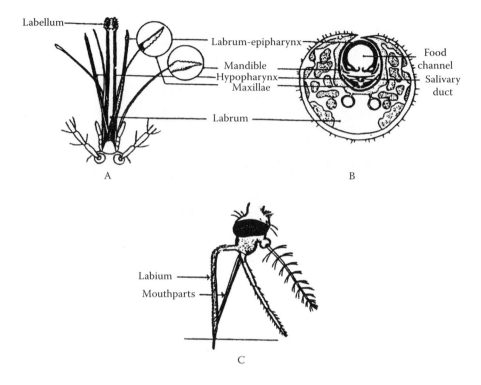

Figure 3
(A) Mosquito mouthparts, (B) cross section of a fascicle, and (C) mouthparts position upon biting. (From U.S. Navy Laboratory Guide to Medical Entomology.)

Fleas (order Siphonaptera) have a pair of stylets composed of maxillary laciniae that pierce host skin. There is also an unpaired stylet formed from the epipharynx. All three structures form the fascicle. The laciniae are bladelike and produce the wound. Blood is then pumped into the pharynx by pharyngeal and cibarial pumps.

Not all flies have piercing–sucking mouthparts, but in those that do, there is considerable variation. Mosquitoes have six stylets (two mandibles, two maxillae, the hypopharynx, and labrum–epipharynx) ensheathed in an elongated, cylindrical labium. This combined structure forms the prominent proboscis of mosquitoes (Figure 3A, Figure 3B, and Figure 4). Upon biting, only the fascicle is inserted into the host's skin; the labium folds up above the skin surface (Figure 3C).

Although relatively short, sand fly mouthparts are efficient organs of penetration.[1] The labrum is broad and tapering to a point and bears a number of small spiny projections. Only the female has mandibles, which take the form of blades bearing numerous small serrations. The maxillary stylets, formed by the galeae, are also bladelike and bear numerous fine denticles on their inner margins and a few additional teeth near the end on their outer margins. The labium is elongated and channeled to carry the stylet and hypopharynx. It has a pair of soft labellar lobes located distally. At the sides of the proboscis are the two maxillary palps, which are longer than the proboscis itself.

Horse flies, deer flies, black flies, and biting midges basically have the same type of mouthparts — scissorlike (Figure 5). Most of the stylets are much flattened compared to those of mosquitoes. In addition, the mandibles are flattened and move in a

Figure 4
Microscopic view of mosquito proboscis.

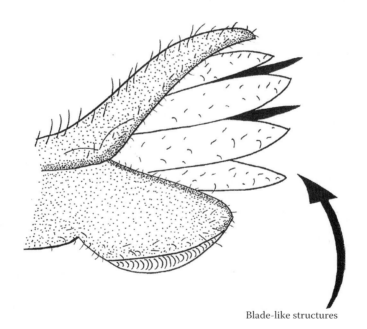

Blade-like structures

Figure 5
Horse fly mandibles are large and bladelike.

transverse manner. Upon feeding, the mandibles move in a scissorlike fashion and the maxillae are thrust in and out of the wound causing pooled blood in the host's tissues. A more specific description of horse fly mouthparts is provided by Snow.[1] The horse fly labrum is relatively soft with a blunt tip and is not used for penetration. However, the mandibles are large and bladelike; these and the styliform maxillary galeae are the piercing organs. Once these mouthparts have made a wound and fluids are flowing out of it, the labium, which does not enter the wound itself but kinks up on the skin's surface, draws up blood by means of pseudotracheae on the labellar lobes.

Stable flies have a hardened, conspicuous proboscis (primarily the labium) that can clearly be seen held out in front of the head. The proboscis has the small labella at its tip composed of everted rasping teeth. With a hard thrust, aided by twisting of the labella, the anterior end of the labium is inserted into the host's skin. The hardened long proboscis enables stable flies to inflict painful bites even through socks, sleeves, and other tight-fitting clothing.

Tsetse flies apply pressure to the skin surface of their hosts with their proboscis. Rasps and teeth on the labellum aid the labium in penetrating the skin. Back-and-forth movements of the fly's head enable the labium to rupture capillaries in the skin. Blood is then rapidly sucked into the food canal of the labrum by the cibarial pump located in the fly's head. During the feeding process saliva enters the wound via the salivary canal of the hypopharynx.

Some noninsect arthropods such as spiders, mites, and ticks have piercing–sucking mouthparts, although the structures are derived from different morphological features than those of insect mouthparts. Mites and ticks (subclass Acari, sometimes also called *acarines*) have a feeding structure, or gnathosoma, that is headlike and may be mistaken for a true head. The gnathosoma, consisting of mouthparts and palps, is arranged to form a tubular structure for obtaining food and passing it into the digestive tract. The cutting–piercing mouthpart structures of acarines are the chelicerae. Chelicerae may be used for tearing, as in the case of scabies mites, or piercing, as in the case of chiggers. In ticks, there is also an anchoring hypostome, which is a very prominent structure and bears teeth on its ventral surface.

III. PATHOLOGY OF INSECT BITES

Arthropod bites basically consist of punctures made by the mouthparts of bloodsucking insects; the actual mechanical injury to human skin is generally minimal. However, in the case of horse flies there may be more tissue damage owing to their *slashing–lapping* feeding method and their large size. By far the most lesions on human skin are produced by the host's immune reactions to the offending arthropod salivary secretions or venom.[2] Sometimes hypersensitivity develops to antigens found in insect saliva.[3] Arthropod saliva is injected while feeding to: (1) lubricate the mouthparts upon insertion, (2) increase blood flow to the bite site, (3) inhibit coagulation of host blood, (4) anesthetize the bite site, (5) suppress the host's immune and inflammatory responses, and (6) aid in digestion. Upon repeated exposure to a particular arthropod's salivary secretions, humans may become allergic to its bites. After becoming sensitized, it is not unusual to have large local reactions to bites by that particular arthropod.

Arthropod bites are often characterized by urticarial wheals, papules, vesicles, and less commonly, blisters. After a few days, or even weeks, secondary infection, discoloration, scarring, papules, or nodules may persist at the bite site.[2] Figure 6 shows secondary infection resulting from scratching a mosquito bite. Complicating the picture further is the development of late cutaneous allergic responses in some atopic individuals. Diagnosis may be especially difficult in the case of biopsies of papules or nodules. Biopsy specimens can reveal a dense infiltrate of a mixture of inflammatory cells; composed variably of lymphocytes, plasma cells, macrophages, neutrophils, or eosinophils. Lesions containing a majority of uniform lymphocytes are occasionally mistaken for a

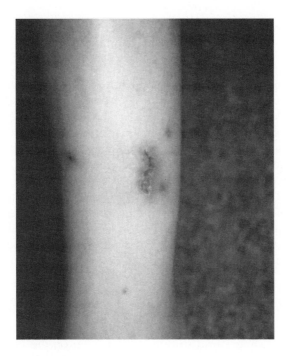

Figure 6
Secondary infection resulting from mosquito bites.

lymphomatous infiltrate. If the infiltrate is predominantly perivascular and extending throughout the depths of the dermis, the lesion might be confused with lupus erythematosus. Eosinophils are frequently seen in papules or nodules resulting from arthropod bites. There may be a dense infiltration of neutrophils, resembling an abscess. Occasionally remains of arthropod mouthparts elicit a foreign body granulomatous response with giant cells. Scabies mites occur in the stratum corneum and can usually be seen upon microscopic examination. New lesions from scabies such as papules or vesicles are covered by normal keratin, whereas older lesions have a parakeratotic surface.[4] There may also be a perivascular infiltrate of lymphocytes, histiocytes, and eosinophils.[4]

Histopathological studies of late cutaneous allergic responses have revealed mixed cellular infiltrates, including lymphocytes, polymorphonuclear leukocytes, and some partially degranulated basophils. A prominent feature of late cutaneous allergic reactions is fibrin deposition interspersed between collagen bundles in the dermis and subcutaneous tissues.

Infectious complications. Secondary infection with common bacterial pathogens can occur in any lesion in which the integrity of the dermis is disrupted, whether by necrosis or excoriation.[5] Infection may result in cellulitis, impetigo, ecthyma, folliculitis, furunculosis, and other manifestations. Three findings may be helpful in making the diagnosis of secondary bacterial infection[5]:

- Increasing erythema, edema, or tenderness beyond the anticipated pattern of response of an individual lesion suggests infection.
- Regional lymphadenopathy can be a useful sign of infection, but it may also be present in response to the primary lesion without infection.
- Lymphangitis is the most reliable sign and suggests streptococcal involvement.

IV. CLUES TO RECOGNIZING INSECT BITES

A. Diagnosis

If a patient recalls no insect or arachnid exposure, arthropod bites may pose frequent difficulty in diagnosis. However, bites should be considered in the differential diagnosis of any patient complaining of itching. Diagnosis of insect bites depends on (1) maintaining a proper index of suspicion in this direction (especially during the summer months), (2) a familiarity of the insect fauna in one's area, and (3) obtaining a good history.[6] It is very important to find out what the patient has been doing lately, e.g., hiking, fishing, gardening, cleaning out a shed, etc. However, even history can be misleading, because patients may present a lesion that they think is an insect bite, when in reality the correct diagnosis is something like urticaria, folliculitis, or delusions of parasitosis. Physicians need to be careful not to diagnose *insect bites* based on lesions alone and should call upon entomologists to examine samples. Other clues that might be helpful in diagnosing insect bites include number of lesions and location on the body (see Section C).

B. Characteristics of Lesions

True bugs. Members of the Hemiptera produce variable lesions upon biting. Bed bug bites are multiple, linear in distribution, and often on the head, neck, and chest. They are characterized by welts, inflammation, and often purpuric spots. Alexander[4] reported that red blotches progressing to form large urticarial wheals are frequent reactions to *Cimex* bites and that bullous lesions are not uncommon. Assassin bugs (including wheel bugs) inflict a bite that is immediately and intensely painful. There may be a small red spot or vesicle at the bite site and considerable swelling; numbness may also occur. The kissing bugs generally have a painless bite, but they may produce four distinct reactions depending on sensitivity: (1) a papule with a central punctum (breathing hole in the skin), (2) small vesicles grouped around the bite site with swelling and little redness, (3) large urticarial lesion with a central punctum and surrounding edema, or (4) hemorrhagic nodular-to-bullous lesions.

Lice. Pubic lice bites produce intense pruritus and discoloration of the skin (bluish gray maculae) if the infestation is long-standing. Other than that, individual lesions are usually not discernable. Head lice bites may be pustular in appearance. Secondary infection such as impetigo contagiosa on the scalp, head, or neck may indicate head lice infestation. The direct effect of body lice feeding is intense irritation, probably due to proteins in their saliva. This leads to widespread excoriation. The usual clinical presentation for body lice is pyoderma in covered areas. Uninfected bites present as small papules or puncta on an erythematous base.[7] In most clinical settings, the patient may already be undressed when the physician gets to see him, and only the ulcerations, infections, and excoriations are present — the acutal lice are to be found in the seams of the patient's clothing. In general, primary sensitization to body lice is attained 3 to 8 months after the original infestation. From that point on, feeding by the lice will cause intense itching. Secondary sensitization may develop some 12 to 18 months after the initial infestation, resulting in a systemic reaction to louse bites characterized by a feeling of malaise and pessimistic frame of mind. The body-lice-infested person becomes apathetic if left alone and irritable if roused (this is the origin of the term *feeling lousy*). In addition, the skin of people who continually harbor body lice becomes hardened and darkly pigmented, a condition known as *vagabond's syndrome.*

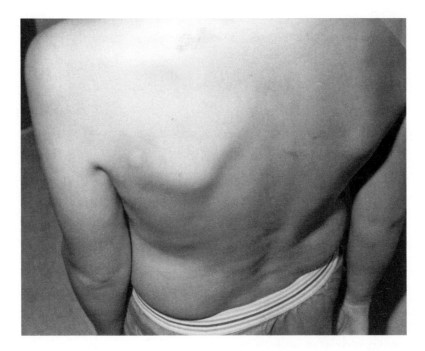

Figure 7
Wheal and flare on shoulder blade caused by mosquito bite 30 min earlier.

Fleas.   Flea bites often occur in irregular groups of several to a dozen or more. The typical flea bite on a human consists of a central spot surrounded by an erythematous ring. There is usually little swelling, but the center may be elevated into a papule, vesicle, or bulla.[4] Papular urticaria is seen in persons with chronic exposure to flea bites. The lesions appear in crops and all stages can be seen simultaneously — fresh wheals, persistent papules, vesicles, scratch marks, exudation, encrustation, and often secondary infection.

Mosquitoes and biting midges.   Mosquitoes and biting midges (Ceratopogonidae) cause itching and welts by their bites. In many individuals the typical wheal and flare reaction is evident approximately 30 min after the bite (Figure 7 and Color Figure 4.4). The size and extent of the lesions are largely a result of the species of insect involved, the sensitivity status of the individual, and age of the individual. Delayed, small but pruritic papules are common after mosquito bites. These generally appear several hours after the bite and usually last 1 to 3 d, but occasionally persist for weeks. During the spring and summer months, children who play outdoors often exhibit numerous mosquito bite lesions of varying ages (Figure 8).

Horse flies and deer flies.   Female horse flies and deer flies (family Tabanidae) have broad, scissorlike mouthparts that can slash a deep painful wound in their host's skin. Bites by tabanids often become secondarily infected, leading to cellulitis. However, in the absence of allergy or secondary infection, there is generally no reddening or swelling.

Black flies.   Black flies often produce reddened, itching papules. Subsequently, multiple nodules may develop. It is not uncommon for the whole affected area to swell markedly. The arms, legs, and face are common sites of attack, but a favorite site is the posterior cervical region, particularly at the base of the hair. In the case of black fly-transmitted onchocerciasis, the condition is extremely pruritic and widespread papules

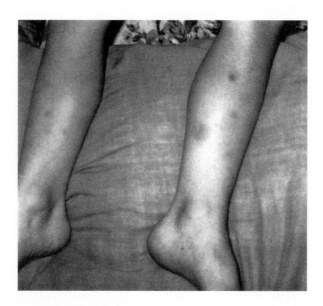

Figure 8
Lesions of varying ages due to mosquito bites.

and lichenification occur, particularly around the shoulders and upper arms, and buttocks and thighs. Microfilariae invade the dermis and destroy the elastic tissue so that the skin may hang in folds, especially in the groin area.

Sand flies.　Sand fly bites may produce itching papules, vesicles, wheals, and a condition closely resembling papular urticaria. Additionally, sand flies may transmit the parasite responsible for cutaneous leishmaniasis. In those cases, exposed skin areas are the most commonly bitten, resulting in papules and nodules that break down, ulcerate, and crust. There is usually healing within a year, but with considerable scarring.

Spiders and centipedes.　Widow spiders differ from violin spiders in regard to reactions to their bites. In the case of widow spiders, there is often a pinprick sensation initially. Two red puncture marks may be present at the bite site, and there may be a developing dull, numbing pain around the site that builds to a peak 1 to 3 h after the bite. Centipedes may also produce two puncture wounds at the bite site, but there is usually redness and swelling present with immediate intense pain. Violin spider bites are rarely, if ever, accompanied by pain initially. However, within 2 to 8 h there is mild to severe pain at the site, erythema, blanching, itchiness, and vesiculation.

Ticks.　Ticks will attach to human skin and remain attached for several days. The actual penetration is usually not painful, except in the case of some species with very long mouthparts. However, inflammation or even hypersensitivity reactions may occur after a few days of tick attachment (Figure 9 and Figure 10; Color Figure 4.5 and Color Figure 4.6). Even after tick removal, a reddened nodule may persist at the bite site for weeks or even months.

Mites.　Chigger bites usually lead to an intensely itchy dermatitis composed of pustules and wheals (Figure 11 and Color Figure 4.7. Lesions occur within 3 to 6 h after exposure to chigger-infested areas. Alexander[4] states that an edematous wheal with a tiny central vesicle may persist for 8 to 15 d.

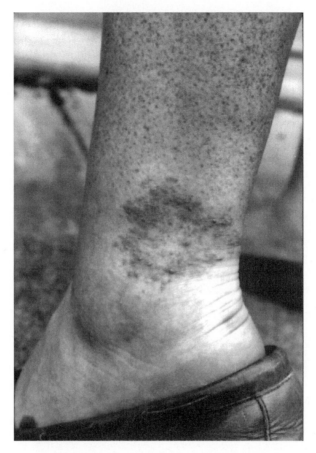

Figure 9
Hypersensitivity reaction resulting from tick bite — presumably a reaction to tick saliva. Tick bite was in the center of the lesion.

Bites by the straw itch mite produce itching a few hours after attack. Often, small urticarial papules with a tiny central vesicle (0.5 mm in diameter) mark the site of skin puncture. Other human-biting mites such as the tropical rat mite produce immediate sharp itching pain upon biting and cause small urticated papules.

Scabies mites have an affinity for the wrists and the areas between the fingers (although they certainly occur in other areas). Accordingly, scabies is often characterized by rashlike lesions that typically occur in tracks in the webbing between the fingers or on the wrists. Alexander[4] described the burrow as a grayish line resembling a pencil mark, about 5 mm long on average. Sometimes, however, scabies lesions may be singular (papules or vesicles) or even bullous. (Note: Rashes may occur over much of the body as a result of scabies infestation, but the rash may not correspond to the location of mite burrows.)

Cheyletiella mites (such as *C. parasitivorax, C. yasguri, C. blakei,* and *C. furmani*) produce scattered or localized urticae, papules, or 2 to 6 mm papulo-vesicles in the areas of skin where direct contact with an infested pet occurs — the flexor sides of arms, breasts, and abdomen.[8]

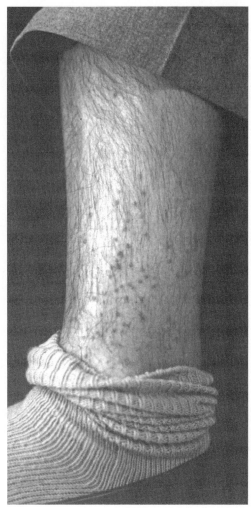

Figure 10
Seed tick bites around ankle, 4 d postattachment. (Photo courtesy of Tom Kilpatrick.)

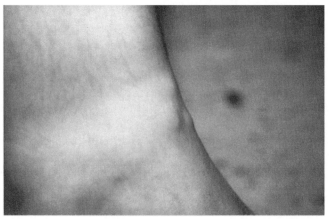

Figure 11
Chigger bites around ankle.

C. *Number of Lesions and Their Location on the Body*

The number of lesions on a patient and their location on the body may also provide clues to the diagnosis of arthropod bites. Table 1 and Table 2 provide information on arthropod bite possibilities based upon these two factors. In addition to the number of lesions and their location, the distribution or pattern of the bites may be diagnostic (Table 3). Residual effects of arthropod bites that occurred days or weeks earlier may also be classified (Table 4).

Table 1

Clues to Biting Arthropods Based on Location of Lesions on the Body

Location of Lesions	Possible Arthropods
Predominantly on left side of body (if right-handed) or right side of body (if left-handed)	Imaginary bugs (see Chapter 7)
Legs or feet	Fleas Mosquitoes Spiders Chiggers Centipedes
Trunk	Chiggers Bed bugs Scabies Ticks Body lice Pubic lice Spiders *Cheyletiella* mites
Genitals[a]	Scabies Chiggers
Arms or hands	Mosquitoes Black flies Mites Biting midges Spiders Fleas *Cheyletiella* mites Centipedes Sand flies Wheel bugs
Head, neck, or face	Mosquitoes Black flies Biting midges Sand flies Head lice Kissing bugs Bed bugs

[a] *The presence of crusted, pruritic papules on the penis and buttocks is highly indicative of scabies.*

Table 2
Clues to Biting Arthropods Based on Number of Lesions on the Body

Number of Lesions	Possible Arthropods
Single	Tick Spider Centipede Wheel bug Kissing bug
Few	Fleas Mosquitoes Stable flies Horse flies and deer flies Kissing bugs Sand flies
Multiple	Mosquitoes Black flies Biting midges (*Culicoides*) Fleas Lice Chiggers Seed ticks Mites Bed bugs Scabies

Table 3
Diagnostic Patterns of Arthropod Bites

Pattern of Bite Lesions	Possible Arthropods
Scattered	Mosquitoes Horse flies and deer flies Black flies Biting midges Head and body lice
Grouped	Fleas Pubic lice Chiggers Scabies
Linear	Bed bugs Chiggers

Source: *Adapted in part from Frazier, C.A., Insect Allergy: Allergic Reactions to Bites of Insects and Other Arthropods, Warren H. Green, St. Louis, 1969, chap. 9. With permission.*

Table 4
Cutaneous Sequelae Resulting from Arthropod Bites

Sequelae	Location on Body	Possible Arthropods
Single nodule	Scalp or trunk	Tick
Multiple nodules	Legs or ankles	Black fly
Purpuric spots	Trunk	Bed bug
Bluish spots	Pubic and perianal area	Pubic lice
Hyperpigmentation	Waistline and genitals	Chigger
Hyperpigmentation	Trunk	Body lice

Source: Adapted in part from Frazier, C.A., Insect Allergy: Allergic Reactions to Bites of Insects and Other Arthropods, Warren H. Green, St. Louis, 1969, chap. 9. With permission.

REFERENCES

1. Snow, K.R., *Insects and Disease*, John Wiley & Sons, New York, 1974, chaps. 4, 5.

2. Frazier, C.A., Diagnosis of bites and stings, *Cutis*, 4, 845, 1968.

3. Moffitt, J.E., Allergic reactions to insect stings and bites, *South. Med. J.*, 96, 1073, 2003.

4. Alexander, J.O., *Arthropods and Human Skin*, Springer-Verlag, Berlin, 1984, p. 251.

5. Kemp, E.D., Bites and stings of the arthropod kind, *Postgrad. Med.*, 103, 88, 1998.

6. Allington, H.V. and Allington, R.R., Insect bites, *JAMA*, 155, 240, 1954.

7. Elgart, M.L., Pediculosis, *Dermatol. Clin.*, 8, 219, 1990.

8. van Bronswijk, J.E.M.H. and de Kreek, E.J., *Cheyletiella* of dog, cat, and domesticated rabbit, *J. Med. Entomol.*, 13, 315, 1976.

9. Frazier, C.A., *Insect Allergy: Allergic Reactions to Bites of Insects and Other Arthropods*, Warren H. Green, St. Louis, 1969, chap. 9.

————————————————————————————*CHAPTER 5*

DERMATITIS, URTICARIA, AND BLISTERING FROM CONTACT WITH ARTHROPODS

TABLE OF CONTENTS

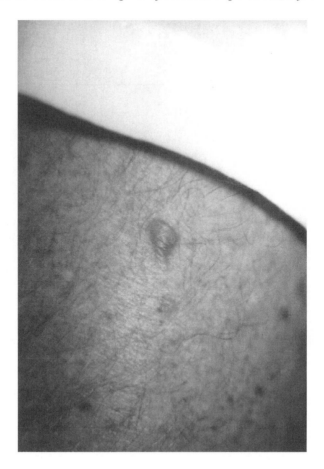

Figure 1
Blister on arm resulting from blister beetle exposure. (Photo courtesy of Ralph Turnbo.)

I. BLISTERING FROM EXPOSURE TO BLISTER BEETLES

A. Medical Significance

Most beetles are harmless and can be handled by humans with no ill effects. However, blister beetles, primarily species in the families Meloidae and Staphylinidae, possess vesicating chemical substances in their body fluids (see also Chapter 12). The blistering fluid in meloids is cantharidin, but may be a somewhat different compound in staphylinids. When the meloid beetles sense danger, they exude cantharidin by filling their breathing tubes with air, closing their breathing pores, and building up body fluid pressure until fluid is pushed out through one or more leg joints. This secretion or "venom" readily penetrates human skin and produces fluid-filled blisters in a few hours (Figure 1).

The blisters that result from exposure to live blister beetles or their dried pulverized body parts are generally few and self-limiting. Alexander[1] states that generally there is no pain or itch associated with the blisters as long as they do not rupture. Extensive blister beetle exposure may irritate the kidneys owing to the penetrating activity of cantharidin.[2] If ingested, the substance can cause nausea, diarrhea, and vomiting; horses have died from ingesting numerous striped blister beetles.[3]

Frazier and Brown[2] reported that a good indication of blister beetle attack is that the blisters are all in the same stage of development and there is no accompanying rash around them. In addition, there may be a line or track of blisters where the beetle crawled over the skin. Alexander[1] reported that areas commonly affected are the face, neck, chest, thighs, and calves, and the buttocks in children.

B.   Contributing Factors and Species

Blister beetle dermatoses are seasonal, with most cases reported in the late summer as the beetles emerge and feed in gardens and field crops. Three common blister beetles in the U.S. are the ash-gray and striped blister beetles in the central and southeastern U.S. and oil beetles (several species) in the southwestern U.S. Contact with the beetles may occur intentionally, such as in the case of a child picking one up, or accidentally, such as the beetle crawling on an unsuspecting victim sitting on the ground, working in the garden, etc. In southern Texas in late summer the author has observed hundreds of black oil beetles crawling over the ground in a random, solitary fashion.

C.   Management and Treatment

Blisters resulting from blister beetle exposure are generally not serious, and reabsorption usually occurs in a few days if the blisters are unruptured. If blisters are ruptured, the skin may flake and there may be a mild reddening of the area before clearing in a week or so.[2] For treatment, exposed areas should be washed, antibiotic ointments applied to prevent secondary infection, and the areas bandaged for protection until the blisters reabsorb.

II.   DERMATITIS AND URTICARIA FROM EXPOSURE TO LARVAL LEPIDOPTERANS

A.   Medical Significance

Caterpillars of many species can produce mild to severe dermatitis, urticaria, nodular conjunctivitis, pain, headache, and even convulsions (rare) via tiny hairs that irritate or inject a venom (Figure 2; see also Chapter 14).[4] According to Alexander,[1] visible lesions are usually preceded by a burning itch. The itch is often followed by the appearance of 2- to 5-mm tiny, rose-red maculopapules or papules. Keegan[5] has given an account of the direct injury produced by caterpillars and moths, and Alexander[1] provides an in-depth discussion of human reactions to lepidopterans. In the scientific literature, distinction is usually made between direct contact with urticating caterpillars (erucism) and contact with scales or hairs of adult butterflies or moths (lepidopterism) (see next section).

Relatively speaking, only a few lepidopterous species are involved in erucism. However, urticating caterpillars can become numerous, resulting in a significant health hazard. In the 1920s the San Antonio schools had to be closed owing to a large outbreak of puss caterpillars and the resulting numerous stings among the school-age children.[6]

Figure 2
Many caterpillars with thick or prominent spines can cause a sting similar to that of wasps.

The caterpillars were so numerous on posts, walls, and bushes that virtually any outdoor activity resulted in a sting.

Even a few urticating caterpillars can be a serious problem if a person has an eye lesion or exhibits acute systemic symptoms when exposed. The puss caterpillar is especially dangerous, sometimes causing cardiovascular and neurological manifestations. Frazier and Brown[2] report a case in which a young woman stepped barefoot on a puss caterpillar. Within minutes she experienced symptoms of shock with respiratory distress; upon arrival at the hospital she had hypotension and bradycardia. She was successfully treated for shock and pain but reported numbness in the foot and leg for several days. Eye lesions may be relatively benign when single pointed hairs are found on the palpebral conjunctiva (they can generally be removed by an ophthalmologist), or they can be extremely serious with loss of vision in the case of numerous deeply embedded hairs or spines that progressively penetrate deeper and deeper owing to eye movement and barbs on the hairs.[1]

B.   Contributing Factors and Species

Both mechanical irritation and injection of venom by the caterpillar spines or hairs contribute to the urticarial response in humans. Many of the hairs, setae, or spines of the offending species are hollow and connected to venom glands. Upon exposure, these structures may be broken, allowing venom to either be injected into the skin or to ooze out onto abraded skin. In addition, in a study of gypsy moth caterpillars, whole first instar larvae contained an average of 17.3 ng of histamine, and the hairlike setae of the fifth instar larvae contained 80 ng of histamine per organism.[7] Accordingly, histamine release, as well as a hypersensitivity reaction, likely play a role in urticaria and dermatitis associated with these caterpillars. Four common species of Lepidoptera that have caterpillars which can sting upon exposure to human skin are the saddleback caterpillar, the IO moth caterpillar, the brown-tail moth caterpillar, and the puss caterpillar. Although not true "stinging" caterpillars, the gypsy moth caterpillar and related groups have been known to cause a dermatitis on human skin upon exposure. Detailed discussions of the biologies of these species are provided in Chapter 14.

C.   Management and Treatment

Treatment mainly involves personal protection and avoidance of the offending species. Embedded spines or hairs can sometimes be removed by "stripping" the lesion with cellophane tape. Acute urticarial lesions may respond satisfactorily to topical corticosteroid lotions and creams (desoximetasone gel has been frequently used), which reduce the intensity of the inflammatory reaction. Oral antihistamines may relieve itching and burning sensations. Systemic administration of corticosteroids has also been used for itching and pain. Pain relievers may also be needed, especially in the case of puss caterpillar stings. Systemic reactions require close monitoring and aggressive treatment (see Chapter 2).

III. INHALANT IRRITATION OR CONTACT DERMATITIS FROM MOTH HAIRS OR SCALES

A. Medical Significance

The term *lepidopterism* refers to urticaria and irritation caused by adult moths and their hairs or scales (see Chapter 26 for further discussion of the species involved). Reports of particularly severe lepidopterism have come from the South American countries of Peru, Brazil, Venezuela, and Argentina. In that area, female moths in the genus *Hylesia* have barbed urticating setae, which are broken off and become airborne as the moths flutter around lights. This results in urticating rashes and upper respiratory irritation. In the U.S., the tussock moths and their relatives (family Lymantriidae) cause rashes, upper respiratory irritation, and eye irritation to forest workers in the Pacific Northwest.[8]

B. Contributing Factors and Species

In most cases of lepidopterism, huge outbreaks of offending moth species lead to environmental air becoming laden with broken hairs or scales. In the U.S., the Douglas fir tussock moth (*Orgyia pseudotsugata*) female covers her egg masses with froth and body hairs. These hairs, along with other hairs from the tips of the female abdomen, become airborne and cause irritation to humans present in and near infested forests. Outbreaks of the Douglas fir tussock moth appear to develop almost explosively, and then subside abruptly after a year or two. During an outbreak, there are literally billions of caterpillars crawling on the ground, trees, brush, and buildings. Accordingly, airborne levels of the offending hairs are quite high. A related species, the white-marked tussock moth (*O. leucostigma*), occurs throughout most of North America and may be involved in cases of lepidopterism, especially in the East. Rothschild et al.[9] also lists moths in the genera *Epanaphe, Anaphe, Gazalina,* and *Epicoma* (family Notodontidae) as being causes of occasional lepidopterism.

C. Management and Treatment

Treatment mainly involves personal protection and avoidance of the offending species. During the season of *Hylesia* moth emergence (South America), it may help to turn off outdoor lights. In affected areas of the U.S., air conditioning and frequent changes of filters may help reduce airborne levels of spines or hairs. Sensitive persons should avoid walking in forests heavily infested with tussock or gypsy moths. Physicians should be aware that cases diagnosed as simple conjunctivitis or keratoconjunctivitis occurring in areas with extreme moth infestation (tussock moths in the Pacific Northwest and gypsy moths in the East) may be due to contact with airborne moth hairs.

Acute urticarial lesions may respond satisfactorily to topical corticosteroid lotions and creams (desoximetasone gel has been frequently used), which reduce the intensity of the inflammatory reaction. Oral antihistamines may relieve itching and burning sensations. In more serious cases, oral prednisone may be indicated. Rosen[10] said systemic administration of corticosteroids in the form of intramuscular triamcinolone acetonide has been remarkably effective in relieving severe itching due to gypsy moth dermatitis.

REFERENCES

1. Alexander, J.O., *Arthropods and Human Skin*, Springer-Verlag, Berlin, 1984, chap. 7.
2. Frazier, C.A. and Brown, F.K., *Insects and Allergy*, University of Oklahoma Press, Norman, OK, 1980, chap. 19.
3. Bahme, A.J., Cantharides toxicosis in the equine, *Southwest. Vet.*, 21, 147, 1968.
4. Diaz, J.H., The epidemiology, diagnosis, and management of caterpillar envenoming in the southern U.S., *J. La State Med. Soc.*, 157, 153, 2005.
5. Keegan, H.L., Some medical problems from direct injury by arthropods, *Intern. Pathol.*, 10, 35, 1969.
6. Foot, N.C., Pathology of the dermatitis caused by *Megalopyge opercularis*, a Texas caterpillar, *J. Exp. Med.*, 35, 737, 1922.
7. Shama, S.K., Etkind, P.H., Odell, T.M., Canada, A.T., and Soter, N.A., Gypsy moth dermatitis, *New Engl. J. Med.*, 306, 1300, 1982.
8. Perlman, F., Press, E., Googins, G.A., Malley, A., and Poarea, H., Tussockosis: reactions to Douglas fir tussock moth, *Ann. Allerg.*, 36, 302, 1976.
9. Rothschild, M., Reichstein, T., Von Euw, J., Alpin, R., and Harman, R.R.M., Toxic Lepidoptera, *Toxicon*, 8, 293, 1970.
10. Rosen, T., Caterpillar dermatitis, *Dermatol. Clin.*, 8, 245, 1990.

CHAPTER 6

MYIASIS (INVASION OF HUMAN TISSUES BY FLY LARVAE)

TABLE OF CONTENTS

I. INTRODUCTION AND MEDICAL SIGNIFICANCE

Fly larvae infesting the organs and tissues of people or animals is referred to as *myiasis*. The condition occurs in several forms that primarily can be classified — at least from their evolutionary roots — as saprophagous and sanguinivorous.[1,2] Not all myiasis involves long-term tissue infestation. For example, the Congo floor maggot in Africa does not embed itself in tissue, but only sucks blood from its host for a short time

CASE HISTORY

HUMAN PARASITES OR MAGGOTS?

A woman sent me several wormlike specimens collected from inside her toilet and on the floor around the toilet. "How are these things getting there?" she asked. "Does this mean these are intestinal parasites?" Upon examination, the specimens were identified as soldier fly larvae, *Hermetia illucians*, common pests inhabiting decaying or putrefying organic matter (Color Figure 21.18). This species has been reported as a cause of intestinal myiasis, but — as in this case — it is difficult to tell if the maggots were excreted into the toilet or were deposited as eggs by the adult fly in the toilet area due to decaying wood and build-up of organic matter at the toilet/floor interface. I recommended that the woman have the toilet taken up, thoroughly cleaned, and a new wax seal installed. She called back saying her husband found many more specimens under the toilet where the flange meets the sewer pipe and even in the wax seal itself.

Comment: Soldier fly larvae are often found in toilets. I have encountered at least ten such events. More than likely, this was a case of mistaken identity — maggots feeding in and around a toilet confused as parasites coming from inside the people using the toilet. Very frequently toilets leak a little around the connection to the sewer, allowing water and organic debris to accumulate — perfect conditions for soldier fly larvae.

(see Chapter 21). Specific cases of myiasis are clinically defined by the affected areas involved. For example, there may be traumatic (wound), gastric, rectal, auricular, and urogenital myiasis, among others. Myiasis can be accidental, when fly larvae occasionally find their way into the human body, or facultative, when fly larvae enter living tissue opportunistically after feeding on decaying tissue in neglected, malodorous wounds. Myiasis can also be obligate, in which the fly larvae must spend part of their developmental stages in living tissue. Obligate myiasis is the most serious form of the condition from a pathogenic standpoint and constitutes true parasitism.

Fly larvae are not capable of reproduction and, therefore, myiasis under normal circumstances should not be considered contagious from patient to patient. Transmission of myiasis occurs only via an adult female fly. See Chapter 21 for discussions of the most common species involved and their biologies.

A. Accidental Myiasis

Accidental enteric myiasis (sometimes referred to as *pseudomyiasis*) is mostly a benign event, but the larvae could possibly survive temporarily, causing stomach pains, nausea, or vomiting. There is some question as to whether or not these cases should be classified as true myiasis because of lack of fly development after the ingested eggs hatch. Numerous fly species in the families Muscidae, Calliphoridae, and Sarcophagidae may produce accidental enteric myiasis. Some notorious offenders are: pomace flies and fruit flies (*Drosophila* spp.); the cheese skipper, *Piophilia casei*, the black soldier fly, *Hermetia illucens*, and the rat-tailed maggot, *Eristalis tenax*. Other instances of accidental myiasis occur when fly larvae enter the urinary passages or other body openings. Flies in the genera *Musca, Muscina, Fannia, Megaselia*, and *Sarcophaga* have often been implicated in such cases.

B. Facultative Myiasis

Facultative myiasis may result in considerable pain and tissue damage as fly larvae leave necrotic tissues and invade healthy ones. Numerous species of Muscidae, Calliphoridae, and Sarcophagidae have been implicated in facultative myiasis. In the U.S., the calliphorid *Phaenicia sericata* has been reported as causing facultative myiasis on several occasions.[3,4] Another calliphorid, *Chrysomya rufifacies*, has been introduced into the U.S. from the Australasian region and is also known to be regularly involved in facultative myiasis.[5] Other muscoid fly species that may be involved in this type of myiasis include *Calliphora vicina, Phormia regina, Cochliomyia macellaria,* and *Sarcophaga haemorrhoidalis.*

C. Obligate Myiasis

Several fly species must develop in the living tissues of a host. This is termed *obligate myiasis* and is caused by species affecting sheep, cattle, horses, and many wild animals. In people, obligate myiasis is primarily due to the screwworm flies (Old and New World) and the human bot fly (Color Figure 6.8). Obligate myiasis is rarely fatal in the case of the human bot fly of Central and South America, but it has led to considerable pathology and death in the case of screwworm flies. Screwworm flies use livestock as primary hosts, but they do attack humans. If, for example, a female screwworm fly oviposits just inside the nostril of a sleeping human, hundreds of developing maggots may migrate throughout the turbinal mucous membranes, sinuses, and other tissues. Surgically, it would be extremely difficult to remove all the larvae. Fortunately, on account of the sterile male release program, screwworm flies have been virtually eliminated from the U.S. and Mexico.

II. CONTRIBUTING FACTORS

A. Accidental Myiasis

Accidental enteric myiasis generally occurs from ingesting fly eggs or young maggots on uncooked foods or previously cooked foods that have been subsequently infested. Cured meats, dried fruits, cheese, and smoked fish are the most commonly infested foods. Other cases of accidental myiasis may occur from the use of contaminated catheters, douching syringes or other invasive medical equipment, or sleeping with the body exposed.

B. Facultative Myiasis

Several fly species lay eggs on dead animals or rotting flesh. Accordingly, the flies may mistakenly oviposit in a foul-smelling wound of a living animal. The developing maggots subsequently invade healthy tissue. Facultative myiasis most often is initiated when flies oviposit in necrotic, hemorrhaging, or pus-filled lesions. Wounds with watery alkaline discharges (pH 7.1 to 7.5) have been reported to be especially attractive to blow flies. Facultative myiasis frequently occurs in helpless semi-invalids who have poor (if any) medical care. Often, in the case of the very elderly, their eyesight is so weak that victims do not detect the myiasis. In clinical settings, facultative myiasis is most likely to occur in incapacitated patients who have recently had major surgery or those having large or multiple uncovered or partially covered festering wounds. However, not all human cases of facultative myiasis occur in or near a wound. In the U.S., larvae of the blow fly, *P. sericata*, have been reported from the ears and nose of healthy patients with no other signs of trauma in those areas.[6]

C.   Obligate Myiasis

Obligate myiasis is essentially a zoonosis; humans are not the ordinary host but may become infested. Human infestation by the human bot fly is very often via a mosquito bite — the eggs are attached to mosquitoes and other biting flies (see Part III for more details); however, human screwworm fly myiasis is a result of direct egg laying on a person, most often in or near a wound or natural orifice. Egg-laying activity of screwworm flies occurs during daytime.

III.   PREVENTION, MANAGEMENT, AND TREATMENT

A.   Prevention

Prevention and good sanitation can avert much of the accidental and facultative myiasis occurring in the industrialized world. Exposed foodstuffs should not be unattended for any length of time, to prevent flies from ovipositing on them. Covering, and preferably refrigerating, leftovers should be done immediately after meals. Washing fruits and vegetables prior to consumption should help remove developing maggots, although visual examination should also be done while slicing or preparing these items. Other forms of accidental myiasis may be prevented by protecting invasive medical equipment from flies and avoiding sleeping nude, especially during daytime. To prevent facultative myiasis, extra care should be taken to keep wounds clean and dressed, especially on elderly or helpless individuals. Daily or weekly visits by a home health nurse can go a long way to prevent facultative myiasis in patients who stay at home. In institutions containing invalids or otherwise helpless patients, every effort should be made to control entry of flies into the facility. This might involve taking such precautions as keeping doors and windows screened and in good repair, thoroughly sealing all cracks and crevices, installing air curtains over doors used for loading and unloading supplies, and installing UV electrocuter (or the newer electronic types) fly traps in areas accessible to the flies but inaccessible to patients. Prevention of obligate myiasis involves avoiding sleeping outdoors during daytime in screwworm-infested areas and using insect repellents in Central and South America to prevent bites by bot fly egg-bearing mosquitoes.

B.   Management and Treatment

Treatment of accidental enteric myiasis is probably not necessary (although there may be rare instances of clinical symptoms), as in most cases there is no development of the fly larvae within the highly acidic stomach environment and other parts of the digestive tract. They are killed and merely carried through the digestive tract in a passive manner. Treatment of other forms of accidental myiasis as well as facultative or obligate myiasis involves removal of the larvae. Alexander[7] recommends debridement with irrigation. Others suggested surgical exploration and removal of larvae under local anesthesia.[6] Care should be taken not to burst the maggots upon removal. Human bot fly larvae have been successfully removed using "Bacon therapy," a treatment method involving covering the punctum (breathing hole in the skin) with raw meat or pork.[8] In a few hours the larvae migrate into the meat and are then easily extracted (see Chapter 21 for more discussion). Maggot infestation of the nose, eyes, ears, and other areas may require surgery if larvae cannot be removed via natural orifices. Because blow flies and other myiasis-causing flies lay eggs in batches, there could be tens or even hundreds of maggots in a wound.

REFERENCES

1. Stevens, J.R. and Wallman, J.F., The evolution of myiasis in humans and other animals in the Old and New Worlds (Part I): phylogenetic analyses, *Trends. Parasitol.*, 22, 129, 2006.

2. Stevens, J.R., Wallman, J.F., Otranto, D., Wall, R., and Pape, T., The evolution of myiasis in humans and other animals in the Old and New Worlds (Part II): biological and life-history studies, *Trends. Parasitol.*, 2006.

3. Greenberg, B., Two cases of human myiasis caused by *Phaenicia sericata* in Chicago area hospitals, *J. Med. Entomol.*, 21, 615, 1984.

4. Merritt, R.W., A severe case of human cutaneous myiasis caused by *Phaenicia sericata*, *Calif. Vector Views*, 16, 24, 1969.

5. Richard, R.D. and Ahrens, E.H., New distribution record for the recently introduced blow fly *Chrysomya rufifaces* in North America, *Southwest. Entomol.*, 8, 216, 1983.

6. Anderson, J.F. and Magnarelli, L.A., Hospital acquired myiasis, *Asepsis*, 6, 15, 1984.

7. Alexander, J.O., *Arthropods and Human Skin*, Springer-Verlag, Berlin, 1984.

8. Brewer, T.F., Wilson, M.E., Gonzalez, E., and Felsenstein, D., Bacon therapy and furuncular myiasis, *JAMA*, 270, 2087, 1993.

CHAPTER 7

DELUSIONS OF PARASITOSIS (IMAGINARY INSECT OR MITE INFESTATIONS)

TABLE OF CONTENTS

I. INTRODUCTION

Some people have delusions of parasitosis (DOP) in which they think they are infested with arthropod parasites. The condition is not uncommon. Any public health or extension service entomologist who has worked with the public has encountered this problem in one form or another. Schrut and Waldron[1] reported seeing over 100 cases in 5 years. Alexander[2] and Slaughter et al.[3] have provided excellent overviews of this subject. Before discussing this phenomenon, some clarifying definitions are in order.

Entomophobia. Entomophobia is often erroneously used as a catchall term to explain abnormal reactions in some people to insects or mites. This word has been loosely used to include several distinct psychic phenomena such as abnormal fear of insects or arachnids, delusions of parasitic mite infestations, and dermatitis caused by "computer mites," "paper mites," or "cable mites" in the workplace. More precisely, however, entomophobia is a terrible fear or dread of insects, mites, or spiders. It may be characterized by hysterical reactions at the sight of the feared arthropods. Entomophobia may be related to, but is distinctly different from, delusions of parasitosis and computer or cable mite dermatitis.

Delusions of parasitosis. DOP is an emotional disorder in which the patient is convinced that tiny, almost invisible insects or mites are present on/in his or her body. Some researchers prefer to label the disorder *pyschogenic parasitosis* because not all patients are delusional in the classic sense.[4] In most cases, DOP is a monosymptomatic hypochondriacal psychosis in which there are no other thought disorders and the delusions are not a result of another psychiatric illness. However, DOP cases can be associated with paranoia or schizophrenia, which can quite possibly endanger the health care provider. One DOP patient attempted to murder her general practictioner.[5] The author has also encountered patients who subsequently began to think that health department officials (including the entomologist!) were "out to get him." The mental and emotional stress from DOP may be severe, leading to destructive behaviors such as quitting jobs, burning furniture, abandoning homes, using pesticides dangerously and repeatedly, etc.

Workplace infestations. Sometimes there is an imaginary insect or mite infestation at the workplace. Workers may complain of cable mites, paper mites, or computer mites biting them and producing lesions. Strictly speaking, this situation may not be DOP in that often there is a real substance in the work environment causing the problem (fiberglass or other insulation, chemicals, paper or cardboard splinters, etc.).

Prior to categorizing persons into any of these three groups, it is imperative that a thorough entomological inspection be carried out at the residence or workplace. This should be done with an objective and open mind because it is possible that there is a real insect or mite problem. A pest control operator (exterminator) could do this, but a university, health department, or extension entomologist would be better as they generally have had more entomological training. Provided in the following text is a brief overview of the insect or mite species that could be the cause of a real infestation (please refer to each of these sections in Part III for further details).

II. ACTUAL ARTHROPOD CAUSES OF DERMATITIS

There are two human-specific mites: the human scabies mite, *Sarcoptes scabei*, and the human follicle mite, *Demodex folliculorum*. The human scabies mite burrows beneath the outer layer of skin, leaving crusted papules and/or a sinuous red trail along the skin for a few centimeters. The infestation causes itching and commonly occurs in the webbing between the fingers and on the wrists and elbows. Skin scrapings of affected areas will usually contain the mites. The human follicle mite occurs in the majority of humans in hair follicles around the nose and eyelids, but generally causes no pathology or discomfort. Physicians should be aware of the follicle mite so that, when detected during a skin scraping, the mites will not be inadvertently reported as the cause of the mysterious biting.

There are a few other mite species (not host specific for humans) that will readily bite people. Usually, these are only opportunistic infestations that are self-limiting. They do not take up residence and reproduce on or in the skin. The tropical fowl mite, *Ornithonyssus bursa* (a parasite of sparrows), the northern fowl mite, *O. sylviarum* (a parasite of chickens), and the tropical rat mite, *O. bacoti* (a parasite of rats) will aggressively bite people, causing pain and itching. In a lot of cases of "mysterious" mite infestation a rat or bird nest can be found some place in the dwelling, and the patient is being bitten by one of these mite species. These mites can be seen (about 1 mm long) and are often captured while biting or crawling on the skin.

Other mites such as the grocer's itch mite, *Glycyphagus domesticus*, the spiny rat mite, *Laelaps echidnina*, the straw itch mite, *Pyemotes tritici*, the baker's mite, *Acarus siro*, and the dried fruit mite, *Carpoglyphus lactis*, may occasionally be involved in cases of dermatitis. However, these cases are usually limited to persons with occupations bringing them into prolonged, close association with these mites. Again, none of these species can live on or in human skin.

Body and head lice, *Pediculus humanus corporis* and *P. humanus capitis*, can cause discomfort and itching. Pubic lice, *Pthirus pubis*, live on humans in the pubic and perianal regions. However, lice are large enough to see, and thus are usually found during a physical examination.

If a patient is elderly or has poor eyesight, fleas may be the cause of their mysterious bites in the home. Slowly pulling a white towel or part of a sheet around in the patient's house should yield several fleas if the house is infested. Incidentally, homeowner flea problems have increased in the last year or two.

Bed bugs also could be the cause of mysterious bites, and their incidence is currently on the rise in the U.S. Contrary to popular opinion, they are not only found where primitive and unsanitary conditions exist, but may also occur in affluent homes, hotels, and institutions. One sign of bed bug infestation is the presence of small specks of blood or dark feces on bedding or skin. Upon examination of the premises, bed bugs can be found (they are plenty big enough to see) beneath loosened wallpaper, in the seams of bedding, or in cracks in the furniture. They are similar in size and appearance to immature cockroaches.

There is a recent report of collembolans (springtails) found in skin scrapings from 18 individuals diagnosed with DOP.[6] However, identification of the insects was made from photographs alone, using special imaging software, and no specimens were recovered and retained as voucher specimens. Currently, results of this paper are controversial and not accepted by a majority of entomologists. More research is required to ascertain the true prevalence, if any, of Collembola in humans.

Again it must be emphasized that cases of imaginary insect or mite infestations must be carefully investigated to rule out actual arthropod causes. Scrapings of skin lesions should be taken for examination for scabies mites. A competent entomologist should examine the patient's workplace and residence for evidence of insects or mites. Careful attention should be given to rat- or bird-infested dwellings, as they may harbor parasitic mites. If repeated collection attempts, skin scrapings, and insecticidal treatments are unsuccessful, and if the patient exhibits symptoms consistent with delusions of parasitosis, then a diagnosis of DOP should be strongly considered.

III.   DELUSIONS OF PARASITOSIS

I usually encounter about 5 to 10 cases of DOP each year (Table 1 is a partial listing of delusory parasitosis cases and workplace infestations I have investigated). The disorder shows patterns of typical behavior.[7]

Sometimes an initial and real insect infestation precedes and triggers the delusion. In a majority of cases the victim is an elderly white female, and the "bugs" may appear and disappear while they are being watched; they enter the skin and reappear and

Table 1

**Breakdown of Selected Imaginary Insect or Mite Infestations
Reported to the Mississippi State Department of Health, 1989–1995**

Case No.	Entity Affected	Age	Sex	Race	Complaint
1.	Individual	~60	M	B	"Parasites in rectum and penis; come out at night to bite"
2.	Individual	Elderly	F	W	"Mites on the skin"
3.	Individual	40	M	B	"Bugs going in and out of skin"
4.	Individual	~60	F	W	"Bugs in my rectum"
5.	Individual	65	F	W	"Bugs biting like fire; making life miserable"
6.	Individual	70	F	W	"Long-standing mite infestation on skin"
7.	Individual	33	M	W	"Mite infestation; appears as white smoke or flakes"
8.	Individual	Elderly	F	W	"Being taken over by bugs"
9.	Individual	75	F	W	"Must be ticks and fleas in my house"
10.	Individual	50	M	W	"Mysterious bug bites for years"
11.	Individual	~60	F	B	"Tiny bugs gliding over my body"
12.	Individual	~60	F	W	"Tiny insects in my ear for 9 months"
13.	Individual	~60	F	W	"Tiny mites biting and making life miserable"
14.	Workplace	25–55	F	W	"Paper mites in medical records"
15.	Workplace	18–50	F	W, B	"Mites in boxes and shipping crates"
16.	Workplace	21–48	F	W, B	"Paper and computer mites in medical records"
17.	Workplace	21–50	F	W, B	"Paper and computer mites in medical records"

invade the hair, nose, and ears (in one case I was investigating, while I was there the woman cried out loudly saying that one of the mites had suddenly run up her nose). Oddly, DOP patients often claim that the bugs are in their rectum or vagina, only coming out at night. The patients claim that the "bugs" are able to survive repeated insecticidal sprays and the use of medicated shampoos and lotions. Frequently there is a history of numerous visits to medical doctors and dermatologists. Lesions may be present, although neurotic excoriation may be the cause.[8] The physician or entomologist is often presented with tissue paper, small bags, or other containers with the presumed pests (Figure 1). However, these usually contain dust, specks of dirt, dried blood, pieces of skin (Figure 2), and occasionally common (nonharmful) household insects or their body parts. Alexander[2] said that presentation of such "evidence," together with the intensity of the patient's belief, is almost pathognomonic of the disorder. Out of desperation the victims may move out of their home, only to report later that the "bugs" have followed them there too. An affected person may be so positive of his infestation and give such a detailed description that other family members may agree with the patient. They may even be "infected" themselves; thus, the delusion has been transferred. Sometimes affected persons become excessively preoccupied with

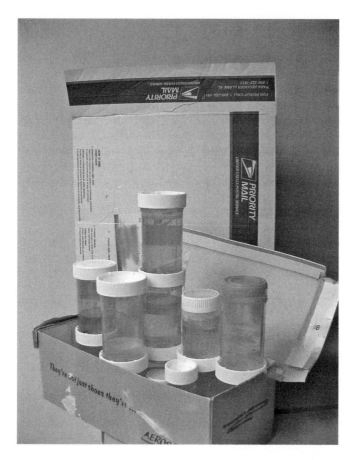

Figure 1
Example of samples sent in to the Mississippi Department of Health for possible "insect infestation."

Figure 2
Pieces of skin from a sample sent in by a DOP patient.

CASE HISTORY

"MITES" IN A WOMAN'S EAR

A 60-year-old white female came to my office after she had been to a family practitioner and three dermatologists to no avail with the problem of "tiny insects living on her skin, but especially inside the ears." Upon being seen, she presented samples she had collected — several coffee cups, small pill bottles, and vials, all containing skin scrapings, ear wax, and/or dried blood. During the interview she stated that the problem began 9 months earlier as an infestation on the skin but, due to repeated pesticide treatments (on her skin), was now in her ears. She also revealed that she regularly forced Kwell® (1% lindane) into her ears via a syringe. Other extreme measures to rid her and her house of the "bugs" included intensive scrubbing of floors and walls, and "rubbing her skin raw."

Careful examination of the patient and the samples she had brought revealed no evidence of insect or mite infestation. Previous examinations by dermatologists had ruled out scabies. Based on the characteristic presentation of "imaginary bugs" in this case, results of the interview and exam, and history of numerous visits to various physicians, a tentative diagnosis of delusions of parasitosis (DOP) was made. The patient was referred back to her family physician with instructions to have him contact the author for further discussion.

Comment. DOP is more common than one might think. During 3 years, 14 cases were reported to the Mississippi Department of Health. Certainly, many more cases never get reported to a state public health agency.

Source: Adapted from *J. Agromed.* 2, 53, 1995. Copyright 1995, the Haworth Medical Press, Binghamton, New York. With permission.

---

cleanliness and spend all available money on cleaners, soaps, disinfectants, insecticides, etc., in order to thoroughly clean, scrub, and sterilize their home.

The following are some excerpts from letters received in my office from patients apparently suffering from DOP.

Patient 1: Elderly Female

The attached package of specimens were (sic) taken from objects in my house when they could be seen — the beds, bathtub, and from myself when I could both see and feel them. I picked them up with a Lysol solution cloth or paper towel … My house has been sprayed and fumigated many times … I have been told that if I have mites, I could not see them … Please let me know how to capture something so small, and mites which cannot be seen. These are making life miserable for me … Is there a possibility that my trouble could be some type of fly larva?

Patient 2: Elderly Female

I couldn't get everything in the (other) containers into this one, and left those that had fallen to the bottom, here. In transferring from 3 containers to this one I'm sure some specimens were left clinging. This is a three day accumulation. There were *definitely* 6–8 live, jumpy specimens. Thank you! I hope you will inspect everything, even scraping the sidewalls (of the container), and top and those that are almost invisible to the naked eye.

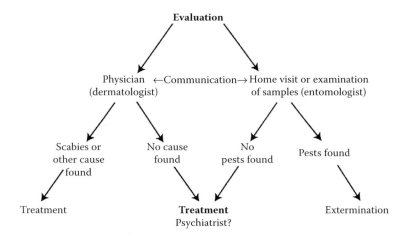

Figure 3
An interdisciplinary approach to treatment of DOP. (From Goddard, J., Imaginary Insect Infesta-tions, Infections in Medicine, 15, 169, 1998.)

These two accounts are quite typical of cases I have investigated. From reading these two letters it seems that affected patients obsessively feel there are mites, invisible to the naked eye, that readily infest humans. It may be a fear of the unknown — the existence of invisible insects or mites that nobody knows much about (even "doctors") — that contributes to this delusion. Thus, whenever these patients feel a tingling sensation or have an itchy place on their skin, they are afraid and later become convinced that mites are on or under their skin.

On the other hand, there may be internal physiological causes of the pruritus, mistakenly believed to be arthropod-produced, such as diabetes, icterus, atopic dermatitis, and lymphoblastomas. At times, the mental disturbance of pellagra takes the form of delusion of parasitosis and disappears with appropriate therapy.[8] Hinkle[9] listed 50 common prescription drugs and their side effects including parathesias, itching, urticaria, and rash. Cases of DOP have also been attributed to cocaine or methamphet-amine abuse.[10,11]

Diagnosis by a physician. DOP is diagnosed as a delusional disorder, somatic type, if symptoms persist more than 1 month.[12] Medical causes may be ruled out by thorough laboratory and neurological evaluation. Schizophrenia and schizophreniform disorders may be eliminated with a detailed patient history and cognitive testing. Physicians should also check for a comorbid psychiatric disorder that may be perpetuat-ing the delusion, because DOP often co-occurs with axis I disorders including major depressive disorder, substance abuse, dementia, and mental retardation. After ruling out actual arthropod causes (see previous section) and underlying medical conditions, physicians may wish to refer the patient to a psychiatrist. But most DOP patients will not see a psychiatrist. Instead, they will seek out another physician, thus starting the whole process over. An interdisciplinary approach, mainly involving family practice physicians, dermatologists, psychiatrists, and entomologists (Figure 3), is needed to help DOP patients. Family practice physicians or general practitioners are usually the first providers who see DOP patients. Physicians need to be careful not to diagnose "insect bites" based on lesions alone and should call upon entomologists to examine samples. Entomologists need to understand the medical complexity of delusions — that there are intensive worries, true delusions, and a host of abnormal personality traits associated with DOP — and avoid any hint of medical evaluation of the patient. Treatment strategies, including antipsychotic drugs, have been proposed.[10,12–16] Many dermatologists report good success in treating DOP patients with pimozide; however, the drug has several adverse effects and should only be used with careful supervision. Recently, risperidone and olanzapine have been proposed instead of pimozide.[10]

IV. WORKPLACE INFESTATIONS

Computer, paper, or cable mite dermatitis in a workplace may not be strictly a delusion of parasitosis. Often, these people are actually experiencing a prickling, tingling, or creeping sensation caused by a real substance in their environment (rock wool or fiberglass insulation, paper fragments, dust, and other debris).[17] This problem is exacerbated by mass hysteria ("Every time I come to work here I get bitten by tiny mites!"). Imaginary mite infestations in the workplace typically involve females who work with old, dusty records or elaborate electronic equipment. I have seen this several times in hospital medical records offices. Contributing factors include inadvertent confirmation by physicians, bites from fleas or mosquitoes at home being scratched at work, and unscrupulous pest control operators willing to spray for computer or cable mites in the absence of a real pest.

Dealing with an imaginary workplace infestation is difficult. First of all, a competent pest control operator should inspect the premises, looking for possible causes of the mysterious biting. He or she should set glue boards (sticky traps) in affected areas and check them at regular intervals. If there is an actual arthropod pest in sufficient numbers to bite employees, they will likely be captured on the monitoring devices. If no actual insect or mite infestation can be identified in the affected area, then environmental factors such as carpets, paper splinters and particles, static electricity, ventilation systems, and indoor air pollutants must be considered. An excellent handout about invisible itches which I use to give affected persons is available from the University of Kentucky Cooperative Extension Service.[18] Sometimes, hiring an "outside" cleaning crew to thoroughly clean the area will eliminate the supposed infestation. If problems continue, an industrial hygienist should be consulted for inspection for these types of irritants.

REFERENCES

1. Schrut, A.H. and Waldron, W.G., Psychiatric and entomological aspects of delusionary parasitosis, *JAMA*, 186, 213, 1963.
2. Alexander, J.O., *Arthropods and Human Skin*, Springer-Verlag, Berlin, 1984, chap. 25.
3. Slaughter, J.R., Zanol, K., Rezvani, H., and Flax, J.F., Psychogenic parasitosis: a case series and literature review, *Psychosomatics*, 39, 491, 1998.
4. Zanol, K., Slaughter, J., and Hall, R., An approach to the treatment of psychogenic parasitosis, *Int. J. Dermatol.*, 37, 56, 1998.
5. Bourgeois, M.L., Duhamel, P., and Verdoux, H., Delusional parasitosis — folie A deux and attempted murder of a family doctor, *Br. J. Psychiatr.*, 161, 709, 1992.
6. Altschuler, D.Z., Crutcher, M., Dulceanu, N., Cervantes, B.A., Terinte, C., and Sorkin, L.N., Collembola found in scrapings from individuals diagnosed with delusory parasitosis, *J. NY Entomol. Soc.*, 112, 87, 2004.

7. Goddard, J., Analysis of 11 cases of delusions of parasitosis reported to the Mississippi Department of Health, *South. Med. J.*, 88, 837, 1995.

8. Obermayer, M.E., Dynamics and management of self-induced eruptions, *Calif. Med.*, 94, 61, 1961.

9. Hinkle, N., Delusory parasitosis, *Am. Entomol.*, 46, 17, 2000.

10. Scheinfeld, N., Delusions of parasitiosis: a case with a review of its course and treatment, *Skinmed*, 2, 376, 2003.

11. Elpern, D.J., Cocaine abuse and delusions of parasitosis, *Cutis*, 42, 273, 1988.

12. Matthews, A. and Hauser, P., A creepy-crawly disorder, *Curr. Psychiatr.*, 4, 88, 2005.

13. Gould, W.M. and Gragg, T.M., Delusions of parasitosis — an approach to the problem, *Arch. Dermatol.*, 112, 1745, 1976.

14. Koblenzer, C.S., *Psychocutaneous Disease*, Grune and Stratton, Orlando, FL, 1987, pp. 116–120.

15. Torch, E.M. and Bishop, E.R., Delusions of parasitosis — psychotherapeutic engagement, *Am. J. Psychother.*, 35, 101, 1981.

16. Driscoll, M.S., Rothe, M.J., Grant-Kels, J.M., and Hale, M.S., Delusional parasitosis: a dermatologic, psychiatric, and pharmacologic approach, *J. Am. Acad. Dermatol.*, 29, 1023, 1993.

17. Possick, P.A., Gellin, G.A., and Key, M.M., Fibrous glass dermatitis, *Am. Ind. Hyg. Assoc.*, January–February Issue, 12, 1970.

18. Potter, M.F., Invisible Itches: Insect and Non-Insect Causes, University of Kentucky, Cooperative Extension Service Publ. No. ENT-58, 1997, 4 pp.

PART II

IDENTIFICATION OF ARTHROPODS AND THE DISEASES THEY CAUSE

CHAPTER 8

IDENTIFICATION OF MEDICALLY IMPORTANT ARTHROPODS

TABLE OF CONTENTS

I. PRINCIPLES OF IDENTIFICATION AND NAMING

Precise identification of an offending arthropod is extremely important in medical settings and may have far-reaching implications. For example, a person bitten by a nonpoisonous spider does not need the expensive and intense treatments that a person bitten by a poisonous one needs. Also, treatment recommendations for pubic lice differ radically from those for nymphal ticks in the pubic region (yet the two are often confused). Obviously, accurate identification is crucial from the outset.

Entomologists have their own jargon and terminology, as do all specialists. However, this terminology can be deciphered once the reader understands the basis for arthropod naming. Generally, arthropods are grouped into classes, orders, families, genera, and species according to shared common morphological structures. Specimens with eight legs and two body regions are placed in one category, specimens with six legs and three body regions are placed in another, etc. Making things easier, the names assigned to these categories are often descriptive of the arthropods in that category. For example, insects with two wings are in the order Diptera, meaning two wings (other common insect orders are provided in Table 1). Species names may further describe the arthropod — *Buggus erythrocephala* would mean having a red head, the term *melanogaster* would mean dark or black

Table 1
Names of Some Common Insect Orders

Insect Order	Common Names
Ephemeroptera	Mayflies
Odonata	Dragonflies, damselflies
Orthoptera	Grasshoppers, crickets
Blattaria	Cockroaches
Isoptera	Termites
Phthiraptera	Lice
Hemiptera	Bugs
Coleoptera	Beetles
Siphonaptera	Fleas
Diptera	Flies
Lepidoptera	Butterflies, moths
Hymenoptera	Ants, wasps, bees

belly, and the name *Calliphora vomitoria* or *C. cadaverina* for a blow fly would give the reader some idea of what kind of things these flies are attracted to or breed in.

However, not all names given to arthropods are descriptive of the organism; some memorialize a specialist in that field. *B. parkeri*, *B. bacoti*, *B. kochi*, *B. blakei*, *B. walkeri*, or *B. smithi* would be named after Drs. Parker, Bacot, Koch, Blake, Walker, or Smith, respectively. The first person to describe a new species (and publish the name) becomes the author of that scientific name and that person's name is often cited with the arthropod, e.g., *Phlebotomus diabolicus* Hall. If subsequently the species is moved to another genus through some kind of taxonomic revision, the author's name is retained but in parenthesis — *Lutzomyia diabolica* (Hall). If two species are subsequently determined to be the same species (this happens frequently), the oldest described species — the first published one — becomes the official name. The other name is then considered a synonym and is discarded. Governing all this is the standard code of rules of nomenclature that has been laid down by the International Commission on Zoological Nomenclature.

II.   BRIEF REVIEW OF ARTHROPOD MORPHOLOGY

A.   Introduction

There is tremendous variety among the arthropods, but identification of the major groups is not difficult if one is familiar with their general morphology. In fact, it is rarely necessary to use a microscope to separate the classes of Arthropoda (insects, spiders, scorpions, etc.). Table 2 presents a summary of the key characteristics of these classes. On the other hand, more specific identification of specimens beyond the *class* can be difficult without the aid of proper literature and, perhaps, special training. Identification to the species level often must be done by specialists who conduct taxonomic research with that particular group. For outside help with identifications, try a local

Table 2
Key Characteristics of Adults of Selected Arthropod Groups

Arthropod Group	Key Characteristics
Insects	• Six legs • Three body regions — head, thorax, abdomen • Some with wings
Spiders	• Eight legs • Two body regions — cephalothorax, abdomen
Mites and ticks	• Eight legs (as adults) • One globose or disk-shaped body region • No true head, mouthparts only
Scorpions	• Eight legs • Broad, flat body region and posterior tail with sting
Centipedes	• Numerous legs — hundred leggers • One pair of legs per body segment • Often dorsoventrally flattened
Millipedes	• Numerous legs — thousand leggers • Two pair of legs per body segment • Often cylindrical

university entomology department (usually at the land-grant universities), a military base (the army and navy employ over 100 professional entomologists), or the entomology department of the Smithsonian Institution.

B.   Characteristics of Insects (Class **Hexapoda***)*

Like all arthropods, insects possess a segmented body and jointed appendages. Beyond that, however, there is tremendous variation: long legs, short legs; four wings, two wings, no wings; biting mouthparts, sucking mouthparts; soft bodies, hard bodies; etc. Compounding identification problems, immature forms of most insects look nothing like the adults. Despite the diversity, adults can at once be recognized as insects by having three pairs of walking legs and three body regions (or tagmata): head, thorax (bearing legs, and wings if present), and abdomen (Figure 1). No other arthropods have wings. However, some insect groups might never have had wings, or may have lost them through adaptation. In particular, several medically important species are wingless (lice, fleas, etc.).

Identification of immature insects presents entirely different problems. In those insects that have simple metamorphosis (grasshoppers, lice, true bugs), the immatures (called *nymphs*) look like the adults, get larger with each molt, and develop wings (if present) during later molts (Figure 2). Identifying nymphs as insects and placing them in their proper orders is generally not a problem. However, in those groups with complete metamorphosis (beetles, flies, bees and wasps, moths and butterflies, fleas, etc.), the immature stage, or larva, looks nothing like the adult (Figure 3). Often, the larva is wormlike. The three body regions are never as distinct as they are in adults, but generally the three pairs of walking legs are evident, although they are often extremely short. Fly larvae (maggots) lack walking legs and, although some (such as mosquitoes)

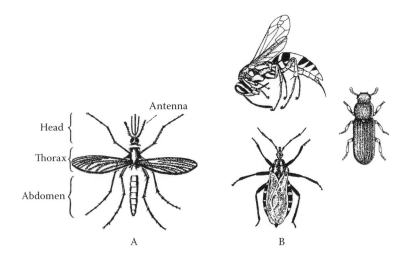

Figure 1
(A) Insect with body regions labeled and (B) several representative insects. (From U.S. DHEW, PHS, CDC Pictorial Keys.)

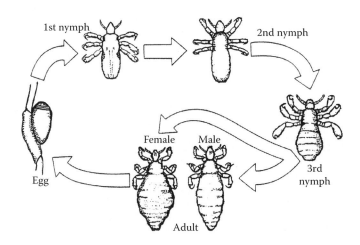

Figure 2
Head lice life cycle — example of simple metamorphosis. (From U.S. DHHS, CDC Publication Number 83-8297.)

have three body regions, others (such as house flies, blow flies, etc.) have no differentiated areas (Figure 4A). Caterpillars and similar larvae often appear to have legs on some abdominal segments (Figure 4D). Close examination of these abdominal *legs* (prolegs) reveals that they are unsegmented fleshy projections, with or without a series of small hooks (crochets) on the plantar surface, and structurally quite unlike the three pairs of segmented walking legs on the first three body segments behind the head.

C. Characteristics of Spiders (Class Arachnida, Order Araneae)

A spider has two body regions: an anterior cephalothorax and posterior abdomen connected by a waistlike pedicel (Figure 5). The former bears the head and thorax

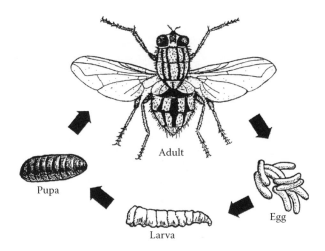

Figure 3
House fly life cycle — example of complete metamorphosis. (From U.S. DHHS, CDC Publication Number 83-8297.)

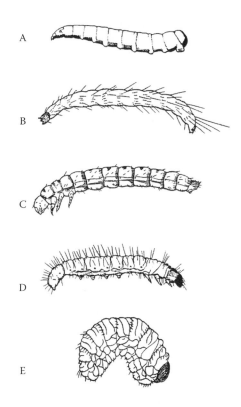

Figure 4
Various types of insect larvae: (A) fly, (B) flea, (C) beetle, (D) moth, and (E) another kind of beetle. (From U.S. DHEW, PHS, CDC Pictorial Keys.)

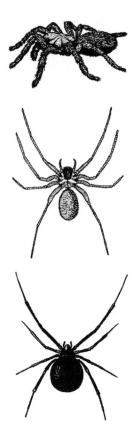

Figure 5
Various spiders: top, tarantula; middle, brown recluse; bottom, black widow dorsal view. (From U.S. DHEW, PHS, CDC Pictorial Keys.)

structures, including the four pairs of walking legs, generally eight simple eyes on the anterior dorsal surface, and mouthparts. The mouthparts, called *chelicerae*, are generally fanglike and are used to inject poison into prey, all spiders being predaceous. Located between the chelicerae and the first pair of walking legs are a pair of short leglike structures called *pedipalpi* that are used to hold and manipulate prey. Pedipalpi may be modified into copulatory organs in males. The abdomen is usually unsegmented and bears spinnerets for web production at the posterior end. Immatures look the same as adults, except that they are smaller.

Harvestmen, or daddy longlegs (order Opiliònes), have many characteristics in common with true spiders; however, they differ in that the abdomen is segmented and is broadly joined to the cephalothorax (not petiolate). Most species have extremely long, slender legs.

D. Characteristics of Mites and Ticks (Class Arachnida, Subclass Acari)

These small arachnids characteristically have only one apparent body region (cephalothorax and abdomen fused), the overall appearance being globose or disk-shaped in most instances (Figure 6). This general appearance quickly separates the Acari from

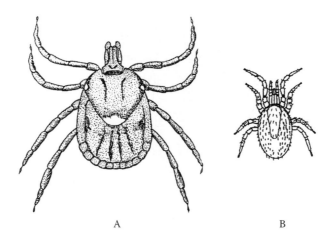

A   B

Figure 6
(A) Tick and (B) mite. (From U.S. DHEW, PHS, CDC Pictorial Keys.)

other arthropods. The body may be segmented or unsegmented with eight walking legs present in adults. Their larvae, which hatch out of eggs, have only six (rarely, fewer) legs, but their single body region readily separates them from insects. They attain their fourth pair of legs at the first molt, and hereafter are called nymphs until they become adults. As in spiders, immature ticks and mites are generally similar in appearance to adults. In general, ticks are considerably larger than mites. Adult ticks are generally pea-sized; mites are about the size of a grain of sand (often even smaller).

E.   Characteristics of Scorpions *(Class* **Arachnida,** *Order* **Scorpiones***)*

The body of a scorpion is easily separable into an anterior broad, flat area and a posterior *tail* with a terminal sting (Figure 7). Although these outward divisions do not correspond to actual lines of tagmatization, they do provide an appearance sufficient to distinguish these arthropods from most others. Like spiders, the mouthparts are chelicerae and the first elongate appendages are pedipalpi. The pedipalpi are modified into pinchers and are used in prey capture and manipulation. Four pairs of walking legs are present. Immatures are similar to adults in general body form.

F.   Characteristics of Centipedes and Millipedes *(Classes* **Chilopoda** *and* **Diplopoda***)*

Centipedes and millipedes bear little resemblance to the other arthropods previously discussed. They have hardened, elongated *wormlike* bodies with distinct heads and many pairs of walking legs. Centipedes — often commonly called *hundred leggers* — are swift-moving predatory organisms with one pair of long legs on each body segment behind the head (Figure 8). Millipedes *(thousand leggers)*, on the other hand, are slow-moving omnivores or scavengers that have two pairs of short legs on each body segment (after the first three segments, which only have one pair each; Figure 8). Immatures are similar to the adults.

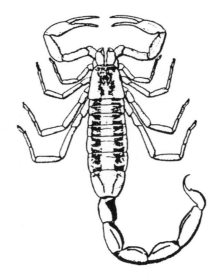

Figure 7
A typical scorpion. (From U.S. DHEW, PHS, CDC Pictorial Keys.)

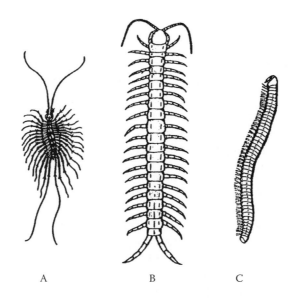

A   B   C

Figure 8
(A) and (B) Centipedes and (C) millipede. (From U.S. DHEW, PHS, CDC Pictorial Keys.)

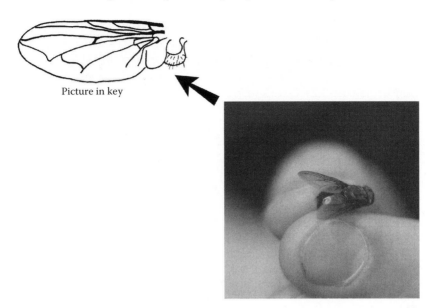

Picture in key

Figure 9
Example of how identification keys work. A specimen is compared to line drawings. Note how the wing veins of the fly in the hand match those in the line drawing.

III.   IDENTIFICATION KEYS

The identification of a particular arthropod to place it in its correct class, order, family, genus, or species is achieved by finding the answers to a paired series of questions set out in the form of keys. Most keys are dichotomous, giving the reader two choices in each couplet, and the reader is then referred to another couplet depending on the answer. This flowchart of sorts continues until an identification is reached. Some keys have good pictures and illustrations to aid in the identification, whereas others have nothing but written descriptions. For example, a key couplet might refer the reader to look at the wing of an insect and compare it to a line drawing of several variations of wings (see Figure 9 for an example). The chart that follows is an updated and revised Centers for Disease Control (CDC) Pictorial Key to the arthropods of medical importance. This chart is provided to enable the nonentomologist reader to identify an arthropod in relation to major group or family level. In addition to this chart, morphological characteristics and helpful identification hints for each of the arthropod groups discussed in this book are provided in each of the respective chapters dealing with those groups.

KEY TO SOME COMMON CLASSES AND ORDERS OF ARTHROPODS OF MEDICAL IMPORTANCE

1

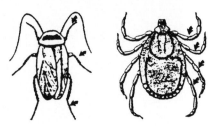 

Three or four pairs of legs
see Set 2

Five or more pairs of legs
see Set 22

2

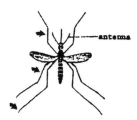 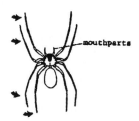

Three pairs of legs, with antennae
Class Insecta
(Insects)
see Set 3

Four pairs of legs, without antennae
Class Arachnida
(Scorpions, Spiders, Ticks, etc.)
see Set 20

3

Wings present, well developed
see Set 4

Wings absent or rudimentary
see Set 12

KEY TO SOME COMMON CLASSES AND ORDERS OF ARTHROPODS OF MEDICAL IMPORTANCE (continued)

4

One pair of wings
Order Diptera
see Set 5

Two pairs of wings
see Set 6

5

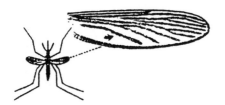

Wings with scales
Mosquitoes

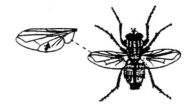

Wings without scales
Other flies

6

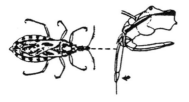

Sucking mouthparts, consisting of an
elongated proboscis
see Set 7

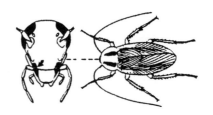

Biting/chewing mouthparts
see Set 8

KEY TO SOME COMMON CLASSES AND ORDERS OF ARTHROPODS OF MEDICAL IMPORTANCE (continued)

7

Wings usually totally covered with scales,
proboscis coiled up under head
Order Lepidoptera
(Butterflies and Moths)

Wings not covered with scales,
proboscis directed backward between
front legs when not in use
Order Hemiptera
(True Bugs and Kissing Bugs)

8

Both pairs of wings membranous and
similar in structure, although they
may differ in size
see Set 9

Front pair of wings leathery or
shell-like, serving as covers for the
membranous hind wings
see Set 10

9

Hind wings much smaller than
front wings
Order Hymenoptera
(Bees, Wasps, Hornets, and Ants)

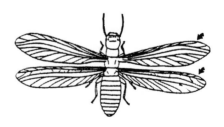

Both pairs of wings similar in size
Order Isoptera
(Termites)

KEY TO SOME COMMON CLASSES AND ORDERS OF ARTHROPODS OF MEDICAL IMPORTANCE (continued)

10

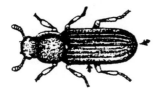

Front wings somewhat hardened
without distinct veins, meeting in
a straight line down the middle
see Set 11

Front wings leathery or paperlike
with a network of veins, usually
overlapping at the middle
Order Orthoptera
(Cockroaches, Crickets, Grasshoppers)
(Note: Cockroaches now placed
in order Blatteria)

11

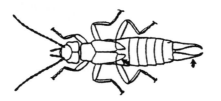

Abdomen with prominent forceps,
wings shorter than abdomen
Order Dermaptera
(Earwigs)

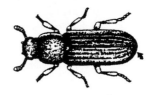

Abdomen without forceps,
wings typically covering abdomen
Order Coleoptera
(Beetles)

12

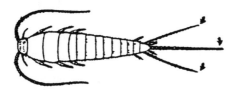

Abdomen with three elongate, tail-like
appendages at tip, body usually
covered with scales
Order Thysanura
(Silverfish and Fire Brats)

Abdomen without three long
tail-like appendages at tip,
body not covered with scales
see Set 13

KEY TO SOME COMMON CLASSES AND ORDERS OF ARTHROPODS OF MEDICAL IMPORTANCE (continued)

13

Abdomen with narrow waist
Order Hymenoptera
(Ants)

Abdomen without narrow waist
see Set 14

14

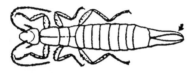

Abdomen with prominent pair of forceps
Order Dermaptera
(Earwigs)

Abdomen without forceps
see Set 15

15

Body strongly flattened from side to
side, antennae small, fitting into
grooves in side of head
Order Siphonaptera
(Fleas)

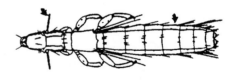

Body not strongly flattened from side to
side, antennae projecting from side
of head, not fitting into grooves
see Set 16

KEY TO SOME COMMON CLASSES AND ORDERS OF ARTHROPODS OF MEDICAL IMPORTANCE (continued)

16

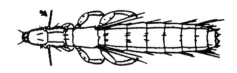

Antenna with nine or more segments
see Set 17

Antenna with three to five segments
see Set 18

17

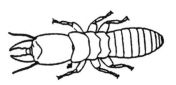

Pronotum covering head
Order Blatteria
(Cockroaches)

Pronotum not covering head
Order Isoptera
(Termites)

18

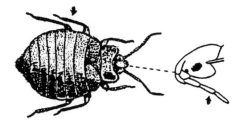

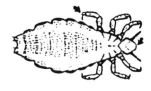

Mouthparts consisting of tubular jointed
beak, tarsi three- to five-segmented
Order Hemiptera
(Bed bugs)

Mouthparts retracted into head
or of the chewing type, tarsi
one- or two-segmented
see Set 19

19

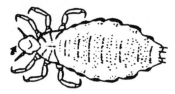 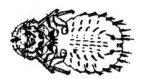

Mouthparts retracted into head, adapted for sucking blood, external parasites of mammals
Order Phthiraptera
(Sucking Lice)

Mouthparts of the chewing type, external parasites of birds and mammals
Order Phthiraptera
(Chewing Lice)

20

Body oval, consisting of a single saclike region
Subclass Acari
(Ticks or Mites)

Body divided into two distinct regions, a combined head-thorax and an abdomen
see Set 21

21

 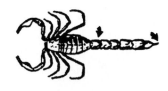

Abdomen joined to head-thorax (cephalothorax) by a slender waist, abdomen with segmentation indistinct or absent, stinger absent
Order Araneae
(Spiders)

Abdomen broadly joined to cephalothorax, abdomen distinctly segmented, ending with a stinger
Order Scorpiones
(Scorpions)

KEY TO SOME COMMON CLASSES AND ORDERS OF ARTHROPODS OF MEDICAL IMPORTANCE (continued)

22

 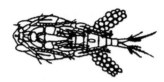

Five to nine pairs of legs in some
species, swimmerets in others, one or two
pairs of antennae present, principally
aquatic organisms
Subphylum Crustacea
(Crabs, Crayfish, Shrimp
and Copepods)

23

Ten or more pairs of legs, swimmerets
absent, one pair of antennae present,
terrestrial organisms, body segments
each with only one pair of legs
Class Chilopoda
(Centipedes)

Ten or more pairs of legs, swimmerets
absent, one pair of antennae present,
terrestrial organisms, body segments
each with two pairs of legs
Class Diplopoda
(Millipedes)

Table 3

Some Questions that May Be Helpful in Determining the Offending Arthropod or Other Causes of Unexplained Dermatitis

Question to Ask Patient	Possible Cause
Preliminary General Questions	
Where do you work? Nature of work?	Occupational exposure to arthropods or irritating fibers, e.g., fiberglass
At work do you sit or stand directly under AC vents, fans, etc.?	Fiberglass or other fibers hitting them
Any curtains or carpets installed or cleaned?	Fibers from such items can cause irritation
What time of year did lesions occur?	Spring and summer — arthropods likely
Patient Remembers Seeing or Removing the Biting or Stinging Bug	
Was it "wormlike"?	Centipede, caterpillar
Was it firmly attached?	Tick, pubic lice
Did it jump greatly?	Flea
Was it "beetlelike"?	Blister beetle
Was it very tiny (pinhead size or smaller)?	Mite, immature tick
Did it fly off?	Biting fly, mosquito, bee, wasp, hornet
Patient Did Not See or Remove Arthropod	
What time of day did these lesions occur?	If night — mosquitoes, if outdoors or unscreened house; bed bugs also possible (indoors)
Where were you when these lesions occurred?	Outdoors — biting flies, mosquitoes, fire ants, chiggers; Indoors — fleas, mites, bed bugs
Do you own pets (especially ones which go outside)?	Fleas, ticks, cheyletiellid mites
Did you go hiking, fishing, or hunting prior to lesion development?	Ticks, chiggers, biting flies, mosquitoes, etc.
Did you work in the yard or garden prior to lesion development?	Biting flies, mosquitoes, fire ants, chiggers, etc.
Did you go berry picking, flower digging, or do some other activity prior to lesion development?	Chiggers, ticks, mosquitoes, biting flies, fire ants
Were you cleaning out an attic, garage, or shed prior to lesion development?	Spiders, scorpions
Are there bats or bird nests in your attic or under the eaves of your windows?	Mites, soft ticks

Note: Not exhaustive; intended as a guide only.

IV. OTHER CLUES TO IDENTIFYING AN OFFENDING ARTHROPOD

When a patient with a lesion supposedly caused by an arthropod comes to a clinic, efforts should be made to identify that organism or at least make an educated guess. Sometimes, pertinent questions about the patient's activities just prior to lesion development are helpful in making this determination. Table 3 is a listing of such questions and the corresponding arthropods. This table is by no means exhaustive and is offered as a guide only. Certainly, in tropical and subtropical areas the list of possible offending arthropods is almost endless.

CHAPTER 9

SIGNS AND SYMPTOMS OF ARTHROPOD-BORNE DISEASES

TABLE OF CONTENTS

I. INTRODUCTION

The human body may react in various ways to disease pathogens or introduced foreign substances such as arthropod venom or saliva. One might think, therefore, that a careful listing of signs and symptoms of arthropod-caused problems would be useful to the practicing clinician in making diagnoses. Unfortunately, these reactions are rarely specific for one particular pathogen or offending arthropod. However, the presence of certain signs and symptoms may alert physicians to diagnostic possibilities. Further confirmation may be aided by specific laboratory findings. The following is an alphabetical listing of common signs and symptoms of arthropod-borne diseases. By no means should it be considered a complete differential diagnosis of any of the symptoms discussed.

II. COMMON SIGNS AND SYMPTOMS

Acute abdomen: Rigid, boardlike abdomen accompanied by intense pain is often a sign of a black widow spider bite. It is caused in this case by the neurotoxic effects of the venom.

Adenopathy: There may be generalized adenopathy in the early stages of African trypanosomiasis, the glands of the posterior cervical triangle being most conspicuously affected (Winterbottom's sign). Also, generalized adenopathy may be seen in the acute stage of Chagas' disease.

Anaphylaxis: Anaphylactic shock is often a result of arthropod stings (less commonly, bites), hypersensitivity to arthropod venom or saliva being the cause in this case.

Anemia: Anemia may be seen in cases of malaria, babesiosis, and trypanosomiasis. Anemia can be especially severe in falciparum malaria. In addition, systemic reactions to brown recluse spider bites (systemic loxoscelism) may result in hemolytic anemia.

Asthma, bronchial: Cockroaches and house dust mites are important causes of nasal allergies and asthma, especially in children. Recently, there has been increased awareness that asthma in urban-dwelling children may be due to (or at least exacerbated by) cockroaches.

Axillary pain, acute: Venoms from arthropod stings may be transported to axillary lymph nodes and cause transient intense pain. This is especially true for some of the stinging caterpillars.

Blister: A blister may occur at arthropod bite sites. Blistering may also occur as a result of blister beetles contacting human skin.

Bulls-eye rash: See **Erythema migrans.**

Chagoma: Indurated, erythematous lesion on the body (often head or neck) caused by *Trypanosoma cruzi* infection (Chagas' disease). May persist for 2 to 3 months.

Chyluria: Presence of chyle (lymphatic fluid) in the urine. Often seen in lymphatic filariasis. Urine may be milky white and may even contain microfilariae.

Coma: Sudden coma in a person returning from a malarious area suggests cerebral malaria. Also, African trypanosomiasis (sleeping sickness) may lead to coma after a long period of increasingly severe symptoms of meningoencephalitis. Rocky Mountain spotted fever and other rickettsial infections may lead to coma.

Conjunctivitis: Chronic conjunctivitis may occur in onchocerciasis and Chagas' disease.

Dermatitis: Several arthropods may directly or indirectly cause dermatitis. Chiggers and other mites may attack the skin, causing a maculopapular rash. Scabies mites may burrow under the skin's surface, making itchy trails or papules. Lice may give rise to hypersensitivity reactions with itchy papules. Chigoe fleas burrow in the skin (especially on the feet), causing local irritation and itching. Macules or erythematous nodules may result as a secondary cutaneous manifestation of leishmaniasis.

Diarrhea: Leishmaniasis (specifically kala-azar) may lead to mucosal ulceration and diarrhea. In falciparum malaria, plugging of mucosal capillaries with parasitized red blood cells may lead to watery diarrhea.

Edema: Edema may result from arthropod bites or stings. Large (but still contiguous) swollen areas or edema from bites or stings are known as *large local reactions*. Loiasis (a nematode worm transmitted by deer flies) may cause unilateral circumorbital edema as the adult worm passes across the eyeball or lid. Passage of the worm is brief, but inflammatory changes in the eye may last for days. Loiasis may also lead to temporary appearance of large swellings on the limbs, known as *Calabar swellings*, at the sites where migrating adult worms occur. Unilateral edema of the eyelid, called *Romaña's sign*, may occur in Chagas' disease. African trypanosomiasis (sleeping sickness) may result in edema of the hips, legs, hands, and face.

Elephantiasis: Hypertrophy and thickening of tissues, leading to an elephant leg appearance, may result from lymphatic filariasis. Tissues affected may include a limb, the scrotum, and the vulva.

Encephalitis: Inflammation of the brain may result from infection by various types of mosquito-borne viruses such as St. Louis, eastern equine, West Nile, and LaCrosse encephalitis viruses.

Eosinophilia: Helminth worms often provoke eosinophilia. Atopic diseases such as rhinitis, asthma, and hay fever also are characterized by eosinophilia.

Eosinophilic cerebrospinal fluid pleocytosis: Cerebrospinal fluid eosinophilic pleocytosis can be caused by a number of infectious diseases (including rickettsial and viral infections), but is primarily associated with parasitic infections.

Epididymitis: Epididymitis, often associated with orchitis, may be an early complication of lymphatic filariasis.

Erythema migrans (EM): Cutaneous lesion that may follow bites of ticks infected with the causative agent of Lyme disease, *Borrelia burgdorferi*. Typically, the lesion consists of an annular erythema with a central clearing surrounded by a red migrating border. Although EM does not always occur, it is virtually pathognomonic for Lyme disease.

Eschar: A round (generally 5- to 15-mm) spot of necrosis may be indicative of boutonneuse fever (a spotted fever group illness) or scrub typhus. The eschar develops at the site of tick or chigger bite.

Excoriation: Lesions produced by excoriation may be a sign of imaginary insect or mite infestations (delusions of parasitosis).

Fever: Fever may accompany many arthropod-borne diseases, including the rickettsioses, dengue, yellow fever, plague, the encephalitides, and others. In some cases there are cyclical peaks of fever, such as in relapsing fever (tick-borne) or malaria. Falciparum malaria is notorious for causing extremely high fever (107°F or higher). Filariasis may be marked by fever, especially early in the course of infection.

Hematemesis: Coffee-ground color or black vomit is a sign of yellow fever.

Hemoglobinuria: Falciparum malaria is the cause of blackwater fever. Sometimes systemic reactions to brown recluse spider bites lead to hemoglobinuria.

Hives: See **Urticaria.**

Hydrocele: Hydrocele is often a result of lymphatic filariasis, developing as a sequel to repeated attacks of orchitis.

Hyponatremia: Sometimes hyponatremia, or low blood sodium level, is a laboratory finding in Rocky Mountain spotted fever.

Itch: Intense itching, especially at night, is a sign of scabies infestation. Itch is also a common result of various insect bites or stings.

Kerititis: Inflammation of the cornea is sometimes a result of ocular migration of *Onchocerca volvulus* microfilariae. The condition may lead to blindness.

Leukocytosis: Leukocytosis may occur in systemic loxoscelism. Systemic loxoscelism is a rare complication of brown recluse spider bites.

Leukopenia: Leukopenia is a prominent sign of ehrlichiosis. It may also occur (3000 to 6000 per mm^3) with a relative monocytosis during the afebrile periods of malaria.

Lymphadenitis: Inflammation of one or more lymph nodes may be a sign of lymphatic filariasis, especially the femoral, inguinal, axillary, or epitrochlear nodes.

Lymphangitis: Lymphangitis may be an early symptom of lymphatic filariasis. The limbs, breast, or scrotum may be involved.

Lymphocytosis: Lymphocytosis is generally seen in Chagas' disease.

Maggots: The presence of fly larvae in human tissues is termed *myiasis*. Various blow flies, bot flies, and other muscoid flies are usually involved.

Meningoencephalitis: Meningoencephalitis may be caused by trypanosomes in the cases of African trypanosomiasis (sleeping sickness) or Chagas' disease (although generally milder). Also, falciparum malaria infection may be cerebral, with increasing headache and drowsiness over several days, or even sudden onset of coma.

Myocarditis: Chagas' disease often leads to myocardial infection. African trypanosomiasis may also cause myocarditis to a lesser extent.

Necrosis: Cell death caused by progressive enzymatic degradation may result from spider envenomization, especially the brown recluse, *Loxosceles reclusa*.

Neuritis: Neuritis may be caused by bee, ant, or wasp venom. Occasionally stings to an extremity result in weakness, numbness, tingling, and prickling sensations for days or weeks. Neuritis may also result from infection with the Lyme disease spirochete.

Nodules, subcutaneous: Tick bites may ultimately result in nodules. Fly larvae in the skin (myiasis) may also present as nodules. Notorious offenders are the human bot fly larva, *Dermatobia hominis*, the Tumbu fly, *Cordylobia anthropophaga*, and rodent bot fly larvae, *Cuterebra* spp.

Onchocercoma: Coiled masses of adult *Onchocerca volvulus* worms beneath the skin enclosed by fibrous tissues may occur in patients living in tropical countries endemic for ochocerciasis.

Orchitis: Orchitis may result from lymphatic filariasis; repeated attacks may lead to hydrocele.

Paralysis: Ascending flaccid paralysis may result from tick attachment. The paralysis is believed to be caused by a salivary toxin injected as the tick feeds.

Proteinuria: Proteinuria, with hyaline and granular casts in the urine, is common in falciparum malaria.

Puncta: A small, point-like pierce mark may mark the bite or sting site of an arthropod. Paired puncta may indicate spider bite or centipede bite.

Rash: Rash may accompany many arthropod-borne diseases such as Rocky Mountain spotted fever, ehrlichiosis, murine typhus, and African trypanosomiasis. The rash may appear to be ring-like, and expanding in the case of Lyme disease (see **Erythema migrans**). An allergic urticarial rash may be seen in the case of bites or stings.

Rhinitis, allergic: Intermittent or continuous rhinitis may be due to exposure to house dust mites or cockroaches.

Romaña's sign: Early in the course of Chagas' disease there may be unilateral palpebral edema, involving both the upper and lower eyelids. This generally occurs when a kissing bug (the vector of the Chagas' organism) bites near the eye.

Shock: Shock may occur from arthropod stings (rarely bites) as a result of hypersensitivity reactions to venom or saliva. Shock may also accompany falciparum malaria.

Splenomegaly: Splenomegaly may result from lymphoid hyperplasia in both African and American trypanosomiasis. It may also occur in visceral leishmaniasis (kala-azar).

Swelling: See **Edema.**

Tachycardia: Both African and American trypanosomiasis may produce tachycardia. In Chagas' disease tachycardia may persist into the chronic stage, where it may be associated with heart block.

Thrombocytopenia: Although not specific for any one disease, thrombocytopenia, or low platelet count, may occur as a result of infection with Rocky Mountain spotted fever organisms.

Ulcers, cutaneous: A shallow ulcer (slow to heal) may be a sign of cutaneous leishmaniasis. In the New World, cutaneous leishmaniasis lesions are most often found on the ear. Also, a firm, tender, raised lesion up to 2 cm or more in diameter may occur at the site of infection in African trypanosomiasis.

Urticaria: Urticaria may result from an allergic or generalized systemic reaction to arthropod venom or (more rarely) saliva.

Winterbottom's sign: In the early stages of African trypanosomiasis, patients may exhibit posterior cervical lymphadenitis.

PART III

ARTHROPODS OF MEDICAL IMPORTANCE

CHAPTER 10

ANTS

TABLE OF CONTENTS

Table 1
Some Ants that Are Known to Sting People

Scientific Name	Common Name	Occurs Where
Solenopsis invicta	Red imported fire ant (IFA)	South America, Southern U.S.
S. richteri	Black IFA	South America, Southern U.S.
S. xyloni	Southern fire ant	California to South Carolina down into parts of Florida
S. geminata	Fire ant	Texas to South Carolina down to South America
Pogonomyrmex barbatus	Red harvester ant	Western U.S.
P. californicus	California harvester ant	Western U.S.
P. occidentalis	Western harvester ant	Western U.S.
P. badius		Eastern U.S.
Myrmica rubra		U.K.
M. ruginodes		U.K.
Paraponera clavata	Bullet ant; Viente-cuatro hora hormiga	South America
Monomorium bicolor		Africa
Myrmecia gulosa	Red bulldog ant	Australia
M. pyriformis	Bulldog ant	Australia
M. forficata		Australia

I.   ANTS IN GENERAL

Ants (family Formicidae) are very successful organisms occurring in tremendous numbers worldwide in terrestrial habitats. They are eusocial insects comprising three castes: queens, males, and workers. Their highly developed social structure makes them an interesting and much-studied group. Two excellent sources of information on the biology and taxonomy of ants are Hölldobler and Wilson[1] and Bolton.[2] All ant species may bite (if they are physically large enough), and some species sting (Color Figure 10.9 and Color Figure 10.10). Ant stings can be significant. The sting of the giant tropical bullet ant, *Paraponera clavata*, is the most painful and debilitating of any known insect.[3,4] Indigenous people say it takes 2 weeks to recover from one sting. In addition, some ant species emit a foul-smelling substance when handled. The imported fire ants, harvester ants, and velvet ants (not actually ants) are discussed in this chapter. Many other species may be of public health importance; a few of them are listed in Table 1.

II.   FIRE ANTS

A.   General and Medical Importance

There are some native fire ants in the U.S., but the imported ones, *Solenopsis invicta* and *S. richteri*, are the worst pests.[5] Imported fire ants (IFAs) sting aggressively and inject a necrotizing venom consisting primarily of alkaloidal compounds, which they

Figure 1
Fire ants bite in order to gain leverage for the act of stinging. (From USDA Yearbook of Agriculture, 1952.)

use to paralyze or kill their prey. The ants characteristically boil out of their mounds in great numbers at the slightest disturbance (Color Figure 10.11). Worker IFAs attach to the skin of their victim with their mandibles and lower the tip of their abdomen to inject the stinger forcefully (Figure 1); therefore, IFAs both bite and sting, but their stings cause the subsequent burning sensation and wheal. Alexander[6] said the characteristic symptom of IFA sting is a burning itch. Within 24 h, a pustule (actually a pseudopustule) develops that may persist for a week or longer.[7] Individuals have sustained up to 10,000 stings without systemic toxic or immunological reactions.[8] However, hypersensitivity to IFA venom may result in severe allergic reactions from just a few stings (see Chapter 2). In addition, some evidence exists that fire ant venom may act as a neurotoxin. One 4-year-old boy in good health had 2 grand mal seizures 30 min after being stung on the foot by 20 IFAs.[9]

B.   General Description

The first, and sometimes second, segment of the abdomen of all ants is nodelike. Adult worker fire ants (Figure 2) have two nodes, are approximately 4 to 6 mm long, and are reddish brown (red form, most widely distributed) or brownish black (black form). Outdoors, fire ants are best recognized by the appearance of their mounds, which are elevated earthen mounds 8 to 90 cm high surrounded by relatively undisturbed vegetation (Figure 3).

IMPORTED FIRE ANTS

Figure 2
Imported fire ant. (From Mississippi's Health, Summer 1990 issue. With permission.)

Importance
Painful stings; allergic RXNs

Distribution
Much of southern U.S.; native to South America

Lesion
Immediate — erythema and central wheal; later — pustules

Disease Transmission
None

Key Reference
Lofgren et al., *Annu. Rev. Entomol.,* 20, 1, 1975

Treatment
Pain relievers, antipruritic lotions for local RXNs; systemic RXNs may require antihistamines, epinephrine, and other measures

Figure 3
Huge fire ant mound.

C. Geographic Distribution

Both IFA species were accidentally introduced into the U.S. (in the Mobile area) in the time period of 1918 to 1940. The red form, *S. invicta*, was brought in from Brazil, and the black form, *S. richteri*, came from Uruguay. The ants may have originally been imported in infested nursery stock or dirt used for ship ballast. Through subsequent unintentional dissemination by people and through dispersal by flooding and mating flights, the red form has spread rapidly, and its range now extends from the southern Atlantic coast westward to California. At least 250 million acres in the U.S. are currently infested. The black form currently occurs only in northeastern Mississippi, northwestern Alabama, and a small portion of southern Tennessee. States currently reporting IFA infestations are Arizona, Alabama, Arkansas, California, Florida, Georgia, Louisiana, Mississippi, New Mexico, North Carolina, Oklahoma, South Carolina, Tennessee, Texas, and Virginia (Figure 4).

D. Biology and Behavior

An IFA colony is usually started by a single fertilized queen with an initial burrow; the first eggs are laid from 24 to 48 h after completion of the burrow. Normal development from egg to worker takes approximately 24 to 30 d depending on temperature and available soil moisture. These first workers open the nest and begin to deepen the tunnel. After about 60 d a few minor workers appear, and at about 5 months a few major workers can be found. The nest is further excavated and the characteristic mound begins to form during the 5-month period after initial colony formation. After 1 year, individuals in the nest may number 10,000 or more, and the colony typically increases from about 30,000 workers after 1 to 1½ years to around 60,000 workers after 2 to 2½

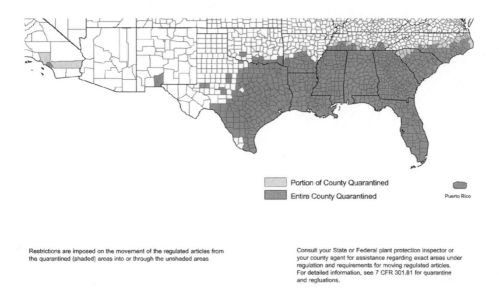

Portion of County Quarantined

Entire County Quarantined

Puerto Rico

Restrictions are imposed on the movement of the regulated articles from the quarantined (shaded) areas into or through the unshaded areas

Consult your State or Federal plant protection inspector or your county agent for assistance regarding exact areas under regulation and requirements for moving regulated articles. For detailed information, see 7 CFR 301.81 for quarantine and regluations.

Figure 4
U.S. Department of Agriculture imported fire ant quarantines, generally representing the distribution of the ants in the U.S. (From USDA, APHIS, updated January 2002.)

years. Major workers are 3 to 5 mm long and live 90 to 180 d; queens are a bit larger, live 2 to 5 years, and can lay 1500 eggs a day. A colony is considered to be mature after 3 years and may contain as many as 200,000 workers or more in some areas. A single mature colony can produce 4500 potential new queens during a year. Some heavily infested areas may contain 50 to 400 or more mature colonies per acre. Reproduction in the red IFA may occur throughout the year but peaks from late May through August. During nuptial flights, newly inseminated queens select a landing site suitable for colony formation, and burrow excavation is begun shortly thereafter.

Fire ants are omnivorous in their feeding habits but prefer insects, spiders, other small arthropods, and earthworms over plant feeding. In heavily infested areas, where food is in short supply, fire ants will eat the germplasm of newly germinated seeds, girdle the stems of small seedlings, and consume hatchling vertebrates. Foraging tunnels 15 to 25 m long are used by workers to collect food for the colony. These tunnels are excavated just below the soil surface and extend outward from the mound in all directions. Foraging workers travel through these tunnels, emerge from an opening, and search for a food source. Once a food source is located, the foraging worker returns to the tunnel laying a trail of pheromone for other worker ants to follow.

E.   Treatment of Stings

Fire ant stings are usually treated in a manner similar to wasp or bee stings. However, the pustules characteristically produced by IFA stings may need to be covered with bandages to avoid excoriation. Ice packs, antibiotic ointments, or hydrocortisone creams are often used for local reactions, and epinephrine, antihistamines, and other supportive measures are indicated for more serious, systemic reactions (see Chapter 2).

HARVESTER ANTS

Figure 5
Worker harvester ant.

Importance
Painful stings; possible allergic RXNs

Distribution
In the U.S., most species are the western type

Lesion
Variable

Disease Transmission
None

Key Reference
Creighton, W.S., *Harvard Bull. Mus. Comp. Zool.*, Vol. 104, 585 pp. 1950

Treatment
Local RXNs — wash wound; pain relievers; systemic RXNs may require antihistamines, epinephrine, and other measures

III. HARVESTER ANTS

A. General and Medical Importance

Harvester ants, *Pogonomyrmex* spp., are dangerous ant species that readily sting people and animals (Figure 5). The reaction to their stings is not always localized; it may spread along the lymph vessels, producing intense pain in the lymph nodes in the axilla or groin long after the original sting pain subsides. Although relatively uncommon, systemic hypersensitive reactions may result from harvester ant stings.

B. General Description

Harvester ants are red to dark brown ants and are 2 to 3 times larger than fire ants; they have large, wide heads (Figure 6). Several species characteristically have many small, parallel ridges on the head. In addition, a good way to distinguish between harvester and fire ants is by the shape and size of their mounds (Figure 3 and Figure 7). Harvester ant mounds are usually flat or slightly elevated and are surrounded by an area of no vegetation 1 to 3 m or more in diameter. Fire ant mounds are distinctly elevated and composed of excavated dirt.

C. Geographic Distribution

There is only one harvester ant species east of the Mississippi River, *Pogonomyrmex badius*, which is distributed throughout the southeastern states. In the western U.S. approximately 20 species occur. Three of the notorious offenders are the red harvester ant (*P. barbatus*), the California harvester ant (*P. californicus*), and the western harvester ant (*P. occidentalis*).

D. Biology and Behavior

These ants are diurnal soil-inhabiting insects that build their nests in open areas. They feed on seeds and store them in honeycomb-like chambers within their nests. Harvester ants clear out a large vegetation-free, craterlike

Figure 6
Microscopic view of harvester ant. Notice large head and mandibles.

Figure 7
Harvester ant mound; actually, a cleared-out area with a centralized entrance. (Photo courtesy of Dr. Chad P. McHugh, Brooks AFB, TX.)

area on the ground surface, and there are one or more holes leading down into the nest. Like other ants, their colonies consist of at least one reproductive queen, several males, and many nonreproductive female workers. The workers are the ones that sting to defend the nest.

E.   Treatment of Stings

Local first aid for harvester ant stings consists of washing the site with soap and water and applying ice packs. Unlike most ant species (and similar to the honey bee),

VELVET ANTS

Figure 8
Female velvet ant.
(From U.S. DHEW, PHS, CDC Pictorial Keys.)

Importance
Painful stings

Distribution
In the U.S., most species occur in South and West

Lesion
Variable

Disease Transmission
None

Key Reference
Mickel, C. E., *Bull. U.S. Natl. Mus.*, No. 143, 1928

Treatment
Local — ice packs; pain relievers systemic — may require antihistamines, epinephrine, and other measures

the sting may be torn off in the wound upon envenomization by *P. californicus*.[10] Allergic reactions may require administration of epinephrine, antihistamines, and other supportive treatment (see Chapter 2).

IV. VELVET ANTS

A. General and Medical Importance

Velvet ants (also sometimes called woolly ants, cow killers, mule killers, or mutillid wasps) are actually wingless female wasps in the family Mutillidae: they only resemble huge ants (Figure 8 and Figure 9). The wingless females have a long stinger and can inflict a painful sting. They are sometimes a problem to barefoot bathers on sandy beaches.

B. General Description

Velvet ants do not have a node on the petiole, which is characteristic of the true ants. Also, females (the ones confused with ants) are generally larger than true ants. Both male and female velvet ants are covered with a bright red, orange, black, or yellow pubescence. There are several species, but the more commonly encountered ones are from 0.75 to 2.5 cm in length.

Figure 9
Velvet ant (CDC photo).

C. Geographic Distribution

Members of this wasp family are widely distributed, but most U.S. species occur in the southern and western areas of the country. One northern species, *Dasymutilla occidentalis*, is common on the sandy beaches of Lake Erie in the summer, causing much distress to barefoot beachgoers.[11]

D. Biology and Behavior

These ants are solitary, diurnal wasps that are often associated with dry, sandy environments. Male mutillids are winged, whereas females are wingless. Their larvae are parasites of bees and other wasps. Females run about in the open searching for a suitable place to lay their eggs.

E. Treatment of Stings

Being solitary wasps (not actually ants), stings by numerous velvet ants are rare. Disinfection of the sting site, application of ice packs, and administration of analgesics are often indicated for normal or local sting reactions. In the event of an allergic reaction, epinephrine, antihistamines, and other supportive measures may be needed.

REFERENCES

1. Hölldobler, B. and Wilson, E.O., *The Ants*, Harvard University Press, Cambridge, MA, 1990, pp. 1–732.
2. Bolton, B., *Identification Guide to the Ant Genera of the World*, Harvard University Press, Cambridge, MA, 1994, pp. 1–478.
3. Schmidt, J.O., Hymenopteran venoms: striving toward the ultimate defense against vertebrates, in *Insect Defenses, Adaptive Mechanisms, and Strategies of Prey and Predators*, Evans, D.L. and Schmidt, J.O., Eds., State University of New York Press, Albany, NY, 1990, p. 387.
4. Schmidt, J.O., Chemistry, pharmacology, and chemical ecology of ant venoms, in *Venoms of the Hymenoptera*, Piek, T., Ed., Academic Press, New York, 1986, p. 425.
5. Taber, S.W., *Fire Ants*, Texas A&M University Press, College Station, TX, 2000, pp. 22.
6. Alexander, J.O., *Arthropods and Human Skin*, Springer-Verlag, Berlin, 1984, chap. 10.
7. Favorite, F., Imported fire ant, *Public Health Rep.*, 73, 445, 1958.
8. Diaz, J.D., Lockey, R.F., Stablein, J.J., and Mines, H.K., Multiple stings by imported fire ants without systemic effects, *South. Med. J.*, 82, 775, 1989.
9. Fox, R.W., Lockey, R.F., and Bukantz, S.C., Neurologic sequelae following the imported fire ant sting, *J. Allerg. Clin. Microbiol.*, 70, 120, 1982.
10. Ebeling, W., *Urban Entomology*, University of California Press, Berkeley, CA, 1978, 351.
11. Frazier, C.A., *Insect Allergy: Allergic Reactions to Bites of Insects and Other Arthropods*, Warren H. Green, St. Louis, 1969, chap. 7.

CHAPTER 11

BEES

TABLE OF CONTENTS

HONEY BEES

Figure 1
Worker honey bee. (From Mississippi's Health, *Summer 1990 issue. With permission.)*

Importance
Painful stings; allergic RXNs

Distribution
Almost worldwide

Lesion
Local swelling; central white spot with erythematous halo

Disease Transmission
None

Key Reference
Michener, C.D., *Annu. Rev. Entomol.*, 14, 299, 1969

Treatment
Local RXNs — ice packs, pain relievers; systemic RXNs — may require antihistamines, epinephrine, and other measures

I. HONEY BEES

A. General and Medical Importance

Honey bees, *Apis mellifera*, are commonly encountered insects and account for numerous stings (and even deaths due to allergy) in the U.S. each year.[1] The problem lies not with their aggressiveness, but with their ubiquity. They are virtually everywhere outdoors during the warm months — in backyard clovers, windowsill flowers, garden vegetable blooms, waste receptacle areas, etc. Unlike most other Hymenoptera, the honey bee worker has a barbed stinger and can sting only once. To escape, the bee must leave its entire stinging apparatus attached to the skin of its victim. Making matters worse, during stinging events honey bees release alarm pheromones associated with the sting gland, causing other bees in the vicinity of a pheromone-marked victim to attack and inflict multiple stings.

B. General Description

Honey bees are the familiar yellow-orange and black-striped bees with two membranous wings that are commonly seen on flowers during spring and summer (Figure 1). They have feathered hairs (when observed under magnification) and no spurs on their hind tibia. Worker honey bees are approximately 15 to 20 mm long, and drones are slightly larger and more robust. There are "races" of honey bees in the U.S.; the gold Italians and the black or gray Caucasian races make up the majority of bees found in this country. A close relative, the Africanized or "killer" bee, has recently entered the U.S. (see Section II).

C. Geographic Distribution

Honey bees are not native to several continents (including North America, north of Mexico), but have been introduced virtually worldwide for pollination and honey production.

D. Biology and Behavior

A. mellifera is a highly social insect. The colony consists of an egg-laying queen, drones to fertilize the queen, and workers to gather food and care for the young. Most honey bee colonies are artificial; however, wild colonies exist, mostly from escaped swarms, and they are usually found in hollow trees. Cells in a honey bee nest are in vertical combs, and two cell layers thick. The colonies are perennial, with the queen and workers overwintering in the hive. Normally, there is only one queen per colony, which may live for several years. When a new one is produced, it may be killed by the old queen, or one of the two may leave with a swarm of workers to build a new nest.

E. Treatment of Stings

Local treatment of honey bee stings consists of removing the stinger, disinfecting the sting site, and applying ice packs to slow the spread of venom. The stinger should be removed as quickly as possible. Conventional advice in scientific literature has emphasized that the sting should be scraped off, never pinched. But one study[2] has shown that the method of removal is irrelevant, and even slight delays in removal caused by concerns about the correct procedure are likely to increase the dose of venom received. Other than stinger removal, Alexander[3] said that nonallergic local reactions seldom require treatment. In the case of a severe or large local reaction, oral antihistamines and topical applications of corticosteroid creams may help, as well as rest and elevation of the affected arm or leg.

Allergic (systemic) reactions to bee stings can be life-threatening events and may require administration of epinephrine, antihistamines, and other supportive treatment (see Chapter 2).

II. AFRICANIZED OR "KILLER" BEES

A. General and Medical Importance

The Africanized honey bee (AHB) is a strain that was brought to South America from Africa to improve honey production. In 1956, 26 swarms of the AHB escaped from experimental

AFRICANIZED HONEY BEES_____

Figure 2
Africanized honey bee worker (left) and European honey bee worker (right). (Photo courtesy of John Kucharski and the USDA Agricultural Research Service.)

Importance
Highly aggressive; painful stings; allergic RXNs

Distribution
South and Central America, Mexico, southern U.S.

Lesion
Same as honey bee sting

Disease Transmission
None

Key Reference
Winston, M.L., *Annu. Rev. Entomol.*, 37, 173, 1992

Treatment
Local RXNs — ice packs, pain relievers; numerous stings (nonallergic RXNs) — treat for histamine overdosage; systemic RXNs — may require antihistamine, epinephrine, and other measures

colonies near Sao Paulo, Brazil, and have subsequently multiplied and spread throughout much of South and Central America.[4] They are now in much of the southern U.S. Winston[5] presents a detailed up-to-date review of the status of the AHB.

The AHB looks almost identical to our domesticated strains of European honey bees (Figure 2), but it is more likely to attack with little provocation, stay angry for a longer time, and exhibit massive stinging behavior in colony defense. Although there are some differences in the venom of AHBs vs. European bees, AHB stings are not more toxic than those of domestic bees. However, death of humans and domestic or wild animals can result from toxic effects of multiple stings (400 to 1000 stings) or from anaphylactic shock from an allergic reaction to very few stings (see Chapter 2).

As the AHB makes its way further into the U.S., numbers of sting-related deaths will likely increase. In Venezuela during 1978 (before AHB) there were 12 deaths attributed to honey bee stings. In 1988 (after AHB) there were 100 bee-sting-related deaths.[6] An increase in mortality of this magnitude may be avoided in the U.S. by better beekeeping practices and prompt medical management of stinging events.

B. General Description

AHBs look identical to the common honey bees (Figure 2). Only a specialist can tell them apart.

C. Geographic Distribution

Africanized bees are native to Africa. Since their accidental introduction in 1956, they have colonized most of South America, Central America, and Mexico. They are now reported from Texas, Louisiana, Arkansas, Florida, parts of New Mexico, Arizona, and California. Their spread into other southern U.S. states is likely.

D. Biology and Behavior

Most biological aspects of AHBs are similar to those of domestic honey bees (see preceding section); however, the mating habits of AHBs are such that Africanized queens tend to mate almost always with Africanized drones. They also reproduce at a high rate and swarm frequently. AHB colonies have more guard bees than European bee colonies; up to 50% of the bees in an AHB colony may be guard bees, which respond to a disturbance.

E. Treatment of Stings

Because the venoms are similar, AHB stings can be treated in the same manner as domestic honey bee stings. One note may be added: In the case of tens or hundreds of stings, extensive histamine release may result from venom action and not necessarily due from an allergic reaction. Accordingly, physicians may need to treat for histamine overdosage.

III. BUMBLE BEES

A. General and Medical Importance

There are several species of bumble bees in the genera *Bombus, Megabombus,* and *Pyrobombus.* Bumble bees, like other stinging Hymenoptera, will attack and sting when their nest is disturbed. A significant number of stinging incidents occur yearly, and some individuals experience allergic reactions. However, bumble bees are neither as aggressive nor as abundant as honey bees, and therefore generally not dangerous.[7]

B. General Description

Bumble bees have feathered hairs on their body and have spurs on their hind tibia. They are robust bees, usually 20 mm or more in length, having black and yellow pubescence on their abdomen (Figure 3 and Figure 4). They are often confused with carpenter bees, which are similar in size and appearance, except that carpenter bees have no pubescence on the abdomen. Carpenter bees rarely sting and are often seen forming galleries in wooden structures such as barns, sheds, stables, etc.

C. Geographic Distribution

Bumble bees occur throughout the U.S. and over much of the world.

D. Biology and Behavior

Most bumble bees are diurnal plant feeders that nest in the ground, usually in loose fibrous

Figure 4
Female bumble bee.

BUMBLE BEES

Figure 3
Bumble bee worker. (From North Carolina Agric. Exp. Stn. Tech. Bull. No. 152.)

Importance
Painful stings; allergic RXNs

Distribution
Almost worldwide

Lesion
Similar to other wasps and bees; central white spot and erythematous halo

Disease Transmission
None

Key Reference
Mitchell, T.B., *North Carolina Agric. Exp. Stn. Tech. Bull.* No. 152, 1962

Treatment
Local RXNs — ice packs, pain relievers; systemic RXNs — may require antihistamines, epinephrine, and other measures

habitats such as mouse nests, insulation, or grass clippings. Their colonies are annual, with only fertilized queens overwintering. These queens start new nests in the spring, and the colony builds to between 100 and 500 bees by late summer. The opening to their nests may appear moundlike from materials excavated by the bees. Bumble bee nest openings that resemble fire ant mounds have been seen.

E.   Treatment of Stings

Local treatment of bumble bee stings involves using ice packs and pain relievers, and washing to lessen the chances of any secondary infection. In the case of a severe or large local reaction, oral antihistamines and topical applications of corticosteroid creams may help, as well as rest and elevation of the affected arm or leg.

For allergic (systemic) reactions, administration of epinephrine, antihistamines, and other supportive treatment may also be required (see Chapter 2).

REFERENCES

1. Harwood, R.F. and James, M.T., *Entomology in Human and Animal Health*, Macmillan, New York, 1979, chap. 17.

2. Visscher, P.K., Vetter, R.S., and Camazine, S., Removing bee stings, *Lancet*, 348(August 3), 301, 1996.

3. Alexander, J.O., *Arthropods and Human Skin*, Springer-Verlag, Berlin, 1984, chap. 10.

4. Michener, C.D., The Brazilian honey bee — possible problem for the future, *Clin. Toxicol.*, 6, 125, 1973.

5. Winston, M.L., The biology and management of Africanized honey bees, *Annu. Rev. Entomol.*, 37, 173, 1992.

6. Gomez-Rodriguez, R., Manejo de la Abeja Africanizada, in Direccion General Desarrollo Ganadero, Caracas, Venezuela, 1986.

7. Biery, T.L., Venomous Arthropod Handbook, U.S. Air Force pamphlet No. 161-43, Brooks AFB, TX, 1977, p. 18.

CHAPTER 12

BEETLES

TABLE OF CONTENTS

I. BEETLES IN GENERAL

A. General and Medical Importance

Beetles belong to the insect order Coleoptera. At least 300,000 species have been described, representing 30 to 40% of all known insects. About 25,000 species of beetles occur in the U.S. and Canada.[1,2] Fortunately, beetles are of minor public health importance. However, some adults may inflict painful bites, several species may secrete irritating chemicals when touched or handled (discussed in the following section), and certain species may act as intermediate hosts for helminths. In addition, beetles such as grain beetles, weevils, and pantry beetles found in stored products can cause inhalational allergies. Some long-horned beetles (family Cerambycidae) have huge jaws and may inflict extremely painful bites if handled. Ladybugs, or lady beetles, have been reported to inflict unprovoked bites or cause stinging sensations by their defensive secretions.[3] Tapeworms, flukes, roundworms, and thorny-headed worms, which infest domestic and wild animals, use beetles as intermediate hosts. Humans may become

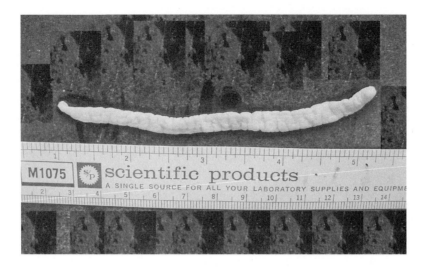

Figure 1
*Acanthocephalan submitted to the Mississippi Department of Health found in the diaper of an
11-month-old baby.*

accidentally infested owing to unhygienic practices or by ingesting the beetles. A case
of acanthocephalan infestation was recently reported to the Mississippi Department
of Health in which an 8-cm worm was found in the diaper of an 11-month-old baby
(Figure 1). Presumably, the infant ate infested beetles or consumed them in flour or
bread products. The child suffered no ill effects from the parasite.

B. Prevention and Treatment of Infestations

Good sanitation is the best way to prevent ingestion of beetles or their body parts.
Food preparation areas should be thoroughly cleaned and sanitized prior to cooking.
Cereal, nuts, candy, flour, and cornmeal can be placed in tight-fitting plastic containers
to prevent beetle infestation. Lady bugs can be kept out of houses by thorough sealing
of exterior cracks and crevices. Once inside, they may be vacuumed.

II. BLISTER BEETLES

A. General and Medical Importance

Blister beetles are plant-feeding insects that contain a blistering agent in their body
fluids. Most are in the family Meloidae, although some are in Staphylinidae. The agent
in meloid beetles, cantharidin (formula $C_{10}H_{12}O_4$), penetrates the skin, producing blisters in a few hours[4] (also see Chapter 5). The blistering agent in staphylinids may be
somewhat different chemically from cantharidin. In any event, handling live beetles
or contact with their pulverized bodies may cause blistering. Cantharidin is poisonous to humans and other animals when ingested and may lead to abdominal pain,
kidney damage, and sometimes death.[5] Horses may die from ingesting blister beetles
in their hay (especially alfalfa).[6,7] On human skin, Lehman et al.[8] reported tingling and
burning prior to blister formation, but other reports indicated an almost symptomless
course except for the blisters.[9] Dermatoses from blister beetle contact are seasonal,

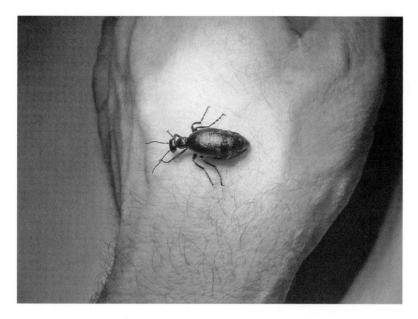

Figure 2
Blister beetle on hand. (Photo copyright 2005 by Jerome Goddard.)

with most cases in the U.S. occurring in July, August, and September. Alexander[10] said that workers harvesting the potato crop, as well as children running around barefoot, are particularly vulnerable.

The *Spanish fly* is a blister beetle in southern Europe that has been used for millennia as a sexual stimulant. Morbidity from its use has been commonly reported. Sexual enhancement is due to inhibition of phosphodiesterase and protein phosphatase activity, and stimulation of beta-receptors, inducing vascular congestion and inflammation.[11]

B. General Description

Blister beetles are elongate, soft-bodied specimens in which the pronotum (section between head and wings, viewed from above) is narrower than the head or wings (Figure 2). Two of the common blister beetle species are both potato beetles: one with orange and black longitudinal stripes (Figure 3A) and another black with gray wing margins (Figure 3B). Members of the genus *Meloe* are called *oil beetles* because they exude an oily substance when disturbed. Oil beetles are approximately 20 to 25 mm long and black, with no hind wings, giving the appearance that their wings are very short (Figure 3C).

C. Geographic Distribution

The ash-gray (*Epicauta fabricii*) and striped blister beetles (*E. vittata*) are common in the central and southeastern areas of the U.S. Oil beetles and other large, rounded species are common in the southwestern U.S. from Texas to southern California. Many other blister beetle species are common in the western U.S. In Europe, three important blister beetles are *Paederus limnophilus, P. gemellus,* and *Lytta vesicatoria* (the Spanish fly).

BLISTER BEETLES

Figure 3
Blister beetles: (A) the striped blister beetle, (B) ash-gray blister beetle, and (C) oil beetle. (Adapted in part from U.S. DHEW, PHS, CDC Pictorial Keys.)

Importance
Contact may produce blisters on skin; ingestion could cause poisoning

Distribution
Many areas of world, including the U.S.

Lesion
Generally painless, large blisters

Disease Transmission
None

Key References
Lehman et al., *Arch. Dermatol.*, 71, 36–38, 1955

Frank, J.H. and Kanamitsu, K., *J. Med. Entomol.*, 24, 155–191, 1987

Treatment
Skin reactions generally not serious; topical antibiotics and bandaging to prevent secondary infection; a poison control center should be consulted in cases of blister beetle ingestion

D.   Biology and Behavior

Blister beetles have a very unusual biology. Larvae of the meloid beetles feed on the eggs of grasshoppers and in the nests of solitary bees. After several larval stages, they pupate and later emerge as adult beetles. Adults are plant feeders that may emerge in large numbers and appear in gardens and field crops quite suddenly.

E.   Treatment of Exposed Areas

Blisters from exposure to the beetles are generally not serious and will be reabsorbed in a few days if unruptured. Even if the blisters are ruptured, there is normally complete clearing in 7 to 10 d. Affected areas should be washed with soap and water, and bandaged until the blisters reabsorb. Antibiotic ointments or creams may help prevent secondary infection. Large blisters, and those occurring on the feet where they might be rubbed, may need to be drained, treated with antiseptics, and bandaged.[12] If the beetles or beetle products are ingested, the cantharidin may cause nausea, diarrhea, vomiting, and abdominal cramps.

REFERENCES

1. Arnett, R.H., Jr., Present and future of systematics of the Coleoptera of North America, in *Systematics of the North American Insects and Arachnids: Status and Needs*, Kosztarab, M. and Schaefer, C.W., Eds., Virginia Polytechnic Institute and State University, Blacksburg, VA, 1990, p. 165.

2. White, R.E., *A Field Guide to the Beetles of North America*, Houghton Mifflin, Boston, MA, 1983.

3. Southcott, R.V., Injuries from Coleoptera, *Med. J. Aust.*, 151, 654, 1989.

4. Harwood, R.F. and James, M.T., *Entomology in Human and Animal Health*, Macmillan, New York, 1979.

5. Tagwireyi, D., Ball, D.E., Loga, P.J., and Moyo, S., Cantharidin poisoning due to "Blister beetle" ingestion, *Toxicon*, 38, 1865, 2000.

6. Helman, R.G. and Edwards, W.C., Clinical features of blister beetle poisoning in equids: 70 cases (1983–1996), *J. Am. Vet. Med. Assoc.*, 211, 1018, 1997.

7. Capinera, J.L., Gardner, D.R., and Stermitz, F.R., Cantharidin levels in blister beetles associated with alfalfa in Colorado, *J. Econ. Entomol.*, 78, 1052, 1985.

8. Lehman, C.F., Pipkin, J.L., and Ressmann, A.C., Blister beetle dermatitis, *Arch. Dermatol.*, 71, 36, 1955.

9. Giglioli, M.E.C., Some observations on blister beetles, family Meloidae, in Gambia, West Africa, *Trans. R. Soc. Trop. Med. Hyg.*, 59, 657, 1965.

10. Alexander, J.O., *Arthropods and Human Skin*, Springer-Verlag, Berlin, 1984.

11. Sandroni, P., Aphrodisiacs past and present: a historical review, *Clin. Auton. Res.*, 11, 303, 2001.

12. Frazier, C.A. and Brown, F.K., *Insects and Allergy*, University of Oklahoma Press, Norman, OK, 1980.

BUGS (THE TRUE BUGS)

TABLE OF CONTENTS

BED BUGS

Figure 1
Adult bed bug, Cimex lectularius.
(From U.S. DHEW, PHS, CDC Pictorial Keys.)

Importance
Nuisance and irritation from bites

Distribution
Worldwide

Lesion
Linear grouping of red blotches, urticarial wheals; sometimes, bullous lesions

Disease Transmission
None

Key References
Usinger, R.L., *Monograph of Cimicidae (Hemiptera-Heteroptera)*, Vol. 7, Thomas Say Foundation, Entomological Society of America, 1966
Ryckman et al., *Bull. Soc. Vector Ecol.,* 6, 93, 1981

Treatment
Antiseptic or antibiotic creams or lotions; antihistamines for urticarial RXNs; eliminate infestation in home

I. BED BUGS

A. General and Medical Importance

The common bed bug, *Cimex lectularius* (Figure 1), has been an associate of humans for thousands of years. The blood-sucking parasites are common in Third World countries, especially in areas of extreme poverty. Bed bugs had nearly disappeared in developing countries until recently, where, in the last 5 to 10 years, they have been making a progressively rapid comeback. The author increasingly receives reports of the parasites inside U.S. hotel rooms. One report of an infested house in Great Britain contained a description of literally thousands of the bugs under and within the bed and in the mattress seams[1] (Figure 2). According to the authors, the area where the bed was against the wall was black with a layer of bed bug excrement, cast skins, and eggs several millimeters thick. Bed bugs have been suspected in the transmission of more than 40 disease organisms such as those causing anthrax, plague, hepatitis, and typhus.[2] However, at this time bed bugs have never been proved to biologically transmit even one human pathogen.[3] Their principal medical importance is the itching and inflammation associated with their bites. Occasionally, hemorrhagic bullae result from bed bug bites.

B. General Description

Adult bed bugs are approximately 5 mm long, oval-shaped, and flattened (Figure 1 and Figure 3). They somewhat resemble unfed ticks or small cockroaches. Adults are reddish-brown (chestnut) in color; the immatures resemble adults but are yellowish-white. Bed bugs have a pyramidal head with prominent compound eyes, slender antennae, and a long proboscis tucked backward underneath the head and thorax. The prothorax (dorsal side, first thoracic segment) bears rounded, winglike lateral horns on each side. Hind wings on bed bugs are absent entirely; the forewings are represented by two small pads.

CASE HISTORY

BED BUGS IN HOTEL ROOM

A physician called asking questions about a strange insect infestation she had encountered in a hotel (one of the major chain hotels) in Denver, CO, during a medical conference. She said that about 11:00 P.M. while reading in bed a small "cockroach-like bug" crawled on her arm. Blood came out when she smashed it. Later, after going to sleep, she was awakened by another one on her arm — it also was full of blood. She captured a few and brought them back for identification. I told her by phone that the bugs were likely bed bugs, but asked to see the specimens. Upon receiving the specimens 2 d later, they were confirmed as *Cimex lectularius*.

Comment. Even though bed bugs are not as common as they used to be, they still may be encountered, even in affluent settings. A traveler may bring in some of the bugs in his or her suitcases or belongings, and soon the bugs take up residence in cracks and crevices, behind baseboards, or under mattresses. Any cockroach-like insects coming out from the mattress or environs to bite are highly suspect (cockroaches have chewing mouthparts and never suck human blood). Get another room — better yet, get another hotel!

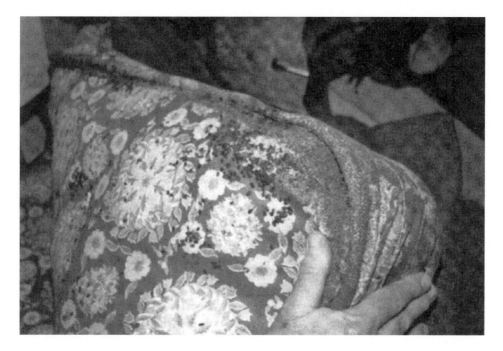

Figure 2
Numerous bed bugs congregating on the underside of a mattress. (Photo courtesy Ian Dick and The London Borough of Islington, Environmental Health Department.)

Figure 3
Bed bug feeding beside the author's wedding ring, showing relative size.

C.   Geographic Distribution

The common bed bug is cosmopolitan, occurring in temperate regions worldwide. Another bed bug species, *C. hemipterous*, is also widespread but is mostly found in the tropics. Several other bed bug species occur on bats and swallows, but they do not usually bite people.

D.   Biology and Behavior

Bed bugs possess stink glands and emit an odor. Homes heavily infested with the bugs have this distinct odor. Bed bugs mostly feed at night, hiding in crevices during the day. Hiding places include seams in mattresses, crevices in box springs, and spaces under baseboards or loose wall paper.

There are five nymphal stages that must be passed before development to adults. Once an adult, the life span is 6 to 12 months. Each nymphal stage must take a blood meal in order to complete development and molt to the next stage (Color Figure 13.12). The bugs take about 5 to 10 min to obtain a full blood meal. Generally, the bite itself is painless. Bed bugs can survive long periods of time without feeding, and when their preferred human hosts are absent they may take a blood meal from any warm-blooded animal.

E.   Treatment of Infestation/Bites

Bed bugs have piercing–sucking mouthparts typical of the insect order Hemiptera. Accordingly, bites from the bugs often produce welts and local inflammation, probably because of allergic reactions to saliva injected via the mouthparts during feeding (see Chapter 4). On the other hand, for many people the bite is nearly undectectable. Bed bug bites are generally self-limiting and require little specific treatment other than antiseptic or antibiotic creams or lotions to prevent infection. Antihistamines may be needed for urticarial reactions. In addition, efforts should be made to eliminate the

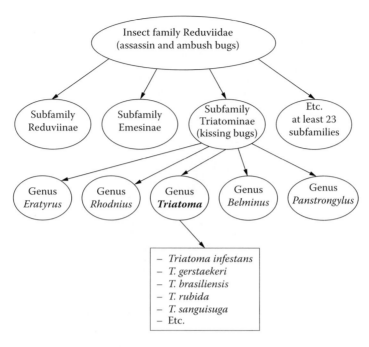

Figure 4
Partial classification scheme for insect family Reduviidae.

source of the bites. Insecticide treatments for bed bugs are effective but should be conducted carefully, as people have prolonged, close contact with the treated areas — beds, couches, etc., where bed bugs live.

II.   CONENOSE BUGS (ASSASSIN AND KISSING BUGS)

A.   General and Medical Importance

Many members of the family Reduviidae have an elongate (cone-shaped) head, and hence the name *conenose bugs*. Most reduviids "assassinate" or kill other insects. Some of the assassin bugs occasionally bite people, producing very painful lesions (Color Figure 13.13). A common offender is *Melanolestes picipes*. There is a report of 2 of these bugs biting an 8-year-old, causing intense pain for hours, swelling at the bite site for days, and later, ulcers.[4] A relatively small but important group of reduviids in the subfamily Triatominae feeds exclusively on vertebrate blood. Notorious members of this group are frequently in the genus *Triatoma*, but not all (see Figure 4 for a classification scheme of this group). Triatomines are called "kissing bugs" because their blood meals are occasionally taken from the area around the human lips (Color Figure 13.14). Other sites of human attack, in order of frequency, are the hands, arms, feet, head, and trunk. Kissing bugs are not able to feed through clothing. Very often, their bites are painless. However, reactions to their bites range from a single papule, to giant urticarial lesions, to anaphylaxis, depending on the degree of allergic sensitivity.[5]

Kissing bugs may transmit the agent of Chagas' disease, or American trypanosomiasis, one of the most important arthropod-borne diseases in tropical America. Chagas' disease is a zoonosis (originally a parasite of wild animals) mostly occurring in Mexico

Figure 5
Approximate geographic distribution of Chagas' disease.

and Central and South America (Figure 5), but a few indigenous cases have been reported in Texas, Tennessee, and California.[6-8] Serological evidence has identified *T. cruzi* as far north as Oklahoma.[9] At present, some 16 to 18 million people are estimated to be infected, with 90 to 100 million people at risk.[10] Chagas' disease has both acute and chronic forms, but is perhaps best well known for its chronic sequelae including myocardial damage with cardiac dilation, arrhythmias and major conduction abnormalities, and digestive tract involvement such as megaesophagus and megacolon.

B. General Description

Assassin bugs are often black or brown in color. They have elongate heads with the portion behind the eyes narrowed and "necklike" (Figure 6A). The beak is short and three-segmented, and its tip fits into a groove in the venter of the thorax. In nontriatomines the beak is thick and curved. The abdomen is often widened in the middle, exposing the lateral margins of the segments beyond the wings. Kissing bugs are similar in appearance to many assassin bugs and may have orange and black markings where the abdomen extends laterally past the folded wings (Figure 6B and Figure 7) and a thin, straight proboscis. In addition, the dorsal portion of the first segment of the thorax consists of a conspicuous triangular-shaped pronotum. Most adult kissing bugs are 1 to 3 cm long and are good fliers. An excellent monograph containing descriptions of the triatomines is provided by Lent and Wygodzinsky.[11]

C. Geographic Distribution

Numerous species of assassin bugs occur essentially worldwide. The wheel bug belongs to this group (see Section III). *Reduvius senilis*, the tan assassin bug, is found in the desert areas of the southwestern U.S. and Mexico. *R. personatus* occurs throughout the U.S. and southern Europe. *Melanolestes picipes* is widely distributed in the U.S.

There are several species of kissing bugs that will attack humans, and some are capable of transmitting *Trypanosoma cruzi*, the causative agent of Chagas' disease. However, only a few species are efficient vectors. The four principal vectors of Chagas' disease in Central and South America are *Panstrongylus megistus, Rhodnius prolixus, Triatoma infestans*, and *T. dimidiata*. The most important vector in Mexico is *T. barberi*. In the southwestern U.S., *T. gerstaeckeri* (Figure 8A) and *T. protracta* (Figure 8B) are important kissing bug species, and *T. sanguisuga* (Figure 8C) occurs throughout much of the U.S. Figure 9 shows the U.S. distribution of five common kissing bugs.

D. Biology and Behavior

Assassin bugs are predaceous on other insects and are often found lying in wait for their prey on various plants or flowers. They, along with the kissing bugs, undergo simple metamorphosis; the developing nymphs look very much like adults, except that they are smaller.

Kissing bugs are nocturnal insects that are able to fly to their hosts with speed and agility. Both sexes bite, and they take their blood meals primarily at night, hiding in any available crack or crevice between feedings (about 36 h). *Triatoma*, as a group, normally feed on a wide variety of small mammals, but will readily feed on humans. Infection rates with *Trypanosoma cruzi*, the causative agent of Chagas' disease, may be as high as 80 to 100% in some adult triatomine bug populations.[12,13] Human kissing bug bites are especially common occurrences in poor, underdeveloped areas with dilapidated or poorly constructed

KISSING BUGS_____

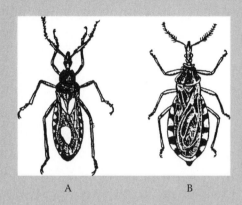

A   B

Figure 6
Adult (A) assassin bug and (B) kissing bug. (From U.S. DHEW, PHS, CDC Pictorial Keys.)

Importance
Blood-feeding on humans — often at night; anaphylaxis has been reported from bites

Distribution
Most medically important species occur in the Americas

Lesion
Variable — papular, nodular, or bullous

Disease Transmission
Chagas' disease

Key Reference
Lent, H. and Wygodzinsky, P., *Bull. Am. Mus. Nat. Hist.* 163 (Art. 3), 123, 1979

Treatment
Local RXNs — antihistamines, Caladryl®; systemic RXNs: antihistamines, epinephrine, and other supportive measures as needed

Figure 7
Kissing bug showing abdomen extending laterally past folded wings.

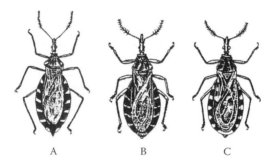

Figure 8
Adult (A) Triatoma gerstaeckeri, *(B)* T. protracta, *and (C)* T. sanguisuga.
(From U.S. DHEW, PHS, CDC Pictorial Keys.)

huts or shacks. Infection is not by the salivary secretions associated with the bite but by fecal contamination of the bite site; however, other routes of transmission may be overlooked. In some communities in Mexico, for example, people believe that bug feces can cure warts or that the bugs have aphrodisiac powers.[12,14] In addition, Mexican children often play with triatomine bugs collected in their houses, and in Jalisco, reduviid bugs are eaten with hot sauce by the Huichol Indians.[12]

E.   Treatment of Infestation and Bites

Treatment for conenose bug bites involves washing the wound with soap and water. Itching lesions may be relieved with topical palliatives such as Caladryl® and oral antihistamines. Mild systemic reactions may require oral or i.m. antihistamines. Anaphylactic shock has been reported.[5] In that case, shock should be treated with epinephrine, antihistamines, and other supportive measures (see Chapter 2). Follow-up for possible development of Chagas' disease may be needed for patients with kissing bug bites in endemic areas as well as for persons in the southern U.S. finding kissing bugs in their bedroom. Physicians with questions about diagnosis and treatment of Chagas' disease may call the Centers for Disease Control, Parasitic Diseases Division, telephone (770) 488-7775.

Figure 9
Approximate U.S. distributions of five commonly encountered kissing bugs.

WHEEL BUGS

Figure 10
Adult wheel bug.
(From U.S. DHEW, PHS, CDC Pictorial Keys.)

Importance
Painful bite

Distribution
U.S.

Lesion
Swollen, inflamed, and indurated at bite site

Disease Transmission
None

Key Reference
Hall, M. C., *Arch. Intern. Med.*, 33, 513, 1924

Treatment
None may be needed; oral analgesics, Caladryl®, topical corticosteroids may help

III. WHEEL BUGS

A. General and Medical Importance

The wheel bug, *Arilus cristatus*, is also in the family of true bugs called Reduviidae. They bite humans only in self-defense. However, the bite is characterized by immediate, intense pain.[15,16] Inflammation may become chronic, producing lesions at the bite site resembling papillomas.

B. General Description

Wheel bugs are gray and approximately 3 cm long (Figure 10). The insects, as their name implies, have a cogwheel-like crest on the dorsal side of their prothorax (Figure 11). This gives the appearance of a half-wheel on their upper body. They also have a small, narrow head and long, piercing–sucking mouthparts tucked under their head and prothorax.

C. Geographic Distribution

The wheel bug is common from New Mexico through the southern and eastern U.S.

D. Biology and Behavior

These insects attack and eat soft-bodied insects. They inject a salivary fluid into their prey via their long beak. Being true bugs (Hemiptera), wheel bugs develop by simple metamorphosis, meaning that immatures are called nymphs and are almost identical to the adults.

E. Treatment of Bites

Bites of the wheel bug, although painful, are usually localized and self-limiting. Specific treatment measures are probably not necessary.

Figure 11
Wheel bug, showing cogwheel-like crest on upper side.

REFERENCES

1. King, F., Dick, I., and Evans, P., Bed bugs in Britain, *Parasitol. Today*, 5, 100, 1989.
2. Burton, G.J., Bed bugs in relation to transmission of human diseases, *Public Health Rep.*, 78, 513, 1963.
3. Goddard, J., Bed bugs bounce back — but do they transmit disease? *Infect. Med.*, 20, 473, 2003.
4. Eads, R.B., An additional report of a reduviid bug attacking man, *J. Parasitol.*, 36, 87, 1950.
5. Moffitt, J.E., Venarske, D., Goddard, J., Yates, A.B., and deShazo, R.D., Allergic reactions to *Triatoma* bites, *Ann. Allerg. Asthma Immunol.*, 91, 122, 2003.
6. Harwood, R.F. and James, M.T., *Entomology in Human and Animal Health*, Macmillan, New York, 1979.
7. Schiffler, R.J., Mansur, G.P., Navin, T.R., and Limpakarnjanarat, K., Indigenous Chagas disease in California, *JAMA*, 251, 2983, 1984.
8. Herwaldt, B.L., Grijalva, M.J., Newsome, A.L., McHee, C.R., Powell, M.R., Nemee, D.G., Steuer, F.J., and Eberland, M.L., Use of PCR to diagnose the fifth reported U.S. case of autothonous transmission of *Trypanosoma cruzi* in Tennessee, *J. Infect. Dis.*, 181, 395, 1998.
9. Bradley, K.K., Bergman, D.K., Woods, J.P., Crutcher, J.M., and Kirchoff, L.V., Prevalence of American trypanosomiasis among dogs in Oklahoma, *J. Am. Vet. Med. Assoc.*, 217, 1853, 2000.
10. Schofield, C.J. and Dolling, W.R., Bed bugs and kissing bugs, in *Medical Insects and Arachnids*, Lane, R.P. and Crosskey, R.W., Eds., Chapman and Hall, London, 1993, chap. 14.
11. Lent, H. and Wygodzinsky, P., Revision of the Triatominae and their significance as vectors of Chagas' disease, *Bull. Am. Mus. Natl. Hist.*, 163, 123, 1979.
12. Schettino, P.M.S., Arteaga, I.D.H., and Berrueta, T.U., Chagas disease in Mexico, *Parasitol. Today*, 4, 348, 1988.
13. deShazo, T., A survey of *Trypanosoma cruzi* infection in *Triatoma* spp. collected in Texas, *J. Bacteriol.*, 46, 219, 1943.
14. Salazar-Schettino, P.M., Customs which predispose to Chagas' disease and cysticercosis in Mexico, *Am. J. Trop. Med. Hyg.*, 32, 1179, 1983.
15. Biery, T.L., Venomous Arthropod Handbook, U.S. Air Force pamphlet No. 161-43, Brooks AFB, TX, 1977.
16. Hall, M.C., Lesions due to the bite of the wheel bug *Arilus cristatus*, *Arch. Intern. Med.*, 33, 513, 1924.

CHAPTER 14

CATERPILLARS (URTICATING)

TABLE OF CONTENTS

I. GENERAL AND MEDICAL IMPORTANCE

Several moth and butterfly families have species whose caterpillars possess urticating hairs or spines that secrete a poison when exposed to human skin[1,2] (see also Chapter 5). Urtication by larval lepidopterans is often termed *eurcism*. Often exposure to urticating caterpillars is accidental or incidental, but in some cases there is deliberate contact (e.g., children playing with caterpillars). In many species there is a severe burning sensation immediately following the sting. This may be followed by swelling, numbness, urticaria, and intense stabbing pain that radiates to a nearby axillary or inguinal region; lymphadenitis may also be present. Usually the effects of these toxic hairs are limited to burning and inflammation of the skin, but they may progress to systemic reactions such as headache, nausea, vomiting, paralysis, acute renal failure, and shock

URTICATING CATERPILLARS

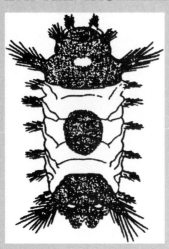

Saddleback caterpillar, *Sibine stimulea*

Importance
Venomous hairs or spines cause stings or irritation

Distribution
Many species involved worldwide

Lesion
Variable — papular eruption, erythema, local swelling

Disease Transmission
None

Key References
Henwood, B. P. and MacDonald, D. M., *Clin. Exp. Dermatol.*, 8, 77, 1983
Rosen, T., *Dermatol. Clin.*, 8, 245, 1990

Treatment
Local RXNs — topical products such as Caladryl®, corticosteroids, and pain relievers; systemic RXNs — may require antihistamines, epinephrine, and other supportive measures

Note: Eye lesions may be particularly serious and should be seen by a specialist

and convulsions (rare).[2,3] Days later, the lesion may show a pattern similar to that of the spines of the offending specimen (Figure 1 and Figure 2). Other caterpillars, such as gypsy moth larvae, are not "stinging caterpillars," but their hairs may cause dermatitis, especially in sensitive persons. Apparently, there has been an increase over the last few decades of reports of dermatologic, pulmonary, and systemic reactions following caterpillar encounters in the southern U.S.[4] Alexander[5] presented an excellent review of the urticating caterpillars with detailed discussion of the nature of their venoms and hair/spine structure.

II. GENERAL DESCRIPTION

Five common urticating caterpillars will be discussed in this section, although there are many others (Table 1). *Automeris io* caterpillars are larvae of the IO moth. Full-grown caterpillars are about 5 to 8 cm long, pale green, with lateral stripes of red or maroon over white running the length of the body (Figure 3A). The brown-tail moth larva, *Euproctis chrysorrhoea*, is a mostly black caterpillar with brown hairs and a row of white tufts on each of its sides. The puss caterpillar, *Megalopyge opercularis*, also sometimes called the opossum bug, asp, Italian asp, or el perrito, is about 3 cm long, tan to dark brown in color, and completely covered dorsolaterally with hair that causes it to resemble small tufts of cotton (Figure 3B). They really do not look like caterpillars. Intermingled among all those fine hairs on the back are clusters of venomous spines. *Sibine stimulea*, the saddleback caterpillar, is 2 to 3 cm long, has a brown sluglike body, and is covered middorsally with markings that resemble a brown or purplish saddle sitting on a green and white saddle blanket (Figure 3C). Gypsy moth larvae, *Lymantria dispar*, are approximately 3 to 5 cm long and gray to brown in color with yellow stripes (Figure 3D). They also may appear to have red or blue spots along the sides and top of the body. As in the other members of the family Lymantriidae, gypsy moth larvae have long tufts of hair along the body.

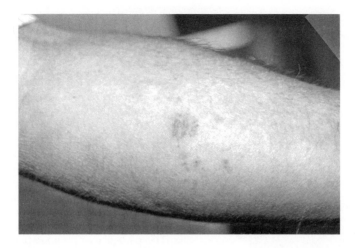

Figure 1
Skin lesion from exposure to stinging caterpillar, 2 d poststing.

Figure 2
Caterpillar responsible for sting lesion depicted in Figure 1.

III.   GEOGRAPHIC DISTRIBUTION

The IO moth is found in the Nearctic region; it is quite common throughout the eastern U.S. Brown-tail moth caterpillars are bothersome pests in Europe and eastern portions of the U.S., where they have been accidentally introduced. The saddleback caterpillar is found in many parts of the world. In the U.S. it is generally distributed southeast of a diagonal line drawn from Massachusetts to the middle of Texas. Puss caterpillars are primarily found in the southern U.S. and southward into Central and South America. They seem to be especially a problem in Texas. Gypsy moths occur in Europe and the eastern U.S.

IV.   BIOLOGY AND BEHAVIOR

IO moth larvae feed on the leaves of a variety of plants, including corn and willow. In most areas they produce only one annual generation, but in the southernmost areas of their distribution there may be a second generation. Therefore, depending on the area of the country, larval stages can be found anytime from spring to fall. The brown-tail

CASE HISTORY

On September 30, 1994, the author was collecting insects at the Copiah County Game Management Area in central Mississippi when his son inadvertently brushed his right forearm against a puss caterpillar. Several of the caterpillars had been seen on a fallen log and some bushes just prior to the sting incident. The offending specimen was collected, identified, and photographed at the site (Figure 1). Within 5 min the patient experienced intense, throbbing pain at the sting site. An erythematous spot approximately 4 cm in diameter developed (Figure 2), containing a few raised papule-like structures, presumably where the poisonous hairs or setae contacted the skin. After 20 min the patient was complaining of severe pain in the right axilla. The patient was photographed during the time of this axillary pain (Figure 3 — actual arm positions, photo not contrived). No other symptoms developed, and the pain subsided within 45 min. The erythematous spot resolved in 24 h.

Figure 1

Puss caterpillar. (Reprinted from Am. Fam. Physician 52, 86, July 1995. Copyright 1995 by the American Academy of Family Physicians. With permission.)

Comment: Many moth families have species whose larvae possess stinging or urticating spines or hairs. One of the most troublesome of these is the puss caterpillar, *Megalopyge opercularis*, a member of the flannel moth family. The puss caterpillar is widely distributed in the southern U.S. extending down into Mexico; members of the species feed on a variety of deciduous trees and shrubs. Interestingly, its appearance is not what one might expect for a caterpillar. Instead of being wormlike, or wormlike with prominent spines, the puss caterpillar is shaped like a teardrop and looks like a tuft of cotton or fur. It may vary in color from light yellow to gray or reddish-brown. Stings from the puss caterpillar may cause immediate intense local burning pain (often referred to as "shooting" pain), headache, nausea,

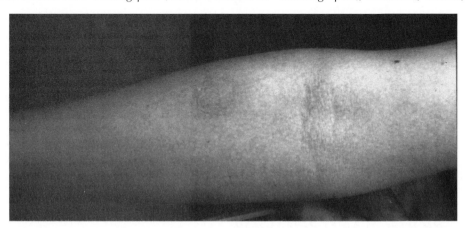

Figure 2

Puss caterpillar sting site 15 min after sting. (Reprinted from Am. Fam. Physician 52, 86, July, 1995. Copyright 1995 by the American Academy of Family Physicians. With permission.)

Figure 3
Axillary pain caused by puss caterpillar sting — actual arm positions, photo not contrived.

vomiting, lymphadenopathy, lymphadenitis, and sometimes shock and respiratory distress. Stings can be especially severe in children.

From reports in entomological and medical literature, this case was fairly typical, except perhaps for the lack of local swelling. Certainly, a person with hypersensitivity to the venom could react in a much more profound manner, possibly leading to shock and respiratory distress. But even in the absence of an allergic reaction, puss caterpillar stings can lead to temporary severe pain and partial immobilization.

Source: Adapted from *Am. Fam. Physician* 52, 86. Copyright 1995, the American Academy of Family Physicians. With permission.

moth is a serious pest of forest and shade trees, as well as many varieties of fruit trees. Caterpillars of this species are most active between April and July.

The saddleback caterpillar may be found feeding on the leaves of a variety of trees, shrubs, and other plants from May to November. The puss caterpillar also feeds on the leaves of a wide range of trees and shrubs. In the southern area of its range, it may have two generations per year. The first generation develops in the spring and early summer, and the second generation develops in the fall. They seem to be especially abundant from September to November. Every few years there are outbreaks of puss caterpillars that lead to numerous stings, especially among children. They are commonly found on the exterior walls of houses, sheds, gates, and fences; thus, risk of human contact is high.

The gypsy moth was introduced into the eastern U.S. in the 1860s and has since become widely distributed throughout New England (and is now spreading southward), where it causes widespread damage to forest trees. Eggs are laid on tree trunks in July and August in masses covered with froth and body hairs from the female. Eggs overwinter and the tiny first-stage larvae emerge the following spring, usually in late

Table 1
Some Species of Lepidoptera Whose Larvae Are Known to Sting or Cause Dermatitis

Species	Condition	Occurs Where
Megalopyge opercularis	Sting	Southern U.S., Central and South America
Several other *Megalopyge* spp.	Sting	South America
Sibine stimulea	Sting	Neotropical, Nearctic
Automeris io	Sting	Nearctic
Several other *Automeris* spp.	Sting	South America
Hemileuca spp.	Sting	Nearctic, South America
Euproctis chrysorrhoea	Sting	Nearctic, Palearctic
E. similis	Sting	Nearctic, Palearctic
E. edwardsii	Sting	Australia
E. flava	Sting	Oriental
Thaumetopoea wilkinsonii	Sting	Palearctic, Afrotropical
T. pityocampa	Sting	Palearctic, Afrotropical
Orchrogaster contraria	Sting	Australia
Several *Hylesia* spp.	Sting and dermatitis	Argentina
Lymantria dispar	Dermatitis	Europe, Eastern U.S.
Orgyia pseudotsugata	Dermatitis	Northwestern U.S.
O. leucostigma	Dermatitis	Nearctic

April or early May. The females can fly, but weakly. Dispersal of the gypsy moth infestation is primarily by young larvae on silken threads traveling tree to tree or limb to limb by "ballooning" in the wind.

V. TREATMENT OF STINGS

Without additional or ongoing contact with the offending caterpillar, lesions resolve at varying times depending on the species involved. Generally, saddleback and IO caterpillar dermatitis subsides within 2 to 8 h, gypsy moth dermatitis within 48 h, and brown-tail moth and puss caterpillar disease (various systemic manifestations) within 7 to 10 d.[6]

Local treatment of urticating caterpillar stings consists of careful repeated stripping of the sting site with adhesive or cellophane tape to remove the spines (if the offending species was stout-spined), application of ice packs, and oral administration of antihistamines to help relieve itching and burning sensation. Acute urticarial lesions may be further relieved by application of topical corticosteroids, which help reduce the intensity of the inflammatory reaction. Allen et al.[7] reported good results with desoximetasone gel applied twice daily to the affected areas. Rosen[6] said systemic administration of corticosteroids in the form of intramuscular triamcinolone acetonide has been remarkably effective in relieving both severe itching due to gypsy moth dermatitis and pain due to puss caterpillar dermatitis.

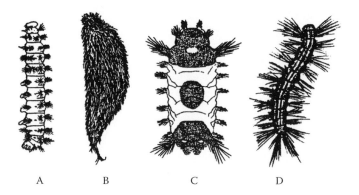

A   B   C   D

Figure 3
Some caterpillars that sting (A–C) or cause dermatitis (D): (A) IO moth larva, (B) puss caterpillar, (C) saddleback caterpillar, and (D) gypsy moth larva. (Adapted from U.S. DHEW, PHS, CDC Pictorial Keys.)

For severe pain associated with the puss caterpillar or IO moth larvae, physicians sometimes administer meperidine HCl, morphine, or codeine[5]; aspirin is reportedly not effective. Systemic hypersensitivity reactions such as hypotension or bronchospasm are usually treated with epinephrine, antihistamines, and other supportive measures (also see Chapter 2). Eye lesions (especially those resulting from forceful contact with caterpillars) may be very serious and should be seen by a specialist.

REFERENCES

1. Maschwitz, U.W. and Kloft, W., Morphology and function of the venom apparatus of insects — bees, wasps, ants, and caterpillars, in *Venomous Animals and Their Venoms*, Vol. 3, Bucherl, W. and Buckley, E., Eds., Academic Press, New York, 1971, p. 38.

2. Keegan, H.L., Some medical problems from direct injury by arthropods, *Intern. Pathol.*, 10, 35, 1969.

3. Gamborgi, G.P., Metcalf, E.B., and Barros, E.J., Acute renal failure provoked by toxin from caterpillars of the species *Lonomia obliqua, Toxicon*, 47, 68, 2006.

4. Diaz, J.H., The epidemiology, diagnosis, and management of caterpillar envenoming in the southern U.S., *J. LA State Med. Soc.*, 157, 153, 2005.

5. Alexander, J.O., *Arthropods and Human Skin*, Springer-Verlag, Berlin, 1984.

6. Rosen, T., Caterpillar dermatitis, *Dermatol. Clin.*, 8, 245, 1990.

7. Allen, V.T., Miller, O., and Tyler, W.B., Gypsy moth caterpillar dermatitis — revisited, *J. Am. Acad. Dermatol.*, 24, 979, 1991.

CHAPTER 15

CENTIPEDES

TABLE OF CONTENTS

I. GENERAL AND MEDICAL IMPORTANCE

Centipedes are long, multisegmented arthropods that characteristically have one pair of legs per body segment (millipedes have two pairs per segment). They are agile, fast-moving creatures that can inflict a painful bite.[1] In fact, some of the larger species can produce extreme pain.[2-4] The venom is a cytolysin-based compound. Most human centipede bites result when a centipede is stepped on, picked up, or otherwise contacts the body. Centipede bites are rarely fatal to humans, but deaths have been reported. Members of the genus *Scolopendra*, which occur in North America, can be 20 to 25 cm long and produce bites with intense burning pain lasting 1 to 5 h. The bite is characterized by two puncture wounds at the site of attack, often red and swollen. Other symptoms may include anxiety, vomiting, irregular pulse, dizziness, and headache.[5] Secondary infections can occur, and superficial necrosis at the bite site may persist for several days. In addition to their bites, large species (such as in the genus *Scolopendra*) have claws that can make tiny punctures if they crawl on human skin.

CENTIPEDES

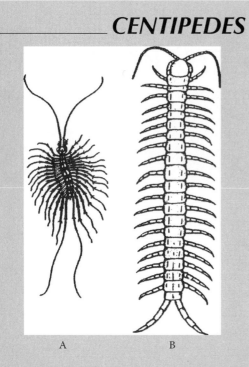

A   B

Figure 1
Centipedes: Scutigera coleoptrata (A) and Scolopendra beros (B) (from U.S. DHEW, PHS, CDC. Pictorial Keys.)

Importance
Painful bites

Distribution
Numerous species worldwide

Lesion
Often, two hemorrhagic punctures

Disease Transmission
None

Key Reference
Keegan, H.L., Some medical problems from direct injury by arthropods, *Intern. Pathol.*, 10, 35, 1969

Treatment
Analgesics; antibiotics; possibly tetanus prophylaxis

II. GENERAL DESCRIPTION

Centipedes are dorsoventrally flattened, have a distinct head, relatively long antennae, along with the one pair of legs per body segment (Figure 1). The first body segment bears a pair of claws that contain ducts for the expulsion of a paralyzing venom contained in a gland at the base of the claw. The number of body segments, and thus number of legs, is variable depending on the species, but it usually ranges from 15 to over 100 pairs. Many species are 3 to 8 cm long, whereas some tropical species may reach 45 cm. The common house centipede, *Scutigera coleoptrata*, is approximately 4 cm long and has long antennae and fragile legs (Figure 1A). From above, the relative lengths of the legs give it an oval appearance.

III. GEOGRAPHIC DISTRIBUTION

Centipede species in the northern U.S. are small and generally harmless to humans, but larger species in the southern U.S. and tropics can inflict a painful bite. The common house centipede, *S. coleoptrata*, occurs throughout southern Europe and the eastern parts of the U.S. and Canada. *Scolopendra polymorpha* occurs in the southwestern U.S., and *S. beros* occurs in southern California. *S. cingulata* is a common species around the Mediterranean and in the Near East. In Asia, *S. morsitans* is very large and can inflict severe bites.

IV. BIOLOGY AND BEHAVIOR

Centipedes lay their eggs in moist soil or vegetation. Development is slow, with about 10 instars. Adult centipedes may live 3 to 5 years. They usually hide during the day under rocks, boards, and bark, or in cracks, crevices, closets, etc. At night, they emerge to hunt for prey such as insects and other small arthropods. The larger species may feed on small vertebrates. This carnivorous habit brings centipedes into close contact with people as they

CASE HISTORY

PAINFUL CENTIPEDE BITE

On July 10, 1997, a man in Jackson, MS, was awakened during the night by a sharp, needle-like bite on his knee. Looking around in bed, he found the offending specimen: a centipede, which he collected for identification (Figure 1). A red and swollen lesion of a few centimeters in diameter appeared on his knee, followed by headache and dizziness which lasted approximately 20 min. The patient self-medicated with Chlortrimeton (4 mg) and all symptoms were completely resolved by morning. No further complications occurred.

Comment: Centipede bites are usually not life-threatening. The arthropods do not actually bite with their mouthparts, but instead with sharpened claws on their modified first pair of legs with which they hold their prey. They have a venom gland situated in the basal segment of the modified legs and inject venom through the claws. Clinical features of centipede bites vary with locality and species, but generally are immediately painful, with some swelling and local tenderness for a few hours. More severe reactions rarely occur, and may include anxiety, vomiting, irregular pulse, dizziness, and headache. This particular centipede was identified as *Hemiscolopendra punctiventris,* a member of the order Scolopendromorpha. This group is principally tropical; in the U.S. they occur mainly in the southern states.

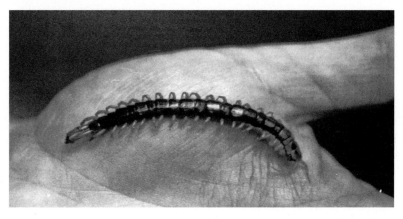

Figure 1
Centipede that bit patient.

frequently enter tents and buildings in search of prey. They inject venom through a pair of powerful claws on the first body segment. The claws are connected to poison glands located in the body trunk.

V. TREATMENT OF BITES

Most centipede bites are uncomplicated and self-limiting.[6] Pain from bites usually subsides in 8 to 36 h. Treatment recommendations include washing the bite site with soap and water, applying ice or cool wet dressings, and taking analgesics for pain. One reference reported pain relief after immersing the bite area in hot water.[3] Alexander[7] said that established inflammatory lesions require appropriate antibiotics.

REFERENCES

1. Remington, C.L., The bite and habits of a giant centipede, *Scolopendra subspinipes*, in the Philippine islands, *Am. J. Trop. Med. Hyg.*, 30, 453, 1950.

2. Acosta, M. and Cazorla, D., Centipede envenomation in a rural village of semi-arid region from Falcon State, Venezuela, *Rev. Invest. Clin.*, 56, 712, 2004.

3. Balit, C.R., Harvey, M.S., Waldock, J.M., and Isbister, G.K., Prospective study of centipede bites in Australia, *J. Toxicol. Clin. Toxicol.*, 42, 41, 2004.

4. Knysak, I., Martins, R., and Bertim, C.R., Epidemiological aspects of centipede bites registered in greater S. Paulo, SP, Brazil, *Rev. Saude Publica*, 32, 514, 1998.

5. Harwood, R.F. and James, M.T., *Entomology in Human and Animal Health*, Macmillan, New York, 1979.

6. Keegan, H.L., Some medical problems from direct injury by arthropods, *Intern. Pathol.*, 10, 35, 1969.

7. Alexander, J.O., *Arthropods and Human Skin*, Springer-Verlag, Berlin, 1984.

CHAPTER 16

COCKROACHES

TABLE OF CONTENTS

I. GENERAL AND MEDICAL IMPORTANCE

Cockroaches are among the most important residential, commercial, institutional, and industrial pests today. Several of the approximately 4000 species in the world have become adapted to living in human habitations (Table 1). These are sometimes referred to as *domestic* or *domiciliary* species and breed in homes, institutions, or industry, where they find food, water, shelter, and warmth (Figure 1). Being indoors, they can remain active throughout the year. They consume any human or animal food or beverage, as well as dead animal and plant materials, leather, glue, hair, wallpaper, fabrics, and the starch in bookbindings.

Cockroaches adversely affect human health in several ways: they sometimes bite feebly, especially gnawing the fingernails of sleeping children; they may enter human ear canals, they contaminate food, imparting an unpleasant odor and taste; and they may transmit disease organisms mechanically on their body parts (Color Figure 16.15). Many species of pathogenic organisms such as bacteria and fungi have been found on cockroach body parts.[1-3] For example, Burgess[4] reported isolation of a strain of

Table 1
Some Cockroach Species Found in Close Association with Humans

Species	Common Name	Occurs Where
Blattella germanica	German cockroach	Cosmopolitan
B. asahinae	Asian cockroach	Far East, Florida
Blatta orientalis	Oriental cockroach	Cosmopolitan
Periplaneta americana	American cockroach	Cosmopolitan
P. fuliginosa	Smoky brown cockroach	Central and southern U.S.
P. brunnea	Brown cockroach	Tropics, parts of U.S.
P. australasiae	Australian cockroach	Tropics, neotropics
Supella longipalpa	Brown-banded cockroach	Tropics, subtropics, parts of temperate zone

Shigella dysenteriae from German cockroaches that was causing a disease outbreak in Northern Ireland. In addition to disease transmission, cockroach excrement and cast skins contain a number of allergens to which sensitive people may exhibit allergic responses (Color Figure 16.16) (also see Chapter 2). Symptoms exhibited by persons with cockroach allergy are similar to those described by Wirtz,[5] which include sneezing, runny nose, skin reactions, and eye irritation. Asthma-related health problems from cockroaches seem to be most severe among children in inner-city areas, but may also be significant in nonurban children.[6] In one study of 476 asthmatic inner-city children, 50.2% of the children's bedrooms had high levels of cockroach allergen in dust.[7] That study also found that children who were both allergic to cockroach allergen and exposed to high levels of this allergen had 0.37 hospitalizations a year, as compared with 0.11 for other children.[7]

Opinions differ regarding the role of cockroaches in disease transmission. Some health officials see no association between cockroaches and disease and think that cockroaches are merely nuisance pests. However, many human disease-causing organisms have been found on the legs, other body parts, or fecal pellets of cockroaches.[8] Several researchers have obtained data indicating that the insects may be most commonly implicated in the transmission of *Salmonella*.[9]

II. GENERAL DESCRIPTION

Cockroaches are dorsoventrally flattened, fast-running, nocturnal insects that seek warm, moist, secluded areas. They have prominent, multisegmented filiform antennae, cerci on the abdomen, and two pairs of wings. The front wings are typically hardened and translucent, whereas the hind wings are membranous and larger. Many species can fly, but the domestic U.S. species rarely do so; however, the imported Asian cockroach in the Florida area both flies frequently and comes to lights.[10] Adult German and brown-banded cockroaches are approximately 15 mm long, whereas the American and Oriental cockroaches are 30 to 50 mm long (Figure 2 and Figure 3). Some tropical species are even longer. Immature cockroaches look similar to adults (except they have no wings), and some of the first nymphal instars are so small as to be confused with ants.

III. GEOGRAPHIC DISTRIBUTION

The German, Oriental, and American cockroaches are cosmopolitan in distribution. The German cockroach is probably the most important overall pest in human habitations worldwide. The American cockroach is an especially severe pest in the tropics and subtropics. Brown-banded cockroaches are a pest in the tropics and subtropics, and they are increasing as a pest in the temperate zone. They probably now infest the entire U.S.

IV. BIOLOGY AND BEHAVIOR

Cockroaches belong to the insect order Blattaria (formerly they were in the Orthoptera) and are closely related to crickets and grasshoppers. They develop by gradual metamorphosis in which the nymphs, when hatched, look similar to the adults, albeit smaller. Some cockroach species live outside and feed on vegetation and other organic matter. However, species that live in buildings are mostly scavengers, feeding on a wide variety of foods including starches, sweets, grease, meat products, glue, hair, and bookbindings. Cockroaches usually choose to live in protected areas that provide a warm and humid environment. American and Oriental cockroaches gather in large groups in protected areas such as wall voids or around steam pipes. The German cockroach spends most of its time hiding in cracks and crevices in dark, warm, and humid areas close to food and water. Brown-banded cockroaches are generally found on ceilings, high on walls, behind picture frames, or in electric appliances (if someone says cockroaches are living inside their telephone or radio, the brown-banded species is probably the culprit). These roaches do not require as close an association with moisture as German cockroaches do.

COCKROACHES

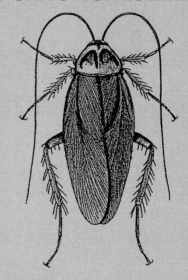

Figure 1
Adult American cockroach; large, commonly encountered specimens.

Importance
Contamination of food; cockroach allergy

Distribution
Numerous species worldwide

Disease Transmission
Mechanical transmission of various bacteria and possibly viruses

Key References
Roth, L.M. and Willis, E.R., *Smithsonian Misc. Coll.,* Vol. 134, No. 10, 1957
Barcay, S.J., Cockroach chapter in *Mallis Handbook of Pest Control,* 9th ed., Hedges, S.A., Ed., GIE Media, Cleveland, OH

Treatment
Avoidance and control; possibly immunotherapy for allergy

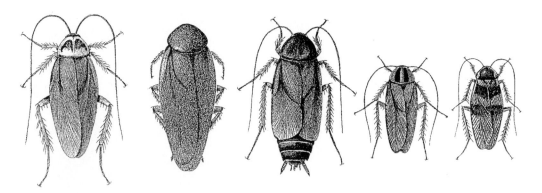

Figure 2
Five common adult cockroaches (L–R): American, smoky brown, Oriental, German, brown-banded.

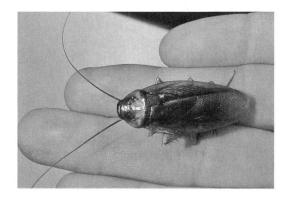

Figure 3
American cockroach. (Photo copyright 2005 by Jerome Goddard.)

V.   TREATMENT OF INFESTATION

Disease transmission or allergic reactions due to cockroaches are best prevented by controlling cockroaches in the residential, institutional, and industrial environment. The best approach to cockroach control involves good sanitation, protecting against new entry, and a combination of least toxic pest control options (baits, dusts, sticky traps, and residuals). In the last decade, numerous cockroach bait products have been marketed that are extremely effective in reducing roach populations. In some cases with proper use of baits, traditional pesticide spraying is no longer needed for cockroach control.

REFERENCES

1. Lemos, A.A., Lemos, J.A., Prado, M.A., Pimenta, F.C., Gir, E., Silva, H.M., and Silva, M.R., Cockroaches as carriers of fungi of medical importance, *Mycoses*, 49, 23, 2006.

2. Tatfeng, Y.M., Usuanlele, M.U., Orukpe, A., Digban, A.K., Okodua, M., Oviasogie, F., and Turay, A.A., Mechanical transmission of pathogenic organisms: the role of cockroaches, *J. Vector Borne Dis.*, 42, 129, 2005.

3. Brenner, R.J., Koehler, P.G., and Patterson, R.S., Health implications of cockroach infestations, *Infect. Med.*, 4, 349, 1987.

4. Burgess, N., Biological Features of Cockroaches and Their Sanitary Importance, in Lectures Delivered at the International Symposium on Modern Defensive Approaches to Cockroach Control, Bajomi, D. and Erdos, G., Eds., The Public Health Commission, Budapest, Hungary, 1982, pp. 45–50.

5. Wirtz, R.A., Occupational allergies to arthropods — documentation and prevention, *Bull. Entomol. Soc. Am.*, 26, 356, 1980.

6. Higgins, P.S., Wakefield, D., and Cloutier, M.M., Risk factors for asthma and asthma severity in nonurban children in Connecticut, *Chest*, 128, 3846, 2005.

7. Rosenstreich, D.L., Eggleston, P., Kattan, M., Baker, D., Slavin, R.G., Gergen, P., Mitchell, H., McNiff-Mortimer, K., Lynn, H., Ownby, D., and Malveaux, F., The role of cockroach allergy and exposure to cockroach allergen in causing morbidity among inner-city children with asthma, *New Engl. J. Med.*, 336, 1356, 1997.

8. Roth, L.M. and Willis, E.R., The medical and veterinary importance of cockroaches, in *Smithsonian Miscellaneous Collection*, Vol. 134 (No. 10), 1957, 56 pp.

9. Rueger, M.E. and Olson, T.A., Cockroaches as vectors of food poisoning and food infection organisms, *J. Med. Entomol.*, 6, 185, 1969.

10. Brenner, R.J., Koehler, P.G., and Patterson, R.S., The Asian cockroach, *Pest Manage.*, 5, 17, 1986.

CHAPTER 17

EARWIGS

TABLE OF CONTENTS

I.   GENERAL AND MEDICAL IMPORTANCE

Earwigs are relatively harmless insects that are occasionally seen inside homes. They are included in this reference because of an old wives' tale that these insects enter human ears, causing much torment (hence the name *earwig*). Earwigs do not enter human ears, or even bite, but some of the larger species may pinch human skin with their abdominal cerci.[1]

II.   GENERAL DESCRIPTION

Earwigs are elongate, slender, flattened insects that are dark-colored and have forceps-like cerci (abdominal pincers). They are generally 4 to 20 mm long, and many species have short, stubby wings (Figure 1). Earwigs have chewing mouthparts and threadlike antennae. Nymphs resemble the adults with differences in abdominal segments and forceps structure. Earwigs may be mistaken for rove beetles (family Staphylinidae).

EARWIGS

Figure 1
Adult earwig.
(From U.S. DHEW, PHS, CDC Pictorial Keys.)

Importance
Harmless — but often mistakenly believed to invade human ears

Distribution
Worldwide

Lesion
None

Disease Transmission
None

Key Reference
Hoffman, K.M., *Proc. Entomol. Soc. Wash.*, 89, 1, 1987

Treatment
None needed

III. GEOGRAPHIC DISTRIBUTION

Numerous species of earwigs occur worldwide. The European earwig, *Forficula auricularia*, and the ring-legged earwig, *Euborellia annulipes*, are cosmopolitan and commonly encountered. Other pest species include the striped earwig, *Labidura riparia*, and the seaside earwig, *Anisolabis maritima*.

IV. BIOLOGY AND BEHAVIOR

Earwigs are mostly nocturnal feeders, feeding on dead and decaying vegetable matter. A few species feed on living plants, and some are predaceous. They may be destructive to garden vegetables, flowers, stored grains, and greenhouse plants. During the day, earwigs hide in cracks and crevices, under bark, or in piles of debris. It is not unusual to find them in private water well buildings and well casings. Accordingly, they can enter the system occasionally and even come out of a faucet. Earwigs lay their eggs in burrows in the ground or among debris, and the female tends them until hatching. This display of maternal care is rare among insects. Earwigs undergo simple metamorphosis and have as many as six nymphal stages. In some areas, they invade homes by the thousands, looking for food and, like cockroaches, become very real pests. For further information the reader is referred to Ebeling,[2] who discusses earwig species and their biology in detail.

REFERENCES

1. Bishopp, F.C., Injury to man by earwigs, *Proc. Entomol. Soc. Wash.*, 63, 114, 1961.
2. Ebeling, W., *Urban Entomology*, University of California Press, Berkeley, CA, 1978, p. 560.

CHAPTER 18

FLEAS

TABLE OF CONTENTS

I. GENERAL AND MEDICAL IMPORTANCE

Fleas are small, laterally flattened, wingless insects that are of great importance as vectors of disease in many parts of the world.[1,2] Public health workers are most concerned with fleas that carry the agents of bubonic plague and murine typhus from rats to people and fleas that transmit plague among wild rodents and secondarily to humans. However, there are other fleaborne diseases. *Rickettsia felis,* a member of the spotted fever group of rickettsiae, has been found in cat fleas and can apparently

FLEAS

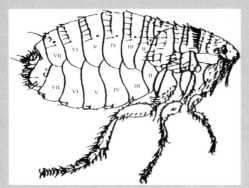

Adult cat flea, *Ctenocephalides felis*

Importance
Biting; annoyance; popular urticaria; vectors of disease

Distribution
Numerous species worldwide

Lesion
Variable depending upon species; in U.S., most lesions irregular, very itchy read wheals; may be papular, vesicular, or bullous

Disease Transmission
Plague; murine typhus; intermediate tape of hosts of dog tapeworm

Key Reference
Lewis R.E., Flea chapter in *Medical Insects and Arachnids*, edited by Lane and Crosskey, Chapman and Hall Co., 1993

Treatment
Topical Corticosteroids; antibiotics if secondary infection; tungiasis (tropics) may require excision of the embedded fleas

infect humans, producing a murine typhus-like illness.[3] Certain rodent fleas are efficient vectors of *Bartonella* organisms.[4] Also, fleas may serve as intermediate hosts for helminths like the dog tapeworm. Despite these disease threats, for many people (especially the lay public), the insidious attacks by fleas on people and domestic animals causing irritation, blood loss, and severe discomfort are equal in importance to disease transmission. Included here are discussions of a few of the more common species, comments as to their medical importance, and notes on their biologies.

The fleabite lesion initially is a punctate hemorrhagic area representing the site of probing by the insect and may have a center elevated into a papule, vesicle, or even a bulla.[5] Lesions may occur in clusters as the flea explores the skin surface, frequently stopping and probing. There is usually formation of a wheal around each probe site with the wheal reaching its peak in 5 to 30 min. Pruritus is almost always present. In most cases there is a transition to an indurated papular lesion within 12 to 24 h. Immunologically, fleabites may produce both immediate and delayed skin reactions.[5,6] In sensitized individuals the delayed reaction appears in 12 to 24 h, persisting for a week or more. The delayed papular reaction with its intense itching is often the reason people present to clinics.

Plague. Plague, a zoonotic disease caused by the bacterium, *Yersinia pestis*, has been associated with humans since recorded history. Few diseases can compare to the devastating effects of plague on human civilization. For example, in the 14th century approximately 25 million people died of plague in Europe. To this day, there are still hundreds of cases occurring annually over much of the world (Figure 1). In the U.S., from 1970–1994 a total of 334 cases of indigenous plague were reported; the peak years were 1983 and 1984, in which there were 40 and 31 cases, respectively.[7] Sylvatic plague, sometimes also called *campestral plague*, is ever-present in endemic areas, circulating among rock and ground squirrels, deer mice, voles, chipmunks, and others. Transmission from wild rodents to

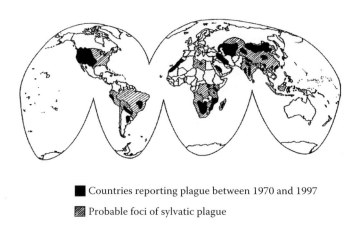

■ Countries reporting plague between 1970 and 1997

▨ Probable foci of sylvatic plague

Figure 1
Status of plague in the world.

humans is rare. *Y. pestis* inflicts damage on the host animal by an endotoxin present on its surface. Hematogenous dissemination of the bacteria to other organs and tissues may cause intravascular coagulation and endotoxic shock, producing dark discoloration in the extremities (thus, the name *black death*). Three clinical forms of plague are recognized: bubonic, septicemic, and pneumonic. The septicemic and pneumonic forms are usually secondary to the bubonic form, and the bubonic form is the most common in the Americas. Pneumonic plague is the most dangerous because of its spread by aerosols (coughing).

There are several methods by which fleas transmit plague. Probably the most important method of infection occurs when fleas ingest plague bacilli along with host (rodent) blood. In the flea stomach, tremendous multiplication of the bacteria takes place (flea feces then also become infected). In some species — especially *Xenopsylla* — further multiplication of the bacteria occurs in the proventriculus, resulting in the flea becoming blocked. When blocked fleas try to feed, there is regurgitation of blood meal products from previous feedings. Blocked fleas become increasingly starved and repeatedly bite in order to get a blood meal. Also, as mentioned, flea feces can be infective, especially when rubbed into abrasions (such as bite wounds) in the skin. Plague bacilli can remain infective in flea feces for as long as 3 years.[8]

Murine typhus. Murine typhus is a rickettsial disease transmitted to humans by fleas and is characterized by headache, chills, prostration, fever, and general pains. There may be a macular rash, especially on the trunk. Infection occurs when infectious flea feces are rubbed into the fleabite wound or other breaks in the skin. Murine typhus is usually mild with negligible mortality, except in the elderly. Severe cases occasionally occur with hepatic and renal dysfunction. The term *murine* indicates that the disease is related to rats. The classic cycle involves rat-to-rat transmission with the Oriental rat flea, *Xenopsylla cheopis,* being the main vector. Murine typhus is one of the most widely distributed arthropod-borne infections endemic in many coastal areas and ports throughout the world.[3] Outbreaks have been reported from Australia, China, Greece, Israel, Kuwait, and Thailand. At one time there were thousands of cases reported annually in the U.S.; from 1931 to 1946 approximately 42,000 cases were reported.[2,9] Since World War II case numbers in the U.S. have fallen drastically to a level of less than 100 per year. Almost all cases in the U.S. are now focused in central and southcentral Texas and Los Angeles and Orange counties in California. However, physicians may encounter murine typhus in returning international travelers. Three recent cases in patients returning to Europe from Indonesia indicate that murine typhus should be considered as a possible cause of imported fever from Indonesia.[10] Interestingly, the ecology of this disease seems to be changing. The classic rat–flea–rat cycle seems to have been replaced in some areas by a peridomestic animal cycle involving free-ranging cats, dogs, opossums, and their fleas.[3]

A. *Cat and Dog Fleas,* Ctenocephalides felis *and* C. canis

Cat fleas are the fleas most often encountered by people in the U.S. (the dog flea is relatively rare in North America). Contrary to their name designation, dog fleas may feed on cats and cat fleas on dogs. In fact, in many areas the predominant flea species infesting dogs is the cat flea. Both species are mainly just pest species. However, the fleas are intermediate hosts of the dog tapeworm, *Diphylidium caninum*, and their bites may produce papular urticaria. Children sometimes become infected via close contact with a flea-infested dog.

B. *Oriental Rat Flea,* Xenopsylla cheopis

This medically important flea is an ectoparasite of Norway rats and roof rats. It is the primary vector of the agent of plague, *Yersinia pestis*, (see previous section) and is involved in the transmission of murine (endemic) typhus organisms, *Rickettsia typhi,* from rat to rat and from rats to people. Murine typhus is one of the most widely distributed arthropod-borne infections, endemic in many coastal areas and ports throughout the world. Almost all cases in the U.S. are concentrated in central and southcentral Texas and in Los Angeles and Orange counties in California.

C. *Human Flea,* Pulex irritans

This flea occasionally becomes abundant on farms, especially in abandoned pigpens. Feingold and Benjamini[6] said *P. irritans* was one of two primary species involved in fleabite allergic reactions in the San Francisco area; however, this species currently is only an infrequent parasite of humans in developed countries. In addition, records of *P. irritans* biting humans in the New World before 1958 may have been due to *P. simulans*, a closely related species.[11]

D. *Chigoe Flea,* Tunga penetrans

This flea, also sometimes called the jigger, nigua, chica, pico, pique, or suthi, burrows into the skin of people in tropical and subtropical regions.[5] In most cases the favorite points of attachment are between the toes and under the toe nails. Unlike most flea species, which spend only a small proportion of their lives on a host animal, *T. penetrans* remains embedded in the skin. The female remains embedded throughout blood engorgement and egg development, which leads to great enlargement of the flea body. Enlarging female fleas on human feet cause intense itching and inflammation and may produce swellings and ulceration. Secondary infection is common. Debilitating sequelae may develop, such as loss of nails and difficulty in walking.[12]

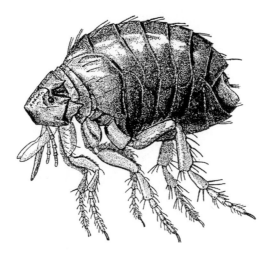

Figure 2
Sticktight flea. (From USDA Bull. No. 248, 1948.)

E. *Northern Rat Flea,* Nosopsyllus fasciatus

The northern rat flea spends most of its adult life on Norway rats and roof rats. This flea is also involved in the transmission of murine typhus organisms among rats and to humans.

F. *Sticktight Flea,* Echidnophaga gallinacea

This flea (Figure 2) is primarily a pest of poultry, but humans are often attacked. As with the chigoe flea, *E. gallinacea* attaches firmly to its host and engorges with blood. It may remain embedded in the integument of the host for some time. Chickens frequently have dark flea-covered patches around the eyes, comb, or wattles.

G. *Sand Fleas*

The lay public uses the term *sand fleas* so much that they are included in this section. In the northern U.S. what people call sand fleas are usually cat or dog fleas found in vacant lots and associated with stray cats or dogs. In the western U.S., cat fleas or human fleas are associated with deer, ground squirrels, stray cats and dogs, or prairie dogs are sometimes termed sand fleas. Cat and dog fleas, and occasionally sticktight fleas, are called sand fleas in the South. In addition, tiny crustaceans in the order Amphipoda occurring abundantly in seaweed along coastal beaches are also called sand fleas or beach fleas. These creatures are not fleas at all.

II. *GENERAL DESCRIPTION*

Adult fleas have laterally compressed bodies, are between 2 to 6 mm long, and are usually brown or reddish-brown with stout spines on their head and thorax (Figure 3). Identification guides are provided by Holland[13] and Benton.[14] Fleas have a short, club-like antenna over each eye. Each segment of their three-segmented thorax bears a pair

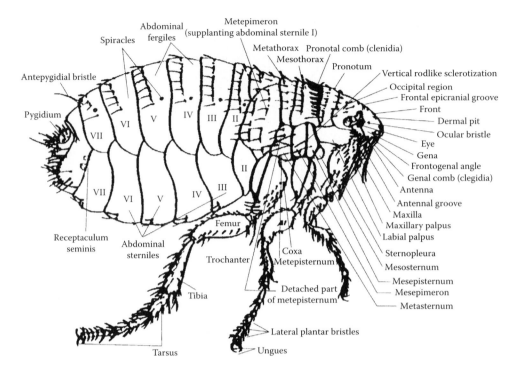

Figure 3
Adult cat flea, Ctenocephalides felis, *with parts labeled. (From U.S. DHEW, PHS, CDC Pictorial Keys.)*

of powerful legs terminating in two curved claws. The chigoe flea is actually quite easily identified as it is so small (1 mm) and has a greatly shortened thorax. Figure 4 is a depiction of the heads of the species discussed here. Most fleas move quickly on skin or in hair and can jump 30 cm or more. They are readily recognized by their jumping behavior when disturbed.

III. GEOGRAPHIC DISTRIBUTION

Most of the 2500 or so species of fleas are seldom seen or encountered, being found only on some small rodents, bats, or birds, often within a restricted range. The distributions of the seven species discussed here follow:

- **Cat and dog fleas** — Worldwide in and around homes with pets. In certain regions one species may occur to the exclusion of the other.
- **Oriental rat flea** — Worldwide wherever *Rattus rattus* is found.
- **Human flea** — Nearly cosmopolitan although there may be large geographic areas apparently free of this species. This seems to be especially true in areas uninhabited by people.
- **Chigoe flea** — Tropical and subtropical regions of North and South America, the West Indies, and Africa. After a 40-year absence, this species has recently reappeared in Mexico.[15] There is one record of the chigoe flea in the U.S. (Texas).[16]

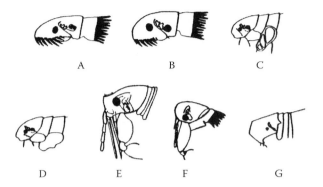

Figure 4
Flea heads: (A) Ctenocephalides felis, *(B)* Ctenocephalides canis, *(C)* Xenopsylla cheopis, *(D)* Pulex irritans, *(E)* Tunga penetrans, *(F)* Nosopsyllus fasciatus, *and (G)* Echidnophaga gallinacea. *(Redrawn from U.S. DHEW, PHS, CDC Pictorial Keys and USDA Misc. Publ. No. 500.)*

- **Northern rat flea** — Widespread over Europe and North America and less commonly in other parts of the world.
- **Sticktight flea** — Widely distributed in the world wherever chickens have been introduced as domestic animals.

IV. BIOLOGY AND BEHAVIOR

Adult fleas have piercing–sucking mouthparts and feed exclusively on blood. The hosts of fleas are domesticated and wild animals, especially wild rodents. If hosts are available, fleas may feed several times daily, but in the absence of hosts, adults may fast for months, especially at low-to-moderate temperatures. Some of the above-mentioned species have specialized life cycles, but in general, the life cycle of most fleas ranges from 30 to 75 d and involves complete metamorphosis.

As cat fleas are a notable pest and seemingly ubiquitous, their life cycle is presented here. Adult female fleas begin laying eggs 1 to 4 d after starting periodic blood feeding. Blood meals are commonly obtained from cats, dogs, and people, but other medium-sized mammals may be affected as well. Females lay 10 to 20 eggs daily and may produce several hundred eggs in their lifetime. Eggs are normally deposited in nest litter, bedding, carpets, etc. Warm, moist conditions are especially favorable for egg production. Eggs quickly hatch into spiny, yellowish-white larvae. Flea larvae have chewing mouthparts and feed on host-associated debris including food particles, dead skin, and feathers. Blood defecated by adult fleas also serves as an important source of nutrition for the larvae. Larvae pass through three instars prior to pupating. Flea larvae are very sensitive to moisture and will quickly die if continuously exposed to less than 60 to 70% relative humidity. Pupating flea larvae spin a loose silken cocoon interwoven with debris. If environmental conditions are unfavorable, or if hosts are not available, developing adult fleas may remain inactive within the cocoon for extended periods. Adult emergence from the cocoon may be triggered by vibrations resulting from host movements.

V. TREATMENT OF BITES AND INFESTATION

The chigoe or burrowing flea, *T. penetrans*, may need to be removed surgically. Alexander[5] said that within the first 48 h of attachment, a sterile needle may be used to accomplish this. During maturation of the eggs, curettage of the contained flea and cautery of the hollow are sufficient. For a mature flea, Alexander recommended total excision of the flea under local anesthesia.[5]

Most other fleas bite but do not remain attached or embedded. Dog or cat fleabites produce reddening, papules, and itching, but generally require no specialized medical treatment. Oral antihistamines may help relieve the itching, and corticosteroids can aid in resolution of the lesions. Scratching of fleabites may produce secondary infection and should be avoided. An antiseptic or antibiotic ointment may be indicated. Of primary concern, however, is to eliminate the source of flea infestation. This may involve sanitation, insecticidal treatment of pets, and spraying of both indoor and outdoor premises.

REFERENCES

1. Bibikova, V.A., Contemporary views on the relationships between fleas and the pathogens of human and animal diseases, *Ann. Entomol. Soc. Am.*, 22, 1, 1977.
2. Harwood, R.F. and James, M.T., *Entomology in Human and Animal Health*, Macmillan, New York, 1979.
3. Azad, A.F., Radulovic, S., Higgins, J.A., Noden, B.H., and Troyer, J.M., Flea-borne rickettsioses: ecologic considerations, *Emerging Infect. Dis.*, 3, 319, 1997.
4. Brown, K.J., Bennett, M., and Begon, M., Flea-borne *Bartonella grahamii* and *Bartonella taylori* in Bank Voles, *Emerging Infect. Dis.*, 10, 684, 2004.
5. Alexander, J.O., *Arthropods and Human Skin*, Springer-Verlag, Berlin, 1984, chap. 11.
6. Feingold, B.F. and Benjamini, E., Allergy to flea bites, *Ann. Allerg.*, 19, 1275, 1961.
7. Craven, R.B., Maupin, G.O., Beard, M.L., Quan, T.J., and Barnes, A.M., Reported cases of human plague infections in the U.S., *J. Med. Entomol.*, 30, 758, 1993.
8. Service, M.W., *Medical Entomology for Students*, Chapman and Hall, New York, 1996, chap. 11.
9. Traub, R., Wisseman, C.L., and Azad, A.F., The ecology of murine typhus: a critical review, *Trop. Dis. Bull.*, 75, 237, 1978.
10. Parola, P., Vogelaers, D., Roure, C., Janbon, F., and Raoult, D., Murine typhus in travelers returning from Indonesia, *Emerging Infect. Dis.*, 4, 677, 1998.
11. Durden, L.A. and Traub, R., Fleas, in *Medical and Veterinary Entomology*, Mullen, G.R. and Durden, L.A., Eds., Academic Press, New York, 2002, p. 103.
12. Feldmeier, H., Eisele, M., Sabola-Moura, R.C., and Heukelbach, J., Severe tungiasis in underprivileged communities: case series from Brazil, *Emerging Infect. Dis.*, 9, 949, 2003.
13. Holland, G.P., The Fleas of Canada, Alaska, and Greenland, in Memoirs of the Entomological Society of Canada, Publ. No. 130, Entomological Society of Canada, Ottawa, 1985.
14. Benton, A.H., *An Illustrated Key to the Fleas of the Eastern U.S.*, Marginal Media, Fredonia, NY, 1983, pp. 1–67.
15. Ibanez-Bernal, S. and Velasco-Castrejon, O., New records of human tungiasis in Mexico, *J. Med. Entomol.*, 33, 988, 1996.
16. Ewing, H.E. and Fox, I., The Fleas of North America, U.S. Department of Agriculture, Misc. Publ. No. 500, 112; 1943, 28.

CHAPTER 19

FLIES (BITING)

TABLE OF CONTENTS

I. BLACK FLIES

A. General and Medical Importance

Black flies (also called *buffalo gnats*, *turkey gnats*, and *Kolumbtz flies*) are small, hump-backed flies that are important as vectors of disease and as nuisance pests.[1,2] In the tropics, black flies are vectors of the parasite, *Onchocerca volvulus*, which causes a chronic nonfatal disease with fibrous nodules in subcutaneous tissues and sometimes visual disturbances and blindness (river blindness). The World Health Organization estimates that about 17.7 million people have onchocerciasis in Africa and Latin America, with approximately 270,000 cases of microfilarial-induced blindness and another 500,000 people with severe visual impairment.[3] In the temperate region black flies are notorious pests, often occurring in tremendous swarms and biting viciously. The bites may be painless at first, but bleed profusely owing to salivary components that prevent clotting.[4] Systemic reactions to black fly bites have been reported, consisting of itching, burning, papular lesions accompanied by fever, leukocytosis, and lymphadenitis. Death can result from anaphylactic shock, suffocation, and toxemia.

B. General Description

Black flies vary in size from about 2 to 5 mm and are thus smaller than mosquitoes. They are black, humpbacked flies with broad wings and stout bodies (Figure 1 and Figure 2). Black flies have short legs and large compound eyes. Their antennae are short (although in 9 to 12 segments) and bare. Crosskey and Howard[5] provide a list of identification keys to black flies by zoogeographic region, and Peterson[6] contains excellent keys to genera of adults, pupae, and larvae of the Nearctic region.

C. Geographic Distribution

There are numerous important species of black flies: *Prosimuliim mixtum* is a serious pest of people and animals in much of the U.S. as well as *Cnephia pecuarum* in the Mississippi Valley. *Simuliim vittatum* and *S. venustum* seriously annoy livestock, fishermen, and campers in U.S. In the Balkans region (former Yugoslavia), there have been severe outbreaks of *S. colombaschense* (the infamous golubatz fly) and *S. erythrocephalum*. Other notorious pests in Europe include *S. equinum, S. ornatum,* and *S. reptans*. In Africa, members of the *S. damnosum* and *S. neavei* complexes are important vectors of onchocerciasis. In Central and South America, vectors of onchocerciasis are *S. ochraceum* and *S. metallicum.*

D.   Biology and Behavior

The larvae of black flies are aquatic and develop in shallow, fast-flowing streams, mainly in upland regions. The filter-feeding larvae have a circlet of numerous radiating rows of tiny hooks located on the tip of the abdomen. These hooks are used for attaching to silk, which the larvae secrete onto solid substrates. Larvae also produce silk for construction of the pupal cocoon. There are usually six to nine larval instars. The densities of larvae can be tremendously high, and in some areas they can cover the entire substrate at the head of stream riffles. Life cycle length varies from 1 year to as little as 2 weeks depending on the species and geographic location. Males form small mating swarms during prenuptial behavior, and mating occurs in the air. Eggs are usually laid in groups of 150 to 600 in or near a water source. Adult black flies are most prevalent during late spring and early summer, when they may be present in large swarms. They are usually encountered near streams or lake outlets, but owing to migration, may be dispersed and found tens of miles from the original water source. Black flies bite in the daytime, mostly in the early morning and near evening.

E.   Treatment of Bites

Black fly bites may be itchy and slow healing. Antiseptic and soothing lotions as well as corticosteroids may relieve pain and itching and help resolve lesions. Systemic reactions characterized by hives, wheezing, fever, leukocytosis, and widespread urticaria require intensive evaluation and treatment. If the reaction is

BLACK FLIES

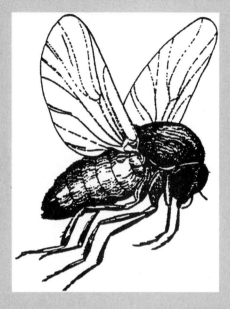

Figure 1
Typical adult black fly.
(From E. Bowles, The Mosquito Book, *Mississippi State Department of Health Publication.)*

Importance
Fierce biters; possible systemic RXNs

Distribution
Numerous species worldwide

Lesion
Variable — often small itching papules, sometimes erythematous wheals and swelling

Disease Transmission
Onchocerciasis

Key References
Gudgel, E. F. and Grauer, F. H., *Arch. Dermatol. Syphiol.,* 70, 609, 1954
Crosskey, R. W., John Wiley, Chichester, U.K., 1990

Treatment
Palliative creams or corticosteroids; oral antihistamines may relieve itching; systemic RXNs may require treatment for shock

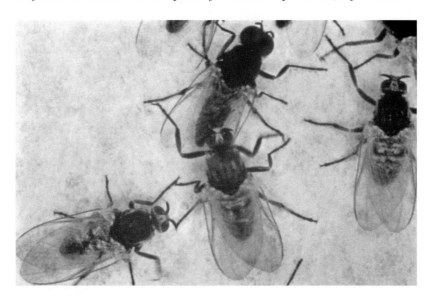

Figure 2
Black flies (CDC photo).

mild, oral antihistamine therapy may suffice, but severe reactions involving shock will probably require epinephrine (see Chapter 2). Persons bitten by black flies in Africa and Latin America may need follow-up for possible development of onchocerciasis.

II.   DEER FLIES

A.   General and Medical Importance

Deer flies belong to the family Tabanidae (the same one as horse flies) but are usually much smaller. Deer flies are extremely annoying to people in the outdoors during summer months, often circling persistently around the head. Like horse flies, deer flies have scissorlike mouthparts and can inflict painful bites. Deer fly bites often become secondarily infected; in hypersensitive individuals they have been known to produce systemic reactions characterized by generalized urticaria and wheezing. In the U.S., the deer fly *Chrysops discalis* mechanically transmits tularemia organisms from rabbits to people by its bites.[7,8] In a recent outbreak, 64% of human cases of tularemia in Wyoming were attributed to deer fly bites.[9] In the African equatorial rain forest, deer flies (particularly *C. silacea* and *C. dimidiata*) transmit the filarial parasite *Loa loa*. Loiasis affects an estimated 2 to 13 million individuals and is characterized by Calabar swellings (localized nonpitting edema mainly on the wrists or ankles, 5 to 20 cm in diameter, lasting from a few hours to a few days), generalized pruritus, arthralgia, fatigue, hypereosinophilia, and sometimes serious central nervous system (CNS) involvement.[10] The pathognomonic symptom of loiasis, subconjunctival migration, is actually uncommon.

B.   General Description

Deer flies (about 8 to 15 mm long) are about half as large as horse flies (Figure 3). They have wings with dark markings and have apical spurs on their hind tibiae (Figure 4).

Many deer fly species are gray or yellow-gray in color with various arrangements of spots on the abdomen.

C. Geographic Distribution

Numerous deer fly species occur almost worldwide. In the U.S., *C. discalis* occurs sparsely in the East but more so in the Southwest. *Chrysops atlanticus* is a troublesome pest in the eastern U.S., and *C. noctifer* is one of the most common western species. *Chrysops silacea* and *C. dimidiata* occur in Africa.

D. Biology and Behavior

Deer flies breed in moist or semiaquatic sites such as margins of ponds, damp earth, or other sites containing various amounts of mud and water. The wormlike carnivorous larvae spend their life in these wet, muddy habitats and then migrate to drier areas of soil to pupate. After the pupal stage, the adult flies emerge. The entire life cycle may take 2 years or more to complete in temperate regions. Females seek a blood meal, whereas males feed on flower and

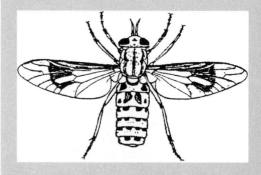

DEER FLIES

Figure 3
Adult deer fly, Chrysops discalis.
(From U.S. DHHS, CDC Publication No. 83–8297.)

Importance
Annoyance; painful bites

Distribution
Numerous species worldwide

Lesion
Deep and painful singular lesions, sometimes leading to cellulitis

Disease Transmission
Tularemia; loiasis (Africa)

Key References
Burger, J.F., *Contrib. Entomol. Int.*, 1, 1, 1995
Foil, L.D., *Parasitol. Today*, 5, 88, 1989

Treatment
Systemic antibiotics if cellulitis; allergic RXNs may require antihistamines and epinephrine; otherwise, soothing ointments, lotions

Figure 4
Deer fly.

HORSE FLIES

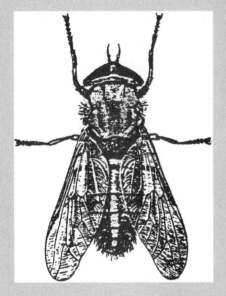

Figure 5
Adult horse fly.
(From USDA Yearbook of Agriculture, *1952.)*

Importance
Annoyance; painful bites; possible systemic RXNs

Distribution
Numerous species worldwide

Lesion
Deep and painful singular lesions sometimes leading to cellulitis

Disease Transmission
None

Key References
Burger, J.F., *Contrib. Entomol. Int.*, 1, 1, 1995
Foil, L.D., *Parasitol. Today*, 5, 88, 1989

Treatment
Systemic antibiotics if cellulitis; allergic RXNs may require antihistamines and epinephrine; otherwise soothing ointments, lotions

vegetable juices. In the temperate zones, deer flies are active in the late spring and summer months, mostly in the early morning and late afternoon. Members of the genus *Chrysops* live in areas having a mixture of forest and open land and feed on various members of the deer family. However, they will aggressively pursue and bite people. Interestingly, in Africa, studies have shown that the presence of wood fires in local villages is attractive to *C. silacea*.[11] Most bloodsucking insects are repelled by the smoke of a wood fire.

E.   Treatment of Bites

Except for secondary infections, which require an appropriate antibiotic, deer fly bites are generally few and self-limiting. Antiseptic and soothing lotions may relieve pain and itching. Allergic reactions characterized by hives, wheezing, or widespread urticaria require intensive evaluation and treatment. If the reaction is mild, oral antihistamine therapy may suffice, but severe reactions involving shock will probably require epinephrine (see Chapter 2).

Follow-up may be prudent for persons bitten by deer flies in the southwestern U.S. (tularemia) or the African equatorial rain forest (*L. loa* filariasis).

III.   HORSE FLIES

A.   General and Medical Importance

Horse flies (also family Tabanidae) are large, robust bloodsucking flies that are notorious pests of horses, cattle, deer, and other mammals. Several species of horse flies will also attack people. Horse fly bites have been known to produce systemic reactions in humans characterized by generalized urticaria and wheezing.

B.   General Description

Horse flies look like giant robust house flies. They are often 20 to 25 mm long and have large prominent eyes (Figure 5). The antennae

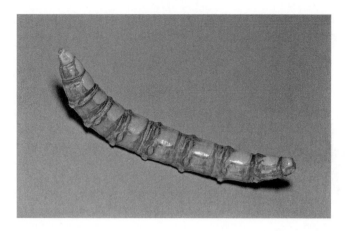

Figure 6
Typical horse fly larva. (From Dr. Blake Layton, Mississippi Cooperative Extension Service, Mississippi State University.)

have only three sections. Their proboscis projects forward, and the female's mouthparts are bladelike for a slashing/lapping feeding method. Horse fly larvae are spindle-shaped and generally white, tan, brown, or even greenish in color.

C. Geographic Distribution

Horse flies are distributed worldwide. Most people-biting species are in the genera *Tabanus, Hybomitra*, and *Haematopota*. Some of the notorious pest species include *Tabanus atratus* (the large black horse fly), *T. lineola,* and *T. similis* (the familiar striped horse flies) in the eastern U.S., and *T. punctifer* and *Atylotus incisuralis* in the West.

D. Biology and Behavior

Horse fly biology is very similar to that of deer flies. Basically, they breed in moist or semiaquatic sites such as pond margins, damp earth, or rotten logs. The grublike larvae spend their life in wet mud, dirt, or shallow water and then pupate in drier patches of soil (Figure 6). Larval development may take a year or more. After a pupal stage, the adult flies emerge. Females seek a blood meal, whereas males feed on nectar from flowers and other vegetable juices. In the temperate region, horse flies are active only in the warmer months of the year.

E. Treatment of Bites

Horse fly bites are generally few and self-limiting. If secondary infection/cellulitis develops, appropriate systemic antibiotics are indicated. Antiseptic and soothing lotions may relieve pain and itching. Allergic reactions characterized by hives, wheezing, and widespread urticaria require intensive evaluation and treatment. If the reaction is mild, oral antihistamine therapy may suffice, but severe reactions involving shock will probably require epinephrine (see Chapter 2).

BITING MIDGES

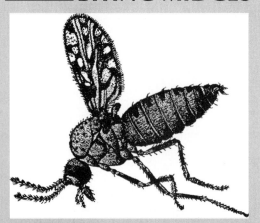

Adult biting midge, *Culicoides furens*

Importance
Annoyance from biting

Distribution
Numerous species occurring worldwide

Lesion
Minute papular lesions with erythematous halo; wheals may occur in sensitized persons

Disease Transmission
Oropouche fever (tropics)

Key References
Kwan, W.H. and Morrison, F.O., *Ann. Soc. Entomol. Quebec*, 19, 127, 1974

Downes, J.A., *Memoirs Entomol. Soc. Canada*, 104, 1, 1978

Treatment
Antipruritic lotions or creams

IV. MIDGES (BITING MIDGES, BITING GNATS)

A. General and Medical Importance

The biting midges are very tiny slender gnats in the family Ceratopogonidae; they are sometimes called *punkies, no-see-ums, gnats,* or *flying teeth* (see Color Figure 19.17). In the Caribbean region and Australia they are referred to as sand flies (not to be confused with psychodid sand flies; see Section V). Adult biting midges are vicious and persistent biters, and some persons have strong reactions to their bites. Their small size allows them to pass through ordinary screen wire used to cover windows and doors. These tiny insects are generally not involved in the transmission of disease agents to humans in the U.S. However, in Africa and South America certain species may be able to transmit filariae, protozoa, and viruses (particularly Shuni and Oropouche viruses in the Simbu group of arboviruses). Oropouche fever, caused by a virus in the Simbu group of Bunyaviridae, is characterized by fever, headache, anorexia, dizziness, muscle and joint pains, and photophobia.[12] It is generally nonfatal, but may be serious enough to lead to prostration. Since it was first isolated from a patient in Trinidad in 1955, this virus has been documented in numerous epidemics in the Amazon region of Brazil.[13] Little is known about the animal reservoirs of Oropouche virus in nature, but some evidence implicates monkeys, wild birds, and sloths. The primary biting midge vector in urban outbreaks is *Culicoides paraensis*.[13]

B. General Description

Biting midges are typically gray in color (although some species may be yellowish), extremely small, 0.6 to 1.5 mm, and delicate with narrow wings that have few veins and no scales. The wings may be clear or hairy, sometimes distinctly spotted (with pigment, not scales as in mosquitoes), and folded scissorlike over the abdomen at rest. The eyes on each side of the head are black and the proboscis protrudes forward and downward. Biting midges somewhat resemble other small species of nonbiting gnats in the family Chironomidae, but they are not as large and mosquito-like as chironomids. Swarms of biting midges are small and inconspicuous. People attacked by this midge will often comment when outdoors, "Something is biting me but I can't see what it is."

C. Geographic Distribution

There are more than 4000 species of Ceratopogonidae in temperate and tropical areas of the world. A worldwide list of the species, subspecies, and varieties in the genus *Culicoides* is provided by Boorman and Hagen.[14] *Culicoides furens* is a vicious biter occurring in salt marshes along the Atlantic and Gulf Coasts, from Massachusetts to Brazil, and along the Pacific Coast, from Mexico to Ecuador. Further inland, *C. paraensis* is a troublesome species throughout much of the eastern U.S. In California, *Leptoconops torrens* and *L. carteri* are very bothersome. *C. pulicaris* is a serious pest of farm workers in Europe, and *Forcipomyia* (*Lasiohelea*) *taiwana* is a severe pest in Taiwan and Japan. In the Near and Middle East, Central Asia, Africa, and Southern Europe, *L. kerteszi* is an avid biter of people. Its bites may become vesicular, resulting in an open lesion that may exude moisture for a week or more. *Austroconops macmillani* is a vicious biter in western Australia, and *L. spinosifrons* is a beach pest in East Africa, Madagascar, India, Sri Lanka, and the Malay Archipelago.

D. Biology and Behavior

Only female ceratopogonids bite people; the males feed on the nectar of flowers. Females feed by the way of small cutting teeth on the elongated mandibles in their proboscis which they use to make a small cut in the skin. A chemical is present in the saliva to prevent blood clotting. Some of the cut capillaries bleed and form a tiny pool of blood that is then sucked up. Larval stages develop in highly organic detritus overlaying the bottom of shallow areas of water or in water-saturated soil high in organic material. They especially seem to be found in salt marshes. The larvae look more like worms than maggots (Figure 7). Female biting midges lay their eggs on the water or on a variety of objects overhanging larval habitats. Most species are active only in the warmer months of the year, but some may be active year round. Lillie et al.[15] found that adult *Culicoides* host-seeking activity was greatest near sunrise and sunset, during the night when the moon was full, and during the afternoon hours in winter and early spring. They usually do not bite during the midday, except when the sky is heavily overcast and the winds are calm.

E. Treatment of Infestation

Alexander[16] said that simple antipruritics (oily calamine lotion or hydrocortisone with added anti-infective agents) are all that is needed for biting midge bites. Probably the best course of action for those affected by biting midges is a combination of avoidance and personal protection measures. Use of finer-mesh screen wire can prevent entry of these flies into dwellings. Repellents containing diethyltoluamide (DEET) and long-sleeved shirts and long pants can provide relief to those persons outside in infested areas. DEET has been shown to be very effective in repelling *Culicoides*.[17]

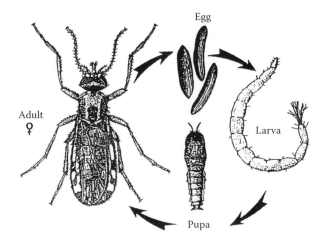

Figure 7
Life stages and life cycle of Culicoides furens. *(Photo courtesy of Florida Medical Entomology Laboratory, IFAS, University of Florida.)*

The bath oil, Avon Skin-So-Soft®, is sometimes used as a repellent, and controlled studies indicate that the product provides some protection from *Culicoides* midges.[17] However, product effectiveness is not because it repels midges, but because the oiliness traps the midges on the skin surface.

V.   SAND FLIES

A.   General and Medical Importance

Sand flies are tiny bloodsucking flies in the family Psychodidae that transmit the causative agents of bartonellosis (Carrión's disease), sand fly fever, and leishmaniasis. Sand fly fever, a viral disease, occurs in those parts of southern Europe, the Mediterranean, the Near and Middle East, Asia, and Central and South America where the *Phlebotomus* vectors exist. Bartonellosis caused by the bacillus, *Bartonella bacilliformis*, occurs in the mountain valleys of Peru, Ecuador, and southwest Colombia. Leishmaniasis occurs in tropical and subtropical areas over much of the world (Figure 8). There has been a recent resurgence of leishmaniasis. It is now found in 88 countries and is increasingly being reported in nonendemic areas. Contributing factors to the increase in leishmaniasis include human settlement in zoonotic foci and urbanization up to the edge of forests. In many cases, the sylvatic cycles have now become peridomestic. The life cycle of *Leishmania donovani* is shown in Figure 9 to illustrate the involvement of sand flies in parasite life cycles.

Clinically, leishmaniasis manifests itself in four main forms: (1) cutaneous, (2) mucocutaneous, (3) diffuse cutaneous, and (4) visceral. The cutaneous form may appear as small and self-limiting ulcers that are slow to heal. When there is destruction of nasal and oral mucosa, the disease is labeled *mucocutaneous leishmaniasis* (Figure 10). Sometimes there are widespread cutaneous papules or nodules all over the body, a condition termed *diffuse cutaneous leishmaniasis*. Finally, the condition in which the parasites invade cells of the spleen, bone marrow, and liver — causing widespread *visceral involvement* — is termed visceral leishmaniasis. There have been reports of

CASE HISTORY

PAINFUL HORSE FLY BITE

On June 20 a woman in Jackson, MS, was awakened during the night by a painful insect bite on her right arm. Looking around, she noticed a large horse fly on the ceiling above the bed. Earlier that evening the horse fly had gotten inside the house and she and her husband had unsuccessfully tried to find it. Apparently, the horse fly later found the victim sleeping and attempted to take a blood meal. Within 8 h the bite site was a 5-cm erythematous indurated plaque, hot to the touch (Figure 1). Other than a persistent itch for a few days, no further complications arose.

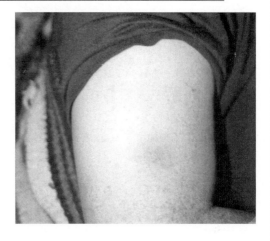

Figure 1
Lesion resulting from horse fly bite, 8 h postbite.

Comment. Horse flies are robust flies known for biting horses, cows, and other large mammals. Some species are almost 2 in long. They do not have tube-like mouthparts for piercing–sucking, but instead, have a complex arrangement of blade-like mandibles and styliform maxillary galeae that cut or slash wounds in their hosts' skin (Figure 2). Horse flies then draw up the blood by means of pseudotracheae on labellar lobes. Blood uptake is quite inefficient, with blood often seen dripping from the wound.

The lesion described previously was warm on palpation because of inflammation and vascular dilatation. Various immune cells attracted to the site (or formed) as part of the immune reaction release histamine, which causes blood vessels to dilate. Induration of a lesion like this is not only from the influx of cells; histamine also causes fluid to be leaked from the small skin blood vessels, making the area feel hard and raised. Other than possible secondary infection, no human diseases are known to be caused by horse flies.

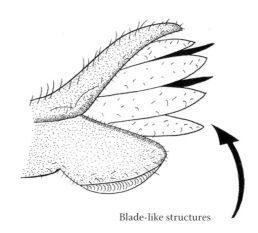

Blade-like structures

Figure 2
Diagrammatic representation of horse fly mouthparts.

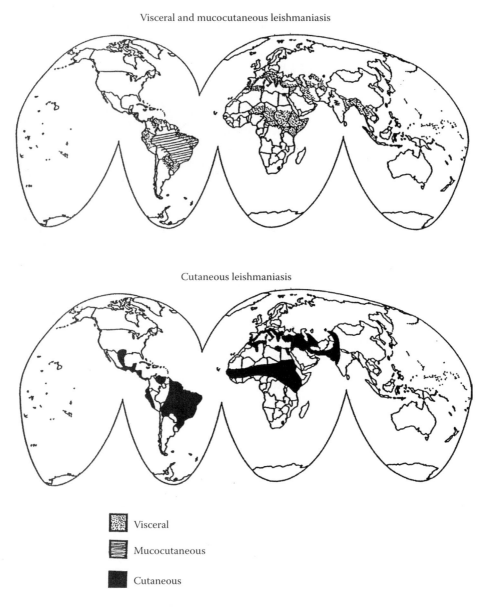

Visceral and mucocutaneous leishmaniasis

Cutaneous leishmaniasis

Visceral

Mucocutaneous

Cutaneous

Figure 8
Approximate worldwide distribution of leishmaniasis.

visceral leishmaniasis in foxhounds in the U.S.,[18] but the only human sand fly-transmitted disease in the U.S. is probably the few cases of cutaneous leishmaniasis diagnosed each year in south Texas.[19–21]

For a long time, these Old World and New World disease forms were correlated with various *Leishmania* species and geographic regions to make a well-defined classification. As is the case in many paradigms in science, this classification is turning out to be not so clear cut. There is apparently a whole spectrum of disease — from cutaneous to visceral — depending on many factors such as species of *Leishmania*, numbers of parasites (parasite burden), and the predominant host immune response. The idea that

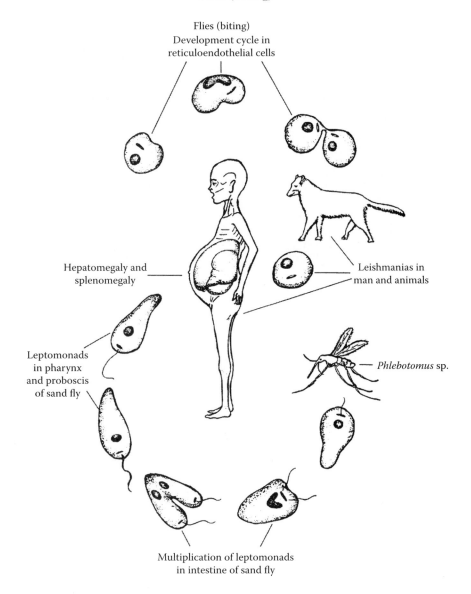

Flies (biting)
Development cycle in
reticuloendothelial cells

Hepatomegaly and
splenomegaly

Leishmanias in
man and animals

Leptomonads
in pharynx
and proboscis
of sand fly

Phlebotomus sp.

Multiplication of leptomonads
in intestine of sand fly

Figure 9
Life cycle of Leishmania donovani. *(Reprinted from Thomas J. Brooks, Jr., Essentials of Medical Parasitology, Macmillan Publishing Company. Copyright 1963, Thomas J. Brooks, Jr. With permission.)*

a few *Leishmania* species each cause a distinct and separate clinical syndrome is no longer valid. For example, a visceral species, *L. chagasi*, has also been isolated from patients with cutaneous leishmaniasis in several Central American countries. However, the particular parasite species and geographic location may still serve as useful epidemiologic labels for the study of the disease complex, and generalized statements can be made. For example, *L. donovani* and *L. infantum* generally cause visceral leishmaniasis in the Old World, whereas *L. tropica* and *L. major* cause cutaneous lesions. In the New World, visceral disease is mainly caused by *L. chagasi*, cutaneous lesions by *L. mexicana* and related species, and mucocutaneous lesions by *L. braziliensis*. Further,

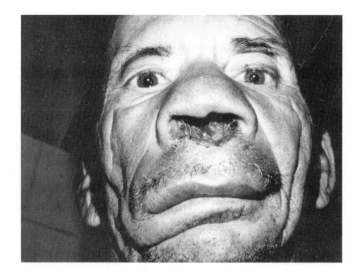

Figure 10
Mucocutaneous leishmaniasis. (Photo courtesy of Armed Forces Institute of Pathology, negative number 74-8873-1.)

by sorting out the species that do not cause human disease in a given area, researchers are more able to focus their studies on the ecology and behavior of those that do.

B. General Description

Sand flies are tiny (about 3 mm long), golden, brownish, or gray long-legged flies (Figure 11 and Figure 12). The females have long, piercing mouthparts adapted for blood sucking. A distinctive feature of these flies is the way they hold their wings V-shaped at rest. They have long, multisegmented antennae and hairs (not scales) covering much of the body and wing margins.

C. Geographic Distribution

Members of the sand fly genera *Phlebotomus* and *Sergentomyia* occur in the Old World, and *Lutzomyia*, *Brumptomyia*, and *Warileya* occur in the tropics and subtropics of the New World. *P. argentipes* is the chief vector of visceral leishmaniasis in many areas of the Old World; *P. langeroni* has recently been incriminated. *Lutzomyia longipalpis* is a major vector in the New World. Old World sand fly vectors of dermal leishmaniasis include *P. caucasicus*, *P. papatasi*, *P. longipes*, and *P. pedifer*. New World vectors of the malady include *L. olmeca*, *L. trapidoi*, *L. ylephiletrix*, *L. verrucarum*, *L. peruensis*, and others. Sand fly fever vectors are primarily *P. papatasi* and *P. sergenti*. Bartonellosis is transmitted by *L. verrucarum*.

As *P. papatasi* is the most important and widespread vector of zoonotic cutaneous leishmaniasis caused by *Leishmania major* in the Old World, its distribution is given in Figure 13.

Lutzomyia anthophora or *L. diabolica*, which generally occur in south and central Texas as well as in parts of Mexico, are thought to be the vectors of human cases of cutaneous leishmaniasis in Texas.

D. Biology and Behavior

Female sand flies are blood feeders; males are not, but will suck moisture from available sources. Despite their medical importance, much of what is known about sand flies is speculative or based on limited field observations. Sand flies are usually found in microhabitats within larger biological communities. Examples of these microhabitats are caves, cavities, tree holes, burrows, pit latrines, animal enclosures, and other buildings. Breeding is believed to take place in these areas, and feeding on hosts occurs there or in the close vicinity. Sand flies in the genus *Brumptomyia* feed on armadillos, whereas *Lutzomyia* spp. feed on both mammals and reptiles. Members of the genus *Sergentomyia* feed on reptiles and amphibians. *Phlebotomus* spp. are exclusively mammal feeders. Sand flies are generally active at night when there is little or no wind, but a few species may feed during the day if disturbed or under cloudy or shaded conditions. They normally rest during the day in their microhabitats. Sand fly development is relatively slow, taking up to several months to complete the life cycle.

P. papatasi, the primary vector of leishmaniasis in the Middle East region, is a nocturnal, desert species flying from sunset through dark. It enters dwellings from animal burrows up to 50 m away. Its activity diminishes rapidly when relative humidity falls below 65%, temperature rises above 80°F, or wind velocity exceeds 5 mi/h.

Figure 12
Adult sand fly. (Photo courtesy of U.S. Armed Forces Pest Management Board.)

SAND FLIES_____

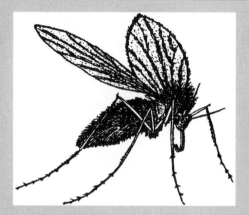

Figure 11
Adult sand fly.
(From E. Bowles, The Mosquito Book, *Mississippi State Department of Health Publication.)*

Importance
Vectors of several important diseases

Distribution
Numerous species almost worldwide

Lesion
Sometimes red papules or urticarial wheals

Disease Transmission
Leishmaniasis, bartonellosis, sand fly fever

Key Reference
Young, D.G. and Perkins, P.V., *J. Am. Mosq. Control Assoc.*, 44(2), Part 2 (Supplement), 1984

Treatment
Except for hypersensitivity RXNs, palliative antipruritic lotions or creams

Figure 13
Approximate geographic distribution of Phlebotomus papatasi.

STABLE FLIES

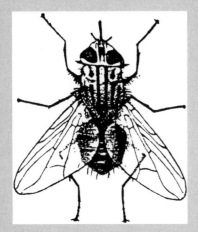

Adult stable fly, *Stomoxys calcitrans*

Importance
Nuisance biting

Distribution
Most of the world

Lesion
Small evanescent papules

Disease Transmission
None

Key Reference
Zumpt, F., Gustave Fischer Verlag
    Publishing Co., Berlin, 1973

Treatment
Except for hypersensitivity RXNs (rare)
or secondary infection, soothing and
antiseptic lotions or creams sufficient

E. Treatment of Bites

Sand flies are relatively uncommon; thus, their bites are generally few and self-limiting. Soothing locations and antiseptic lotions may relieve itching and prevent secondary infection. Follow-up of patients with sand fly bites or persons with exposure to areas endemic with the several sand fly-transmitted diseases is needed to diagnose and treat these maladies. In the case of leishmaniasis, diagnosis is delayed by physicians unfamiliar with the clinical features and various forms of the disease. Cutaneous leishmaniasis is often initially described as a *boil* or *ulcer that fails to heal*. In addition, cutaneous leishmaniasis needs to be differentiated from other skin lesions such as those resulting from leprosy, tuberculosis cutis, sporotrichosis, histoplasmosis (*Histoplasma duboisii*), molluscum contagiosum, and cutaneous sarcoidosis. The most current treatment recommendations for leishmaniasis can be obtained by consultation with the Centers for Diseases Control and Prevention, Parasitic Diseases Division, phone number (770) 488-7775.

VI. STABLE FLIES

A. General and Medical Importance

The stable fly, *Stomoxys calcitrans*, is a significant medical and veterinary pest.[22] People who say "a house fly bit me" are usually mistakenly referring to the stable fly. The flies (sometimes also called *dog flies*) are fierce biters of people, pets, and livestock, and are a major pest in some seacoast areas, impeding development. Because of their bloodsucking habits, the flies have been suspected of transmitting a number of human diseases by mechanical action, but proof is lacking.

B. General Description

Stable flies are 5 to 6 mm long, have a dull gray thorax with four dark longitudinal stripes, and have a dull gray abdomen with dark spots (Figure 14). They look very similar to house flies, but they are slightly larger and have a rigid proboscis projecting forward in a bayonet-like fashion (Figure 15). In contrast, house flies have sponging mouthparts that project downward.

C.   Geographic Distribution

The stable fly occurs in most parts of the world. There is a related species, *S. nigra*, that occurs throughout Africa and attacks horses, donkeys, cattle, and people.

D.   Biology and Behavior

Both male and female stable flies are vicious biters, both taking blood meals. It only takes 3 to 4 min for them to engorge to full capacity. The female lays her eggs in plant waste, cut grass, old hay stacks, piles of fermenting seaweed, or manure (if there is sufficient straw or hay mixed in it). Larval development takes 8 to 30 d, depending on temperature. The larvae (maggots) look almost identical to those of house flies. Stable flies may overwinter as pupae, but normal pupal development takes about 1 to 3 weeks. Overall, under optimum conditions, the total time for development of stable flies from eggs to adults is about 35 d. Adults remain alive for a month or two. They bite people and animals during the day.

E.   Treatment of Bites

Stable fly bites are generally few and self-limiting. Antiseptic and soothing lotions may relieve pain and itching. Allergic reactions characterized by hives, wheezing, and widespread urticaria are rare but require intensive evaluation and treatment.

VII.   TSETSE FLIES

A.   General and Medical Importance

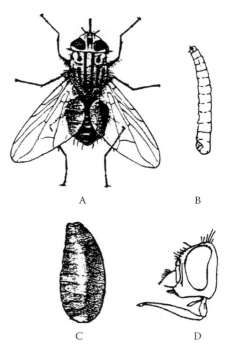

A   B

C   D

Figure 14
Stable fly life cycle. (From Alabama Cooperative Extension Service.)

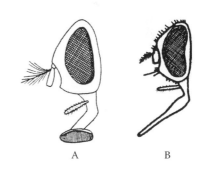

A   B

Figure 15
(A) House fly and (B) stable fly mouthpart comparison. (From U.S. DHHS, CDC Publication No. 83-8297.)

There are over 20 species of flies in the genus *Glossina* that are called tsetse flies. Most of the species are vectors of trypanosomes of people and animals (Figure 16); however, at least six species are of primary importance as vectors of African trypanosomiasis, caused by subspecies of the protozoan, *Trypanosoma brucei*. The disease is

TSETSE FLIES

Adult tsetse fly

Importance
Vectors of sleeping sickness

Distribution
Tropical Africa

Lesion
Small punctate hemorrhages unless sensitive to saliva (in which case, itchy wheals result)

Disease Transmission
African trypanosomiasis

Key References
Buxton, P.A., H.K. Lewis and Co. Publishers, London, 1955

Willet, K. C., *Annu. Rev. Entomol.*, 8, 197, 1963

Treatment
Soothing lotions and antiseptic lotions or creams

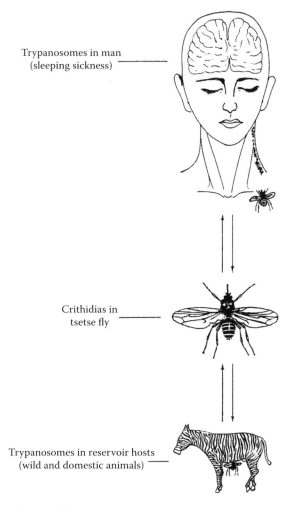

Trypanosomes in man (sleeping sickness)

Crithidias in tsetse fly

Trypanosomes in reservoir hosts (wild and domestic animals)

Figure 16
Life cycle of Trypanosoma rhodesiense *and* I. gambiense. *(Reprinted from Thomas J. Brooks, Jr., Essentials of Medical Parasitology, Macmillan Publishing Company. Copyright 1963, Thomas J. Brooks, Jr. With permission.)*

called *sleeping sickness* because there is often a steady progression of meningoencephalitis, with increase of apathy, fatigability, confusion, and somnolence. The patient may gradually become more and more difficult to arouse and finally becomes comatose. The disease is re-emerging and the WHO estimates that 350,000 to 500,000 cases of African trypanosomiasis occur each year. Countries affected the most include the Democratic Republic of Congo, Angola, and Sudan. Making matters worse,

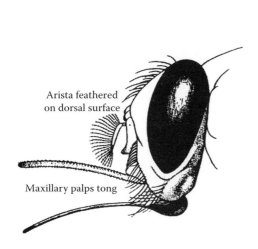

Arista feathered
on dorsal surface

Maxillary palps tong

Hatchet cell

Figure 17
*Side view of tsetse fly head. (From U.S. Navy
Laboratory Guide to Medical Entomology.)*

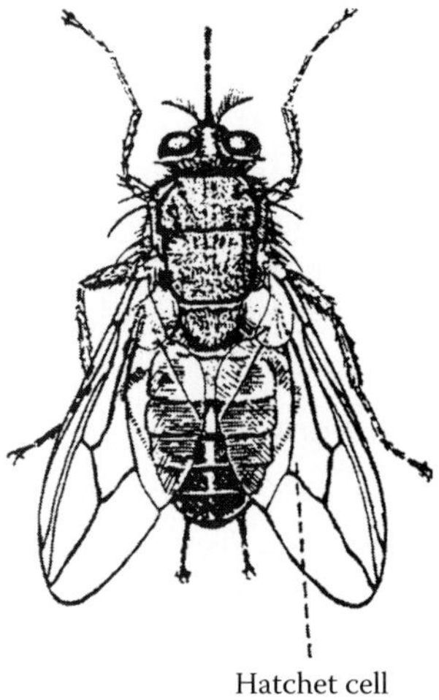

Figure 18
*Adult tsetse fly. (From USDA Yearbook of
Agriculture, 1952.)*

there is rising resistance to melarsoprol, the only widely available drug for CNS involvement in African trypanosomiasis. Nash[23] and Willet[24] provided very good reviews of
the complicated African trypanosomiasis problem. Briefly, the chief vectors of *Trypanosoma brucei gambiense*, the cause of the Gambian form of sleeping sickness, are
Glossina palpalis, G. fuscipes, and *G. tachinoides.* Cases of Gambian sleeping sickness
occur in western and central Africa and are usually more chronic. In eastern Africa, the
Rhodesian (or eastern) form, which is virulent and rapidly progressive, is caused by *T.
brucei rhodesiense.* The primary vectors of the Rhodesian form are *G. morsitans, G.
swynnertoni,* and *G. pallidipes.* The eastern form is commonly contracted by travelers
on game safaris or ecovacations. Other than the possibility for sleeping sickness transmission, the bites of tsetse flies are of minor consequence. However, some individuals
become sensitized to the saliva, and subsequent bites produce welts.

B.   General Description

Tsetse flies are 7 to 13 mm long and yellow, brown, or black. They fold their wings
scissorlike over their back at rest, and this, along with other body features, makes them
appear wasp- or honey bee-like. The arista arising from the short, three-jointed antennae has rays that are branched bilaterally (Figure 17). Tsetse flies have a long, slender
proboscis that is held out in front of the fly at rest. The discal cell (first M) looks like a
meat cleaver or hatchet (Figure 18).

C.   Geographic Distribution

Tsetse flies are generally confined to tropical Africa between 15° N and 20° S latitude.
Glossina morsitans is a bush species found in wooded areas and brush country in eastern Africa. In western and central Africa, where members of the *G. palpalis* group are
the principal vectors, the flies are predominantly found near the specialized vegetation
lining the banks of streams, rivers, and lakes.

D. Biology and Behavior

Tsetse flies feed on a wide variety of mammals and a few reptiles; people are not their preferred hosts. Both sexes feed on blood and bite during the day. Tsetse flies have a life span of about 3 months (less for males), and females give birth to full-grown larvae on dry and loose soil in places like thickets and sandy beaches. The larvae burrow a few centimeters into the substrate and pupate. The pupal stage lasts 2 weeks to a month, after which the adult fly emerges to continue the life cycle.

E. Treatment of Bites

Tsetse fly bites are generally self-limiting. Antiseptic and soothing lotions may relieve pain and itching. Persons sensitive to tsetse fly bites may require antihistamine and other treatment to control the immune response. Persons bitten in endemic areas for sleeping sickness should be followed up for development of the disease. There are treatment strategies (quite complex) for African sleeping sickness. For up-to-date information and recommendations, contact the Parasitic Diseases Division, Centers for Disease Control and Prevention, telephone (770) 488-7775.

REFERENCES

1. Harwood, R.F. and James, M.T., *Entomology in Human and Animal Health*, Macmillan, New York, 1979.
2. Nelson, G.S., Onchocerciasis, *Adv. Parasitol.*, 8, 173, 1970.
3. WHO, Onchocerciasis and Its Control, World Health Organization Technical Report Series No. 852, Geneva, 1995, pp. 1–103.
4. Cupp, E.W. and Cupp, M.S., Black fly salivary secretions: importance in vector competence and disease, *J. Med. Entomol.*, 34, 87, 1997.
5. Crosskey, R.W. and Howard, T.M., A new taxonomic and geographical inventory of world black flies, Special Publ., The Natural History Museum, London, 1, 1997.
6. Peterson, B.V., Simuliidae, in *An Introduction to the Aquatic Insects of North America*, 3rd ed., Merritt, R.W. and Cummins, K.W., Eds., Kendall/Hunt Publishers, Dubuque, IA, 1996, p. 591.
7. Hopla, C.E., The ecology of tularemia, *Adv. Vet. Sci. Comp. Med.*, 18, 25, 1974.
8. McDowell, J.W., Scott, H.G., and Stojanovich, C.J., Tularemia, in Centers for Disease Control Training Guide, U.S. Public Health Service, Washington, D.C., 1964, 65 pp.
9. CDC, Tularemia transmitted by insect bites — Wyoming, 2001–2003, *MMWR*, 54, 170–173, 2005.
10. Pinder, M., *Loa loa* — a neglected filaria, *Parasitol. Today*, 4, 279, 1988.
11. Caubere, P. and Noireau, F., Effect of attraction factors on sampling of *Chrysops silacea* and *C. dimidiata*, vectors of *Loa loa* filariasis, *J. Med. Entomol.*, 28, 263, 1991.
12. Mellor, P.S., Boorman, J., and Baylis, M., *Culicoides* biting midges: their role as arbovirus vectors, *Annu. Rev. Entomol.*, 45, 307, 2000.
13. Mullen, G.R., Biting midges, in *Medical and Veterinary Entomology*, Mullen, G.R. and Durden, L.A., Eds., Academic Press, New York, 2002, p. 163.
14. Boorman, J. and Hagen, D.V., A name list of world *Culicoides*, *Int. J. Dipterol.*, 7, 161, 1996.
15. Lillie, T.H., Kline, D.L., and Hall, D.W., Diel and seasonal activity of *Culicoides* spp. near Yankeetown Florida, monitored with a vehicle-mounted insect trap, *J. Med. Entomol.*, 24, 503, 1987.

16. Alexander, J.O., *Arthropods and Human Skin*, Springer-Verlag, Berlin, 1984.

17. Schreck, C.E. and Kline, D.L., Repellency determinations of four commercial products against six species of ceratopogonid biting midges, *Mosq. News*, 41, 7, 1981.

18. Duprey, Z.H., Steurer, F.J., Rooney, J.A., Kirchoff, L.V., Jackson, J.E., Rowton, E.D., and Schantz, P.M., Canine visceral leishmaniasis, United States and Canada, 2000–2003, *Emerging Infect. Dis.*, 12, 440, 2006.

19. Furner, B.B., Cutaneous leishmaniasis in Texas: report of a case and review of the literature, *J. Am. Acad. Dermatol.*, 23, 368, 1990.

20. Grimaldi, G., Jr., Tesh, R.B., and McMahon-Pratt, G., A review of the geographic distribution and epidemiology of leishmaniasis in the New World, *Am. J. Trop. Med. Hyg.*, 41, 687, 1989.

21. McHugh, C.P. and Kerr, S.F., Isolation of *Leishmania mexicana* from *Neotoma micropus* collected in Texas, *J. Parasitol.*, 76, 741, 1990.

22. Newson, H.D., Arthropod problems in recreational areas, *Annu. Rev. Entomol.*, 22, 333, 1977.

23. Nash, T.A.M., A review of the African trypanosomiasis problem, *Trop. Dis. Bull.*, 57, 973, 1960.

24. Willet, K.C., African trypanosomiasis, *Annu. Rev. Entomol.*, 8, 197, 1963.

CHAPTER 20

FLIES (NONBITING)

TABLE OF CONTENTS

FILTH FLIES

Adult house fly, *Musca domestica*

Importance
Nuisance; contamination; myiasis

Distribution
Numerous species wordwide

Lesion
Non-biting, no lesion

Disease Transmission
Several agents mechanically

Key Reference
B. Greenberg's Two Vol. *Flies and Disease,* Princeton University Press, 1971

I. HOUSE FLIES AND OTHER FILTH FLIES

A. General and Medical Importance

The term *filth flies* refers to members of the families Muscidae, Sarcophagidae, and Calliphoridae, which are domestic nonbiting flies commonly seen in and around human dwellings. Although other species or groups could be considered filth flies, this topic will be restricted to those fly families (mentioned earlier) representing the house flies, flesh flies, and blow flies. These flies do not bite, but are medically important in the mechanical transmission of disease agents from feces or dead animals to foodstuffs or food preparation areas. Throughout the world, these flies serve as carriers of organisms causing diseases such as typhoid, diarrhea, bacillary amoebic dysentery, cholera, giardiasis, pinworm, and tapeworm. They may spread these agents via their mouthparts, body hairs, or sticky pads of their feet, as well as through their vomitus or feces.[1] Controlled studies have shown that fly management in communities leads to a reduction in the incidence of diarrhea.[2,3] The egg-laying habits of these insects may also lead to another human malady, myiasis, which is the infestation of people by the maggots of flies (see Chapter 6).

B. General Description

House flies are about 5 to 8 mm long, with a dull gray thorax and abdomen (not shiny). The thorax has four longitudinal dark stripes (Figure 1). Mature house fly larvae are 10 to 13 mm long and usually creamy white in color (Figure 2). Overall, the larvae have a conical shape with two dark-colored mouth hooks at the narrow end and two oval spiracular plates at the broad posterior end. The three sinuous slits in the spiracular plate are diagnostic features for this species (Figure 1).

Flesh flies look like house flies but are generally larger (11 to 13 mm long); they have three dark longitudinal stripes on their thorax, a checkerboard pattern of gray on the abdomen, and sometimes a reddish-brown tip on the abdomen (Figure 3). The larvae of flesh

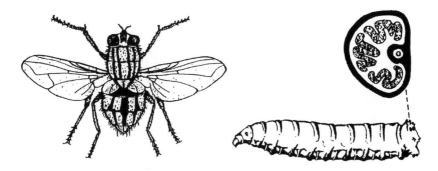

Figure 1
Adult (left) and larval (right) house fly. (From U.S. DHHS, CDC Publication No. 83-8297.)

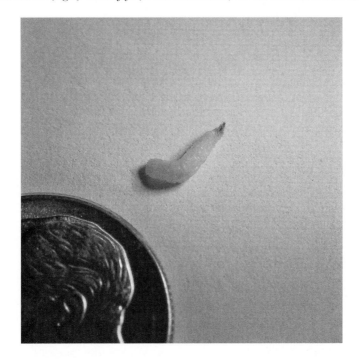

Figure 2
House fly larva (maggot), showing relative size.

Figure 3
Adult (left) and larval (right) flesh fly. (Redrawn from USDA, ARS, Agriculture Handbook No. 655, February 1991.)

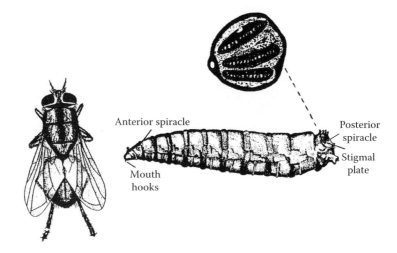

Figure 4
Adult (left) and larval (right) blow fly. (From U.S. Navy Laboratory Guide to Medical Entomology.)

flies are similar to those of house flies, except that they have straight spiracular slits and often an incomplete ring around the spiracular plate (Figure 3).

Blow flies (also known as green flies or bluebottle flies) are about the same size as flesh flies, although some of the bluebottle flies (genus *Calliphora*) are larger and more robust (Figure 4). Blow flies, with the exception of the cluster fly, are metallic bronze, green, black, purplish, or blue colored. They are the commonly encountered green flies seen on flowers, dead animals, and feces or, occasionally, indoors. Blow fly maggots resemble both house fly and flesh fly maggots but have straight spiracular slits and very often a complete sclerotized ring around the spiracular plate (Figure 4).

C. Geographic Distribution

House flies, *Musca domestica*, occur worldwide in association with human dwellings. There are numerous blow fly species occurring over most regions of the world. *Phaenicia sericata* is a very common blow fly that is cosmopolitan in distribution. *Calliphora vicina* is one of the most common bluebottle species in Europe and North America. Various species of flesh flies also occur worldwide. One of the most common is *Sarcophaga haemorrhoidalis*, which is virtually worldwide in distribution and common in the U.S.

D. Biology and Behavior

Like all Diptera, these flies exhibit a complete life cycle, having egg, larva, pupa, and adult stages (Figure 5). All three filth fly groups discussed in this section have similar biologies. Females lay their eggs, singly or in clusters, on or adjacent to an appropriate medium for larval development. However, flesh fly females deposit living larvae rather than eggs. One to a few days later the eggs hatch and the first-stage larvae emerge, entering the food source where they feed and develop. As the outer skin of a larva is nonliving and cannot grow, this skin (exoskeleton) must be periodically shed as the larva grows. Molting occurs three times in these filth fly families. Eventually, larval growth is completed and a pupa is formed inside the last larval skin. The larval

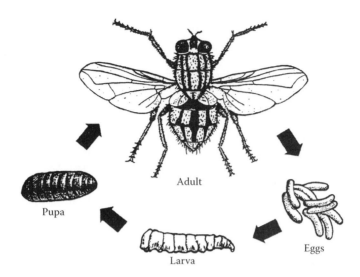

Adult

Pupa

Larva

Eggs

Figure 5
Life cycle of the house fly. (From U.S. DHHS, CDC Publication No. 83-8297.)

growth period ranges from 2 d to 3 weeks depending on temperature, and the pupal stage takes a similar length of time. The entire house fly life cycle can occur in 8 to 10 d under summer conditions. Emerging adults are active, moving from one attractant to another throughout most of the daylight hours. House flies are strongly attracted to feces, decaying organic material, and foodstuffs. Garbage and pet feces are almost always the important breeding sources of house fly problems in urban settings. Flesh flies breed in decaying meat or animal excreta. Most blow flies lay their eggs on animal carcasses, and the developing maggots quickly dispose of the carrion. However, some blow fly species breed in dog manure or decaying organic matter. The entire blow fly life cycle requires 9 to 25 d or more.

E. Treatment of Infestation

Filth flies are best controlled by a combination of good sanitation, mechanical exclusion, ultraviolet light traps, and chemical control (when necessary). Good sanitation includes emptying and steam-cleaning dumpsters on a regular schedule and otherwise eliminating breeding sites. Mechanical exclusion methods include air curtains and properly fitted doors and screens. Ultraviolet light traps (some electrocute the flies, others stun them and trap them on glue boards) work well indoors, and the newer models are quite safe. Insecticidal treatments are best performed by competent pest control operators.

II. EYE GNATS

A. General and Medical Importance

Members of the fly family Chloropidae are variously known as eye gnats, grass flies, eye flies, and fruit flies (although other flies are the true fruit flies). There are about 55 genera and about 270 described species in the Nearctic region.[4] Members of the

EYE GNATS

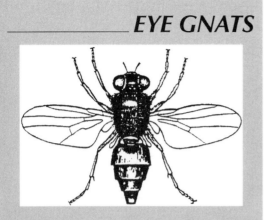

Figure 6
Typical eye gnat.
(From U.S. DHHS, CDC Publication No. 83-8297.)

Importance
Mechanical disease transmission

Distribution
Numerous species almost worldwide

Lesion
Nonbiting; sometimes cause minute scratch lesions on eyeball

Disease Transmission
Yaws and pinkeye

Key Reference
B. Greenberg's Two Vol. *Flies and Disease*, Princeton University Press, 1971

genus *Hippelates* are troublesome pests that are strongly attracted to the eyes of humans and other animals where they feed on the fluids of the eye surface and tear canal. Other species are attracted to serous discharges from sores, as well as excrement. They do not bite but they do have sponging mouthparts that are "spined"; these spines may cause eye irritation.[5] The lesions (minute scratches on the eyeball) that result from these spines are thought to be entry points for disease agents such as that of pinkeye and yaws.

B.   General Description

These flies are small, bare flies about 2 mm long that are often shiny black (Figure 6). Most *Hippelates* eye gnats have a large, black, curved spur on the hind tibia. Their antennae are short with bare aristae. The gnats are frequently seen on dogs around the eyes and anal–genital area.

C.   Geographic Distribution

This is a large and widespread group composed of hundreds of species; numerous species occur almost worldwide.

D.   Biology and Behavior

Eye gnats are common in agricultural areas, meadows, and other places where there is considerable grass. Larvae of most species feed in grass stems, decaying vegetation, and excrement. *Hippelates* larvae can be found feeding on decaying organic matter or roots of plants in the soil. Eggs hatch in about 3 d, the larval stage lasts about 2 weeks, and the pupal stage follows; adults emerge a few days later.

E.   Treatment of Infestation

Because eye gnats do not actually bite, treatment involves only the incidental diseases transmitted, e.g., pinkeye and yaws. Avoidance of the gnats is important. Most insect repellents work reasonably well in minimizing their nuisance presence around humans (although it should be emphasized that repellents should not be sprayed on or near the eyes).

III.   NONBITING MIDGES

A.   General and Medical Importance

Nonbiting midges (Chironomidae) are often confused with mosquitoes, but they are innocuous insects except during times when they are unusually abundant. They do not bite, but massive emergences may produce traffic hazards, bother livestock, and cover residences.[6,7] In addition, there may be an economic impact from damage to machinery, paint finishes, and airplanes.[8] Sometimes it is difficult to keep swarms out of the eyes or to avoid inhaling them. Persons with allergies may develop hypersensitivity to midge body parts or hemoglobin.[9]

B.   General Description

Midges are delicate insects, approximately the same size as mosquitoes, but they have a short proboscis that is not adapted for piercing (Figure 7). They have six-segmented antennae (possibly more, but no less), and in the males the antennae are plumose. Chironomids often have tiny hairs on the wing veins and on the membrane of the wing itself. Adult specimens are variously colored (usually brown or gray) and range in size from barely visible to mosquito-sized. Chironomids hold their wings rooflike over the abdomen at rest. They form huge swarms near water.

Chironomid larvae are wormlike, having a hardened shell-like head capsule. Mature larvae have 12 body segments and prolegs on their first and last body segments. Some chironomid larvae are red and are frequently called *bloodworms*.

C.   Geographic Distribution

Numerous species of chironomids occur worldwide. Three notorious pest species include

NONBITING MIDGES

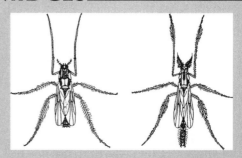

Adult nonbiting midges, Chironomidae

Importance
Massive emergences; airborne body parts induce hypersensitive RXNs in some people

Distribution
Numerous species almost worldwide

Disease Transmission
None

Key Reference
Ali, A., *J. Am. Mosq. Control Assoc.*, 7, 260, 1991

Treatment
Sensitive persons can avoid infested areas in spring and summer; stay indoors during emergences; wear dust masks when outdoors; see allergist for treatment

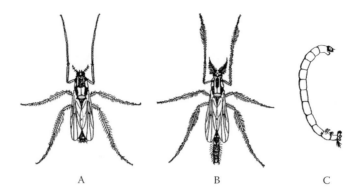

A   B   C

Figure 7
(A) Adult female, (B) male, and (C) larval nonbiting midge (Chironomidae).

Tanytarsus lewisi in northeastern Africa, *Glyptotendipes paripes* in the southeastern U.S., and *Chironomus decorus* in the northeastern U.S.

D.   Biology and Behavior

Midges are extremely abundant around standing water (especially freshwater ponds, lakes, and reservoirs), as the larvae are aquatic. The life stages of chironomids include the egg, four larval instars, pupa, and adult. Depending on the species and time of year, development from egg to adult is 1 month to 2 years. The larva is the only feeding stage; larval food items are varied depending on species, and include such items as small particulate organic matter, plants, or even other insects or tiny crustaceans. The larvae generally inhabit the bottom sediments in water sources. Most adult midges rest during daytime and fly at night. They are attracted to light in great numbers and sometimes huge swarms hover in the air, producing a humming sound.

E.   Treatment

Persons sensitive to airborne chironomid body parts or emanations can sometimes move away from the source of the chironomid midges, e.g., a large freshwater reservoir. Efficient air-conditioning can help keep the indoor environment free of allergen. Sensitive persons may also gain protection from the offending particles by wearing a dust or particle mask when outdoors during chironomid emergences. In addition, sensitive persons should consult an allergist for treatment.

REFERENCES

1. Frazier, C.A. and Brown, F.K., *Insects and Allergy*, University of Oklahoma Press, Norman, OK, 1980, p. 107.
2. Chavasse, D.C., Shler, R.P., Murphy, O.A., Huttly, S.R.A., Cousens, S.N., and Akhtar, T., Impact of fly control on childhood diarrhea in Pakistan: community-randomized trial, *Lancet*, 353, 22, 1999.
3. Watt, J. and Lindsay, D.R., Diarrheal disease control studies. I. Effect of fly control in a high morbidity area, *Public Health Rep.*, 63, 1319, 1948.

4. Sabrosky, C.W., Chloropidae, in *Manual of Nearctic Diptera*, 2, McAlpine, J.F., Ed., Monograph No. 28, Research Branch, Agriculture Canada, Ottawa, 1987, p. 1049.

5. Harwood, R.F. and James, M.T., *Entomology in Human and Animal Health*, Macmillan, New York, 1979, p. 251.

6. Bay, E.C., Anderson, L.D., and Sugarman, J., The batement of a chironomid nuisance on the highways at Lancaster, California, *Calif. Vector Views*, 12, 29, 1965.

7. Beck, E.C. and Beck, W.M. Jr., The Chironomidae of Florida, II. The nuisance species, *Fla. Entomol.*, 52, 1, 1969.

8. Ali, A., Perspectives on management of pestiferous Chironomidae, an emerging global problem, *J. Am. Mosq. Control Assoc.*, 7, 260, 1991.

9. Cranston, P.S., Allergens of non-biting midges: a systematic survey of chironomid haemoglobins, *Med. Vet. Entomol.*, 2, 117, 1988.

CHAPTER 21

FLIES (THAT MIGHT CAUSE MYIASIS)

TABLE OF CONTENTS

I. HUMAN BOT FLIES

A. General and Medical Importance

Some fly larvae develop in living flesh (also see Chapter 6); the human bot fly is one of them. This fly, *Dermatobia hominis* (Figure 1), is a parasite of humans, cattle, swine, cats, dogs, horses, sheep, other mammals, and a few birds in Mexico and Central and South America.[1,2] The larvae (Figure 2) burrow into the host's tissues, feeding and eventually emerging to drop to the ground and pupate. In people, the larvae have been recovered from the head, arms, back, abdomen, buttocks, thighs, and axilla. Human infestation is often characterized by painful discharging cutaneous swellings on the body. The condition is rarely fatal, except possibly in very young children (less than 5 years old); the larvae infesting the scalp penetrate into the incompletely ossified skull and enter the brain.[3] Although the parasite does not occur in the U.S., cases are occasionally seen in travelers to endemic areas. One such case was reported from Ohio in which a local physician submitted a second-stage larva to the Ohio Department of Health for identification.[4] The larva had been removed from a patient who had recently returned from Brazil. The Mississippi Department of Health has recently been involved in two cases of people returning from Belize.

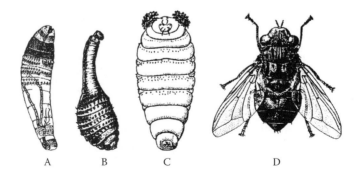

A   B   C   D

Figure 1
Human bot fly: (A) first-stage larva, (B) second-stage larva, (C) third-stage larva, and (D) adult. (From James, M.T., USDA Misc. Publ. No. 631, Washington, D.C., 1947, 175 pp.)

B.   General Description

D. hominis resembles a bluebottle fly (a large bluish member of the family Calliphoridae), is approximately 15 mm long, has a yellowish or brown head and legs, and has a plumose arista (a bristlelike branch off the antennae).

C.   Geographic Distribution

D. hominis does not occur in the U.S., but is common in parts of Mexico and Central and South America. Sancho[5] reports its distribution from the northern provinces of Mexico (Taumalipas, bordering southern Texas) to the northern Argentine provinces of Misiones, Tres Rios, Corrientes, and Formosa — roughly between latitudes 25° N and 32° S. Vacationers may acquire this parasite in tropical America, and return home before completion of maggot development.

D.   Biology and Behavior

Human bot flies have a unique and almost unbelievable life cycle. The adult flies catch various bloodsucking flies and attach eggs to their sides with a quick-drying glue. Common egg carriers are day-flying mosquitoes (*Psorophora* spp.) and muscid flies (*Sarcopromusca, Stomoxys,* and *Synthesiomyia* spp.).[6] These carrier flies later feed on humans (or other hosts) at which time the newly hatched bot fly larvae penetrate the host skin. The larvae feed inside a subdermal cavity of the host for 5 to 10 weeks. For respiration, they maintain a hole open to the external air. When mature, the larvae emerge, drop to the soil, and pupate. Interestingly, the fully fed larvae leave their hosts during the night or early morning, never during the afternoon, presumably to avoid desiccation.[5] After a month or so, the adult flies emerge to mate and begin the life cycle over.

E.   Treatment of Infestation

Alexander[7] said attempts to express *D. hominis* larvae by pressure on or around the lesion are fruitless and painful. The recommended therapeutic procedure for all forms of myiasis

HUMAN BOT FLY____

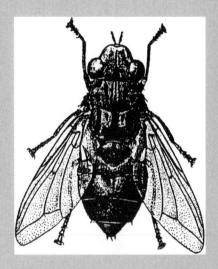

Human bot fly, *Dermatobia hominis*

Importance
Larvae cause myiasis in humans

Distribution
Mexico, Central and South America

Lesion
Superficial, painful swelling with central opening

Disease Transmission
None

Key Reference
Sancho, E., *Parasitol. Today,* 4, 242, 1988

Treatment
Excision of larvae; bacon therapy
(See *JAMA,* 270, 2087, 1993)

Figure 2
Human bot fly third-stage larva.

is direct removal of the maggots and treatment to prevent or control secondary infection. Secondary infection may result if the larva is ruptured or killed within its cavity and not removed. A traditional Central American remedy is occlusion of the punctum (breathing hole in the skin) with raw meat or pork, a treatment sometimes called *Bacon therapy*. Apparently, the larva will almost always migrate into the applied piece of meat. An American group modified this approach and was able to successfully extract 10 larvae after 3 h of Bacon therapy.[8]

II. SCREWWORM FLIES

A. General and Medical Importance

Two important species of screwworm flies that feed in living tissues are *Chrysomya bezziana*, the Old World screwworm, and *Cochliomyia hominivorax*, the New World screwworm. Larvae of these calliphorid flies are obligate parasites of living flesh (humans and domestic and wild mammals), feeding during their entire larval period inside a host. Screwworm infestation causes damage to tissues that is compounded by two factors: (1) screwworms may travel through living tissues and not just remain subdermal like the Tumbu fly or others, and (2) eggs are laid in batches of 100 to 400 resulting in multiple infestation within the host.

Female flies most often oviposit on or near a wound. However, human infestations have resulted from the flies ovipositing just inside the nostril while a person sleeps during the day, especially if there is a nasal discharge. Upon hatching, the larvae begin feeding, causing extensive destruction of tissue and a bloody discharge. Tissues around the lesion become swollen and pockets may be eaten out beneath the skin. Infested persons may die from tissue destruction. Human cases are uncommon, but do occur in areas where screwworm infestations occur in livestock. In 1935 there were 55 reported cases during a large outbreak among livestock in Texas.[9] Because the New World screwworm fly has been eradicated from the U.S., human cases are rare and due entirely to foreign travel. An imported case of traumatic screwworm myiasis was reported in a soldier wounded during military action in Panama.[10] Cases of myiasis caused from the Old World screwworm fly are common in Asia but relatively rare in Africa.[11]

B. General Description

Adult *Chrysomya bezziana* are green to blue in color and have the base of the stem vein (radius) ciliate above, but they have at most only two narrow longitudinal thoracic stripes (Figure 3A). They are approximately 8 to 12 mm long. Full-grown *Chrysomya bezziana* larvae have the usual 12 segments with belts of dark spines encircling them (Figure 3B). Adult *Cochliomyia hominivorax* are bluish green, medium-sized flies with a yellowish orange face and three dark stripes on the thorax (Figure 3C). The base of the stem vein (radius) bears a row of bristlelike hairs on the dorsal side. Full-grown larvae are about 2 cm long, appear pinkish in color, and have prominent rings of spines around the body (Figure 3D).

C. Geographic Distribution

The distribution of the New World screwworm once extended from the U.S. to southern Brazil, but it now has been limited to South and Central America due to an eradication program involving the sterile male release technique. The Old World screwworm occurs in the Afrotropical and Asian regions, extending south into Indonesia, the Philippines, and New Guinea (but not Australia).

D. Biology and Behavior

Females of both screwworm species are attracted to wounds in mammals and lay eggs at the edge of the wounds. Eggs are deposited in batches of 150 to 500 and hatch approximately 15 h later. The larvae feed while embedded

SCREWWORM FLIES

Adult Old World screwworm fly,
Chrysomya bezziana

Importance
Human myiasis — sometimes fatal

Distribution
Central and South America; Oriental and Afrotropical regions; part of South Pacific region

Lesion
Pocketlike sinuses in tissues

Disease Transmission
Generally none

Key Reference
James, M.T., USDA Miscellaneous Publication No. 631, Washington, D.C., 1947, 175 pp.

Treatment
Removal of larvae by irrigation or surgical means

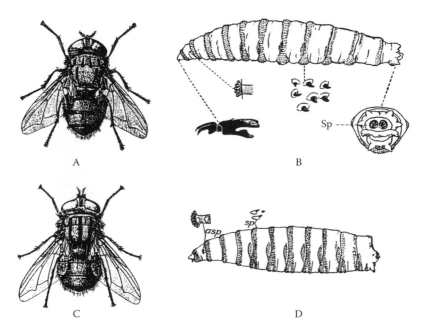

Figure 3
Screwworm flies: (A) adult female Chrysomya bezziana, *(B) larval* C. bezziana, *(C) adult female* Cochliomyia hominivorax, *and (D) larval* C. hominivorax. *(From James, M.T., USDA Misc. Publ. No. 631, Washington, D.C., 1947, 175 pp.)*

inside living tissue; sometimes, however, the peritremes (the plate surrounding the breathing tubes) are visible. The larvae emerge from the host as prepupae 4 to 7 d later and fall to the ground, where they pupate for a week or more. The entire life cycle from egg to egg takes about 24 d under optimum conditions. Adult screwworm flies are active all year round but only fly during daylight.

E. Treatment

Treatment of screwworm fly myiasis involves removal of the larvae. This may be done in several ways, but often there is surgical exploration and removal of the larvae under local anesthesia.[12] Alexander[7] advocated irrigation with chloroform or ether under local or general anesthesia. Infestation of the nose, eyes, ears, and other areas may require surgery if larvae cannot be removed via natural orifices. Because screwworm flies lay eggs in batches, there could be tens or even hundreds of maggots in a wound.

III. CONGO FLOOR MAGGOT

A. General and Medical Importance

The Congo floor maggot, *Auchmeromyia senegalensis (= luteola),* is a blow fly. Its larvae suck human blood while their hosts are sleeping on the floor in huts or other dilapidated, infested dwellings. Although this blood-feeding behavior may not actually be myiasis in the strictest sense, this species is discussed in the following subsections. Congo floor maggots are not known to transmit disease agents. The bite of the larva is generally felt as a pinprick, but sensitive persons may experience pain, swelling, and irritation.

B.   General Description

Adult *A. senegalensis* are 8 to 13 mm long, yellow-brown flies. They are similar in appearance to the Tumbu fly (next section) but have the second visible abdominal segment longer than the third (Figure 4 and Figure 5). Males of the Congo floor maggot fly have widely separated eyes. Full-grown larvae are up to 18 mm long.

C.   Geographic Distribution

The Congo floor maggot is an African calliphorid occurring throughout Africa south of the Sahara Desert and in the Cape Verde islands.

D.   Biology and Behavior

Congo floor maggot larvae occur in the earthen floor of huts or other similar dry-soil situations. They feed periodically at night by sucking blood from individuals sleeping on the ground. The larvae take about 20 min to feed on their host, and then drop to the ground. They may feed daily if given the opportunity. The larvae go through three stages, which may include 6 to 20 blood feedings. They eventually pupate in the soil for about 2 weeks, after which the adults emerge. The entire life cycle is approximately 10 weeks under optimal conditions, and development continues round the year with no diapause. The adults feed on feces, fallen fruit, and other fermenting materials. The adults mate and then the female lays her eggs in dry, dusty, or sandy soil, often in a hut. This fly has also been reported to be associated with warthogs, aardvarks, and hyenas.

E.   Treatment of Bites

Blood feedings by the Congo floor maggot are generally not serious but may cause considerable discomfort. Because the maggots cannot climb vertically, avoidance is possible by sleeping in beds at least 10 cm off the ground.

CONGO FLOOR MAGGOT

Figure 4
Auchmeromyia senegalensis.

Importance
Nocturnal blood-feeding maggots; nuisance

Distribution
Sub-Sahara Africa; Cape Verde islands

Lesion
Generally, trivial pinprick lesions

Disease Transmission
None

Key Reference
James, M.T., USDA Miscellaneous Publication No. 631, Washington, D.C., 1947, 175 pp.

Treatment
Generally none needed

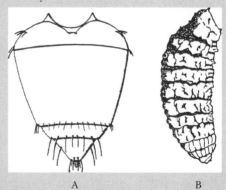

A   B

Figure 5
Auchmeromyia senegalensis: *(A) abdomen of male showing exceptionally long second segment and (B) larval stage. (From James, M.T., USDA Miscellaneous Publications No. 631, Washington, D.C., 1947, 175 pp.)*

TUMBU FLY

Figure 6
Cordylobia anthropophaga *adult female.*

Importance
Furuncular type of myiasis

Distribution
Tropical Africa

Lesion
Red or shiny nodules on skin; lesion very similar to those caused by *Dermatobia hominis*

Disease Transmission
None

Key Reference
James, M.T., USDA Miscellaneous Publication No. 631, Washington, D.C., 1947, 175 pp.

Treatment
Direct removal of larvae

IV.   TUMBU OR MANGO FLY

A.   General and Medical Importance

The Tumbu fly, *Cordylobia anthropophaga*, is a blow fly (family Calliphoridae) whose larvae can burrow into human subcutaneous tissues producing a furuncular type of myiasis (see also Chapter 6). The lesion resembles a boil and often has a serous exudate. The first symptom is a slight itching, which occurs in the first day or two of the infestation. In a few days a reddish papule develops, and then the typical boil-like lesion develops. With multiple infections the lymph glands may become enlarged and fever may be present.[13] Children are most commonly affected, and lesions usually occur on areas of the body normally covered with clothing, as the adult flies tend to oviposit on soiled clothing. In an imported case seen in Maine, lesions appearing as boils approximately 25 mm in diameter contained second-stage Tumbu fly larvae.[13]

B.   General Description

The adult Tumbu fly is similar in appearance to the adult Congo floor maggot as well as other flies in related genera; precise identification is therefore difficult. The body of both sexes is yellow-brown and about 6 to 12 mm long (Figure 6). The second and third visible abdominal segments are of equal length. Males of this species have nearly contiguous eyes. (Congo floor maggot males have widely separated eyes.) Full-grown larvae are 13 to 15 mm long.

C.   Geographic Distribution

Tumbu flies occur in the Afrotropical region.

D.   Biology and Behavior

Tumbu flies cause cutaneous myiasis in several mammalian hosts, including humans. Dogs are the most common domestic hosts, and several species of wild rats are the preferred field hosts. The adult flies feed on plant

CASE HISTORY

MAGGOT IN URINARY TRACT

A large regional hospital sent in a fly maggot supposedly recovered from a urine sample (Figure 1). Laboratory personnel apparently did not think such an event was possible. "Was it really a maggot?" they asked. "What should we do?" "Are there more in the urinary tract?" Upon examination, the maggot was identified as a syrphid fly larva. Although not usually involved in urinary myiasis, the species was from a group of flies known to occasionally cause other forms of myiasis in people. Accordingly, the case was considered *urinary tract myiasis* and the physicians acted accordingly.

Comment: The maggots are generally expelled in cases of accidental urinary myiasis by noninvasive species. Therefore, specific treatment may not be needed.

Source: Adapted from *Lab. Med.,* 25, 369, 1994. Copyright 1994, The American Society of Clinical Pathologists. With permission.

Figure 1
Syrphid fly larva recovered from urine sample; suspected case of accidental urinary myiasis.

juices, rotting plant or animal matter, and feces. They are most active in the early morning (7:00 to 9:00) and late afternoon (4:00 to 6:00). Adult flies lay their eggs in shady areas of sand contaminated with excrement or sometimes on soiled clothing. Improperly washed diapers placed in the shade to dry are a common oviposition site. The flies do not oviposit directly on the skin or hair of a host. Eggs hatch and the larvae remain buried in the dirt or sand until a host approaches. In the presence of a host the larvae actively search for and get on the host. Penetration of the skin is rapid and ordinarily unnoticeable. The larva develops in the subcutaneous tissues for about 10 d. As the larva develops the lesion begins to look boil-like and a clear fluid, occasionally stained with blood or larval feces, may ooze from the boil. Larvae (now prepupae) then exit the lesion and pupate in the ground. About 10 to 14 d later the adults emerge, mate, and begin laying new batches of eggs. The adult female only lives 2 to 3 weeks. People are most commonly parasitized during the rainy season, but fly development continues year round.

E.   Treatment of Infestation

As in other types of myiasis, the only known therapeutic procedure, other than applying local palliatives, is direct removal of the maggots (surgically, if necessary) and treatment to prevent or control secondary infection. Service[14] says a standard method for Tumbu larvae removal is to cover the small hole in the swelling with medicinal liquid paraffin. This prevents the larva from breathing through its posterior spiracles and forces it to wriggle a little further out of the swelling to protrude the spiracles. In doing so, it lubricates the pocket (in the skin) and the larva can then be extracted by gently pressing around the swelling.

V.   SARCOPHAGID FLIES (WOHLFAHRTIA SPP.)

A.   General and Medical Importance

Two important sarcophagid flies that cause myiasis are *Wohlfahrtia magnifica* and *W. vigil*. *W. magnifica* larvae produce traumatic myiasis throughout their host's tissues, much like a screwworm fly. Human fatalities due to this species have been reported.[2] *W. vigil* larvae are not as invasive; they are usually limited to dermal tissues, producing a furuncular or boil-like lesion.[2]

B.   General Description

The *Wohlfahrtia* spp. look very similar to flesh fly adults (*Sarcophaga*), but instead of the checkerboard pattern on their abdomen they have clearly defined spots (Figure 7). The abdomen of *W. vigil* is almost entirely black (Figure 8B).

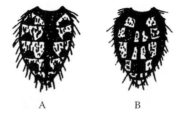

A   B

Figure 7
Comparison of (A) Wohlfahrtia *and*
(B) Sarcophaga *abdomens.*

C. Geographic Distribution

W. magnifica occurs in the southern Palearctic region. *W. vigil* occurs in northern North America, primarily Canada and the northern U.S.

D. Biology and Behavior

W. magnifica is an obligate parasite in the wounds and natural orifices of warm-blooded animals, including humans. It never develops in carrion or rotting materials. Females of both *W. magnifica* and *W. vigil* are larviparous, i.e., the adult fly deposits living larvae on potential hosts. Both tiny skin lesions (tick bite, scratch, etc.) and mucous membranes of natural orifices are used as entry points by the larvae. The developing larvae feed for 5 to 7 d within a host, after which they emerge, fall to the ground, and pupate. *W. vigil* is also an obligate parasite, but it rarely penetrates deeper than dermal tissues. Dogs, cats, rodents, rabbits, mink, foxes, and humans are hosts of this pest. The larvae form a boil-like swelling under the skin with a circular opening through which the larva can be partially seen. After maturing, the larvae emerge, drop to the ground, and pupate. Multiple lesions are common on patients infested with *W. vigil*, with the average number being 12 to 14.[2]

E. Treatment

W. vigil infestations are relatively easy to treat, because the larvae are easily seen and removed. Unlike other forms of furuncular myiasis, expression seems easy and uncomplicated.[7] *W. magnifica* infestations are much more serious, because they often involve numerous, voraciously feeding larvae within the nose or in existing wounds. Irrigation or possible surgical exploration to remove the larvae will be needed.

WOHLFAHRTIA FLIES

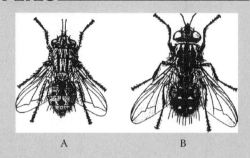

A   B

Figure 8

Adult female (A) Wohlfahrtia magnifica *and (B)* W. vigil. *(From James, M.T., USDA Miscellaneous Publications No. 631, Washington, D.C., 1947, 175 pp.)*

Importance

W. magnifica — myiasis similar to screwworm fly; *W. vigil* — furuncular myiasis

Distribution

W. magnifica — southern Palearctic; *W. vigil* — northern U.S. and Canada

Lesion

W. magnifica — deep lesions in the nose or in existing wounds; *W. vigil* — papule- or pustule-looking lesions containing larvae

Disease Transmission

None

Key Reference

James, M.T., USDA Miscellaneous Publication No. 631, Washington, D.C., 1947, 175 pp.

Treatment

Removal of the larvae by expression, direct picking, or irrigation

CASE HISTORY

ANIMAL BOT FLY LARVA IN CHILD

A 3-year-old boy complained of being "stung" on his side and neck (approximately 3 cm under the lower jaw bone) while watching television early one morning. His mother said that within 5 min, typical stinglike "welts" occurred at the places the child pointed out. Within a day or so, a line of vesicles extended away from the lesions — presumably where the larvae were migrating within the skin. The lesion on his side extended upward in a sinuous fashion about 10 cm, ending in a small papule. No further development occurred at the side lesion (apparently the developing larva died). The larva on his neck migrated about 4 cm laterally and began to enlarge. After about 14 d the dermal tumor was inflamed and contained an opening in the center about 3 mm in diameter, through which apparently the larva obtained air. The child cried often and complained of severe pain. Despite numerous trips to physicians, the myiasis was not diagnosed until almost 4 weeks after the initial stinging incident. An emergency room (ER) physician expressed the larva, which was ultimately forwarded to the Health Department for identification. Upon examination, the specimen (Figure 1) appeared to be a second-stage larva of a fly in the genus *Cuterebra* (the rabbit and rodent bot flies). Figure 2 is a drawing of the posterior spiracular slits of the specimen.

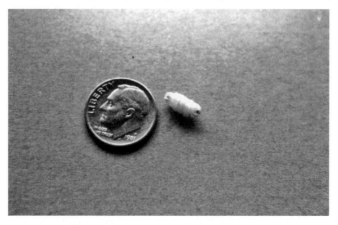

Figure 1
Cuterebra fly larva removed from neck of 3-year-old child.

VI.   OTHER FLIES THAT MAY OCCASIONALLY CAUSE MYIASIS

A.   General and Medical Importance

Larvae of flies in the genus *Cuterebra*, often found in squirrels (called *wolves*) and rabbits, may rarely parasitize humans, forming a warblelike dermal tumor.[15,16] In a case the author consulted on (see Case History above), a 3-year-old boy had two *Cuterebra* larvae — on his side and neck — forming boil-like lesions. Several health care providers examined the boy and either diagnosed the lesions as boils or larval migrans (from dog hookworm), because there were short migration trails visible in the skin. To everyone's surprise, one physician finally recognized the myiasis and expressed a larva from the neck lesion.

Other flies may occasionally cause myiasis in humans (Figure 9 to Figure 11). See Chapter 6. This behavior is termed *facultative myiasis* (see Color Figure 21.18). In some

Comment: These fly larvae commonly infest squirrels, chipmunks, other wild rodents, and rabbits. Occasionally, they also infest dogs and cats. Human infestations are extremely rare. Although most reports in the scientific literature say that the eggs or larvae are not deposited directly on the hosts (being laid instead in pathways, dens, etc., of the hosts), there is evidence of direct deposition in this case. Physicians should carefully examine boils or other skin lesions for fly larvae, especially if the lesion is refractory to standard treatments.

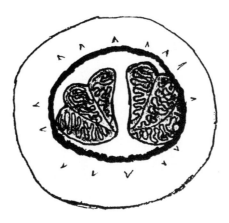

Figure 2
Drawing of posterior spiracles from Cuterebra *larva infesting child.*

Figure 9
Typical blow fly. They are usually shiny blue or green in color.

of these cases the larvae enter living tissues after feeding in neglected, malodorous wounds. In other cases, ports of entry include natural orifices such as the ears, urinary opening, or anus. Table 1 provides a list of some species occasionally involved in

Table 1
Some Fly Species Not Ordinarily Causing Myiasis but May Opportunistically Be Found in Living Tissues or Hosts

Type of Myiasis	Fly Species	Fly Distribution	Comments	Reference
Urinary	*Fannia canicularis*	Cosmopolitan	Breeds in animal feces or decaying vegetable matter	James[2]
	F. scalaris	Cosmopolitan	Found outdoors more than *F. canicularis*; also known as the latrine fly	James[2]
	Musca domestica	Cosmopolitan	Breeds in animal feces, garbage, etc.	
	Muscina stabulans	Cosmopolitan	Breeds in animal feces and decaying vegetable matter	
	Teichomyza fusca	Nearly cosmopolitan	Breeds in excrement, sewers, and drains	
	Megaselia scalaris	Nearly cosmopolitan		Singh and Rana[18]
	Eristalis tenax	Cosmopolitan	Larvae called rat-tailed maggots	Mumcuoglu et al.[19]
Cutaneous or traumatic	*Phaenicia sericata*	Cosmopolitan	Very common saprophagous flies attracted to fresh carrion	Anderson and Magnarelli,[12] Greenberg,[20] and Merritt[21]
	P. cuprina[a]	Nearctic	Breeds in garbage and dog feces	
	Phormia regina	Holarctic	One of the most numerous flies in the southern U.S.	Hall[22] and Hall and Townsend[23]
	Cochliomyia macellaria	Nearctic and neotropical regions	Very abundant carrion breeder	Alexander[7]

Site	Species	Distribution	Notes	Reference
	Calliphora vicina	Holarctic	Active mostly in the cool-weather months of the year	Alexander[7]
	Sarcophaga haemorrhoidalis	Nearly cosmopolitan	Breeds in carrion and excrement	James[2]
	Chrysomya rufifacies	Oriental and neotropical	Known as the hairy maggot blow fly	
	Hermetia illucens	Nearly cosmopolitan	Adult known as black soldier fly	Adler and Brancato[24]
	Cuterebra spp.	New World	Rodent bots	Rice and Douglas[16]
Rectal	*Eristalis tenax*	Cosmopolitan; most common in holartic	Larvae called rat-tailed maggots	
	F. scalaris	Cosmopolitan		
	F. canicularis	Cosmopolitan		
	Musca domestica	Cosmopolitan		Goddard[25]
	Muscina stabulans	Cosmopolitan		
	Phaenicia sericata	Cosmopolitan		Davies[26] and Keller and Keller[27]
Aural	*P. cuprina*[a]	Nearctic		Goddard[28]
	Sarcophaga citellivora			
	Phormia regina	Holarctic		Damsky et al.[29]

[a] *Only meaning the Nearctic form here (= P. pallescens of Shannon).*

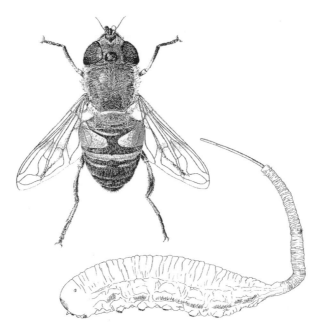

Figure 10
Drone fly, Eristalis tenax, *adult and larva; often reported as a cause of accidental or facultative myiasis. (From USDA, ARS, Agriculture Handbook No. 655, February 1991.)*

Figure 10a
Rat-tailed maggot. (From Dr. Blake Layton, Mississippi State University, Cooperative Extension Service. With permission.)

facultative myiasis and notes on their biologies. Sometimes these fly larvae can be identified by looking at the shape and pattern of their posterior spiracles (Figure 12 and Figure 13).

One of the most commonly implicated fly groups is the Calliphoridae (blow flies). Several species of blow flies, and especially *Phaenicia sericata* and *Phormia regina,*

Fannia canicularis

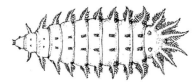

Fannia scalaris

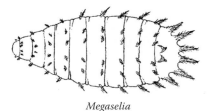

Megaselia

Figure 11
Fly larvae frequently involved in urinary (facultative) myiasis. (From USDA, ARS, Agriculture Handbook No. 655, February 1991.

have been reported to cause facultative myiasis in humans. The discussion below will be limited to those two species.

B.   General Description

Phaenicia sericata is a typical-looking blow fly — shiny green or coppery green (Figure 14A). *Phormia regina*, also called *the black blow fly*, is more slender and is olive-colored or nearly black (Figure 14C). The pubescence surrounding the anterior spiracle of *Phormia regina* is orange.

C.   Geographic Distribution

Phaenicia sericata is nearly cosmopolitan in distribution and is probably the most abundant species of *Phaenicia* in North America. *Phormia regina* is Holarctic in distribution. Goddard and Lago[17] found *Phormia regina* to be the most common blow fly in Mississippi, with specimens encountered throughout the year (peak numbers occurring in April and September).

D.   Biology and Behavior

The general life history of blow flies is included in Chapter 20 on nonbiting filth flies. Other specific comments on these two species are presented here. *Phaenicia sericata* is a species that is quick to appear at fresh carrion, although it is also attracted to a wide range of decaying substances. In many cases, facultative myiasis cases seem to be the result of *Phaenicia sericata* being attracted to festering and malodorous wounds with subsequent invasion of healthy human tissue. *Phormia regina* is very common throughout the U.S. during the warmer months (April to September). It is also abundant near a wide variety of decaying substances, especially carrion.

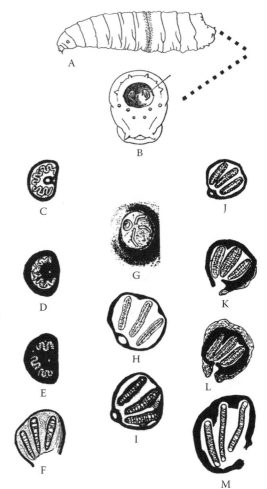

Figure 12

Posterior spiracles of various fly larvae occasionally involved in myiasis: (A) mature muscoid fly larva, (B) rear view of fly larva, (C) Musca domestica, (D) Musca sorbens, *(E)* Musca crassirostris, *(F)* Wohlfahrtia vigil, *(G)* Muscina stabulans, *(H)* Cynomyopsis cadaverina, *(I)* Phaenicia caeruleiviridis, *(J)* Phaenicia sericata, *(K)* Chrysomya chloropyga, *(L)* Chrysomya albiceps, *(M)* Sarcophaga crassipalpis. *Note: D, E, K, and L occur in the Old World only. (Adapted in part from James, M.T., USDA Miscellaneous Publication No. 631, Washington, D.C., 1947, 175 pp.)*

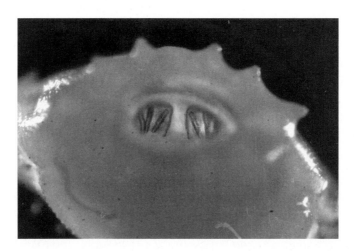

Figure 13

Posterior end of fly larva removed from the ear of a 4-month-old child. Note size and shape of spiracular slits. (Photo courtesy of Dr. Alan Causey, University of Mississippi Medical Center, Jackson.)

CASE HISTORY

TRAUMATIC MYIASIS

On February 17, 2000, a physician from a local hospital submitted a vial of insect larvae for identification. The tiny, wormlike specimens had been removed from a middle-aged woman found lying in the woods with a large open wound in her scalp. Upon microscopic examination, the larvae were found to be second-stage blow fly (insect family Calliphoridae) larvae.

Comment: This was a classic case of traumatic myiasis. It is not unusual for blow flies to lay eggs in wounds of people or animals, especially when the wound is neglected or malodorous. In nature, these flies oviposit on dead and rotting carcasses, so one can see how they might occasionally choose a festering wound on a living animal. However, this is still "accidental" or "opportunistic" myiasis and not purposeful, obligate myiasis.

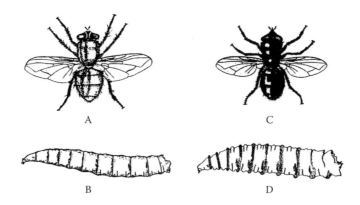

Figure 14
(A) Adult and (B) larval Phaenicia sericata; *(C) adult and (D) larval* Phormia regina. *(From U.S. DHHS, CDC Publication No. 83-8297 and James, M.T., USDA Miscellaneous Publication No. 631, Washington, D.C., 1947, 175 pp.)*

E. Treatment of Infestation

Therapeutic measures for blow fly myiasis include removal of the maggots (surgically, if necessary) and treatment for secondary bacterial and fungal infections should they occur.

REFERENCES

1. Guimaraes, J.H. and Papavero, N., A tentative annotated bibliography of *Dermatobia hominis*, *Arq. Zool.*, 14, 223, 1966.
2. James, M.T., The Flies That Cause Myiasis in Man, USDA Misc. Publ. No. 631, Washington, D.C., 1947, 175 pp.

3. Rossi, M.A. and Zucoloto, S., Fatal cerebral myiasis by the tropical warble fly, *Am. J. Trop. Med. Hyg.*, 22, 267, 1973.

4. Anonymous, Surprise in a returning traveler, *Ohio Vector News*, 10, 2, 1991.

5. Sancho, E., *Dermatobia* the neotropical warble fly, *Parasitol. Today*, 4, 242, 1988.

6. Catts, E.P. and Mullen, G.R., Myiasis, in *Medical and Veterinary Entomology*, Mullen, G.R. and Durden, L., Eds., Academic Press, New York, 2002, p. 317.

7. Alexander, J.O., *Arthropods and Human Skin,* Springer-Verlag, Berlin, 1984.

8. Brewer, T.F., Wilson, M.E., Gonzalez, E., and Felsenstein, D., Bacon therapy and furuncular myiasis, *JAMA*, 270, 2087, 1993.

9. Harwood, R.F. and James, M.T., *Entomology in Human and Animal Health*, Macmillan, New York, 1979.

10. Mehr, Z., Powers, N.R., and Konkol, K.A., Myiasis in a wounded soldier returning from Panama, *J. Med. Entomol.*, 28, 553, 1991.

11. Zumpt, F., *Myiasis in Man and Animals in the Old World*, Butterworths, London, 1965.

12. Anderson, J.F. and Magnarelli, L.A., Hospital acquired myiasis, *Asepsis*, 6, 15, 1984.

13. Rice, P.L. and Gleason, N., Two cases of myiasis in the U.S. by the African Tumbu fly, *Cordylobia anthropophaga, Am. J. Trop. Med. Hyg.*, 21, 62, 1972.

14. Service, M.W., *Medical Entomology for Students*, Chapman and Hall, New York, 1996.

15. Goddard, J., Human infestation with rodent bot fly larvae: a new route of entry? *South. Med. J.*, 90, 254, 1997.

16. Rice, P.L. and Douglas, G.W., Myiasis in a man by *Cuterebra, Ann. Entomol. Soc. Am.*, 65, 514, 1972.

17. Goddard, J. and Lago, P.K., An annotated list of the Calliphoridae of Mississippi, *J. GA Entomol. Soc.*, 18, 481, 1983.

18. Singh, T.S. and Rana, D., Urogenital myiasis caused by *Megaselia scalaris*: a case report, *J. Med. Entomol.*, 26, 228, 1989.

19. Mumcuoglu, I., Akarsu, G.A., Balaban, N., and Keles, I., Eristalis tenax as a cause of urinary myiasis, *Scand. J. Infect. Dis.*, 37, 942, 2005.

20. Greenberg, B., Two cases of human myiasis caused by *Phaenicia sericata* in Chicago area hospitals, *J. Med. Entomol.*, 21, 615, 1984.

21. Merritt, R.W., A severe case of human cutaneous myiasis caused by *Phaenicia sericata*, *Calif. Vector Views*, 16, 24, 1969.

22. Hall, D.G., *The Blow flies of North America*, Thomas Say Foundation, Washington, D.C., 1948.

23. Hall, R.D. and Townsend, L.H.J., The Blow flies of Virginia, Virginia Polytechnic Institute and State University, Research Bull. No. 123, 1977, 42 pp.

24. Adler, A.I. and Brancato, F.P., Human furuncular myiasis caused by *Hermetia illucens, J. Med. Entomol.*, 32, 745, 1995.

25. Goddard, J., Unpublished data, in Mississippi Department of Health notes and records, 1, 1989.

26. Davies, D.M., Human aural myiasis: a case in Ontario, Canada and a partial review, *J. Parasitol.*, 62, 124, 1976.

27. Keller, A.P.J. and Keller, A.P.I., Myiasis of the middle ear, *Laryngoscope*, 80, 646, 1970.

28. Goddard, J., Unpublished data, in Mississippi Department of Health notes and records, 1987.

29. Damsky, L.J., Baur, H., and Reeber, E., Human myiasis by the black blow fly, *MN Med.*, 59, 303, 1976.

CHAPTER 22

LICE

TABLE OF CONTENTS

BODY LICE

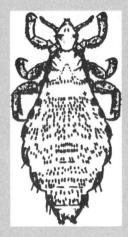

Figure 1
Adult body louse.
(From U.S. DHEW, PHS, CDC Pictorial Keys.)

Importance
Irritation; disease transmission

Distribution
Essentially worldwide

Lesion
Red papules, 3 to 4 mm in diameter; purpuric halo may be present

Disease Transmission
Epidemic typhus; trench fever; epidemic relapsing fever

Key References
Kim, K.C., Pratt, H.D., and Stojanovich, C.J., *The Sucking Lice of North America*, Pennsylvania State University Press, University Park, PA, 1986
Witkowski, J.A. and Parish, L.C., What's new in the management of lice, *Infect. Med.*, 14, 287, 1997

Treatment
Treatment of all clothing and bedding of infested individuals; pediculicidal shampoos and lotions

I. BODY LICE

A. General and Medical Importance

The body louse, *Pediculus humanus corporis*, is a blood-feeding ectoparasite of humans. Body and head lice look almost identical, but head lice remain more or less on the scalp and body lice on the body or in clothing. Body lice are relatively rare among affluent members of industrial nations, yet they can become a severe problem under crowded and unsanitary conditions such as war or natural disasters. Body lice may transmit the agent of epidemic typhus, *Rickettsia prowazeki*, and there have been devastating epidemics of the disease in the past. Typhus is still endemic in poorly developed countries where people live in filthy, crowded conditions. Besides louse-borne typhus, body lice transmit the agents of trench fever and epidemic relapsing fever. Trench fever, caused by *Bartonella quintana*, is still widespread in parts of Europe, Asia, Africa, Mexico, and Central and South America, mainly in an asymptomatic form. However, it is now recognized as a reemerging pathogen among homeless populations in the U.S. and Europe, where it is responsible for a wide spectrum of conditions such as chronic bacteremia, endocarditis, and bacillary angiomatosis.[1] Epidemic relapsing fever occurs primarily in Africa; there were 4972 cases with 29 deaths reported worldwide in 1971.[2] Aside from the possibility of disease transmission, body lice may cause severe skin irritation. The usual clinical presentation is pyoderma in covered areas. Characteristically, some swelling and red papules develop at each bite site. There are intermittent episodes of mild to severe itch associated with the bites. Compounding this, some individuals become sensitized to antigens injected during louse biting, leading to generalized allergic reactions. Subsequent excoriation of the skin by the infested individual may lead to impetigo or eczema. Alexander[3] noted that long-standing infestations may lead to a brownish-bronze pigmentation of the skin, especially in the groin, axilla, and upper thigh regions.

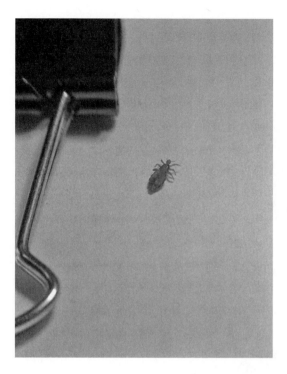

Figure 2
Body louse next to metal clamp.

B. General Description

An authoritative identification manual of all sucking lice in North America is available.[4] Human body lice are tiny (2 to 4 mm long), elongate, soft-bodied, light-colored, wingless insects (Figure 1 and Figure 2). They are dorsoventrally flattened, with an angular ovoid head and a nine-segmented abdomen. The eggs are small (about 1 mm), oval, white or cream-colored objects with a distinct cap on one end. The head bears a pair of simple lateral eyes and a pair of short five-segmented antennae. Body lice are usually about 15 to 20% larger than head lice. They have modified claws enabling them to grasp tightly to hair shafts or clothing while they feed via piercing–sucking mouthparts. Body lice eggs are primarily laid in clothing.

C. Geographic Distribution

Body lice occur essentially worldwide.

D. Biology and Behavior

The body louse lives primarily on the clothing of infested individuals and moves to adjacent body areas to feed. The eggs are attached to fibers of clothing with a strong glue-like substance; they seem to be especially located along the seams inside of underwear or other places where clothing contacts the body. Eggs hatch in 5 to 11 d, depending on the temperature, and the young (called *nymphs*) begin feeding. Developing nymphs

HEAD LICE

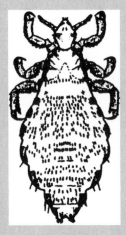

Figure 3
Adult head louse.
(From U.S. DHEW, PHS, CDC Pictorial Keys.)

Importance
Irritation; itching; secondary infection

Distribution
Essentially worldwide

Lesion
Itchy scalp, often secondarily infected as lesions of impetigo contagiosa

Disease Transmission
None

Key References
Kim, K.C., Pratt, H.D., and Stojanovich, C.J., *The Sucking Lice of North America*, Pennsylvania State University Press, University Park, PA, 1986

Witkowski, J.A. and Parish, L.C., What's new in the management of lice, *Infect. Med.*, 14, 287, 1997

Treatment
Pediculicidal shampoos

feed and molt several times before reaching the adult stage. The egg-to-adult cycle is about 3 to 5 weeks long. Body lice can survive in clothing even if it is habitually removed at night, but usually will die if the clothing is not worn for several days. They seem to especially prefer woolen clothing.

E. Treatment of Infestation

Because body lice infest both the patient and his or her clothing, control strategies involve frequently changing clothing, washing infested garments in very hot water or having them dry-cleaned, and using pediculicidal lotions or shampoos.[5] The primary element of control is to ensure that all clothing and bedding of infested persons is sanitized or treated. Clothing and bedding can also be disinfected by spraying with pyrethrin preparations or other approved insecticides or dusts.[1] When mass treatments are indicated (such as in a war or natural disaster), insecticide dusts are applied directly to the body.

II. HEAD LICE

A. General and Medical Importance

The appearance and behavior of the head louse, *P. humanus capitis*, are somewhat similar to the body louse, but it is confined to the scalp. Head lice have been pests of humans throughout the ages, as evidenced by the discovery of ancient hair combs dating back to the first century B.C. containing the remains of lice and their eggs. Head lice generally pose no significant health threat to infested persons, although heavily infested persons often have occipital and cervical adenopathy and occasionally a generalized morbilliform eruption. The primary negative effects of lice infestation are embarrassment and social sanctioning.

B. General Description

Head lice are tiny (1 to 3 mm long), elongate, soft-bodied, light-colored, wingless insects

Figure 4
Microscopic view of head louse. Note strong claws on two front legs especially suited for holding hair shafts.

(Figure 3). They are dorsoventrally flattened, with an angular ovoid head and a nine-segmented abdomen (Figure 4). The head bears a pair of simple lateral eyes and a pair of short five-segmented antennae. Head lice possess specially modified claws that enable them to grasp tightly to hair shafts while they feed through specially modified piercing–sucking mouthparts. Head lice look almost identical to body lice. However, they vary distinctly in behavior; head lice occur chiefly on the head, whereas body lice occur on the body and clothing. Head lice eggs (nits) are about 1-mm-long oval objects with a distinct cap on one end. They are white to cream-colored when viable and firmly attached to the hair. Under examination with the naked eye, they can be confused with dandruff, globules of hair oil or hairspray, and other substances, but they are easily identified under moderate magnification.

C. Geographic Distribution

Head lice occur on humans essentially worldwide.

D. Biology and Behavior

Head lice live on the skin among the hairs on the patient's head. Nits are laid on the shaft of the hair, near the base, and attached with a strong gluelike substance, most commonly behind the ears and at the nape of the neck (Figure 5). As long as adult lice remain on the scalp, they can live for about a month. Because they require the warmth and blood meals afforded by the scalp, it is generally reported that lice can survive only about 24 h if removed. But in one study involving hundreds of head lice removed

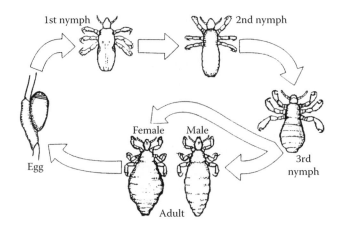

Figure 5
Head lice life cycle. (From U.S. DHEW, PHS, CDC Pictorial Keys.)

from children, none survived longer than 15 h; most died between 6 to 15 h.[6] Lice eggs hatch after 5 to 10 d, and the young (called *nymphs*) begin feeding immediately. Each feeding lasts several minutes. As the lice develop, they feed frequently, both day and night, particularly when the patient is quiescent. In as little as 18 to 20 d they mature through three molts, mate, and the females are ready to lay eggs to start the next generation. The egg-to-adult cycle averages 3 weeks.

E. Treatment of Infestation

Management of head lice infestations requires three general steps[7]: (1) delousing infested individuals through application of pediculicidal shampoos or cream rinses, with retreatment as necessary; (2) removing nits from the hair as thoroughly as possible; and (3) delousing personal items (clothes, hats, combs, pillows, etc.).

An important principle of head lice management is to treat all infested members of a family concurrently. If an infested school-aged child is the only family member treated, he or she may be quickly reinfested by a sibling or parent who is also unknowingly infested. Individuals with lice should be treated with one of the approved pediculicidal shampoos or cream rinses. There are several over-the-counter shampoos (RID®, A200®, R&C Shampoo®, and others) containing pyrethrins that are generally effective. However, because these products are not sufficiently ovicidal, successful management may require 1 or 2 repeat applications at 7- to 10-d intervals to kill newly hatching lice. There is one over-the-counter product containing the synthetic pyrethroid, permethrin, that is ovicidal (Nix®), which is applied as a creme rinse after shampooing. Because Nix® has significant ovicidal activity in addition to its pediculicidal effect, a single treatment may sometimes be sufficient for total eradication of lice and viable eggs.[7] However, many state health departments still recommend a second treatment 7 to 10 d later — even with Nix®. Another product, KWELL® (containing lindane, an organochlorine), is available by prescription as a shampoo. However, inappropriate use of lindane, such as frequent, repeated applications or ingestion, can result in neurotoxicity. Brandenburg et al.[8] have shown that single treatments with the synthetic pyrethroids are more effective than treatments with lindane. Recently, the FDA reapproved malathion (Ovide® lotion) as a prescription drug for the treatment of head lice infestation. Any of these products, even lindane, are generally safe when used as directed. No matter which product they choose, patients should be instructed to follow the precautions and instructions on the product label.

Concerning the removal of nits, it is advisable to remove as many as possible for two reasons. First, reducing the number of viable ova in the hair reduces the chance of persistent infestation and treatment failure. Second, removal of nits lessens concern about continued infestation and risk of spread to others, thus allowing resumption of normal daily routine and school. Removal of all nits is probably impossible. For this reason, persons should be instructed to use the specially designed combs provided with the pediculicides and to follow the product instructions. Combing out nits is more successful if the hair is damp.

Finally, efforts should be made to delouse personal belongings of infested individuals. Washable clothing, hats, bedding, and other personal items should be washed properly and dried in a clothes drier for at least 20 to 30 min. Nonwashable clothes should be dry-cleaned. Other personal items such as combs and brushes should be thoroughly washed in one of the pediculicidal products or soaked in hot water (130°F or more) for 5 to 10 min. Upholstery exposed to potential infestation should be vacuumed thoroughly. Use of certain appropriately labeled pyrethrum aerosols on furniture, carpeting, and other places where lice are suspected may kill a stray louse or two, but as the lice do not willingly leave their host and do not survive long off the host, these treatments would serve primarily as a psychological adjunct to other treatments and generally are not recommended.

Lice resistant to pediculicides. During recent years, there has been an increase in reports of head lice treatment failures. These failures are most often the result of reinfestation or lack of adherence to already-established treatment protocols. However, there does appear to be some development of resistance in certain lice populations to conventional agents such as permethrin, pyrethrins, and organochlorine insecticides.[9-11] A study of field-collected lice from Buenos Aires showed definite resistance to permethrin.[9] However, it is unlikely that any particular population of lice is resistant to all pediculicides. The various lice treatment products have different active ingredients in various classes of insecticide — for example, KWELL® has lindane (chlorinated hydrocarbon); Ovide® has malathion (organophosphate); NIX® has permethrin (synthetic pyrethroid); A200®, R&C®, and RID® have pyrethrins (botanical). Patients with suspected populations of resistant head lice should switch and be treated with a product containing a different class of active ingredient, again paying careful attention to label directions. For example, treatment failures with Nix® may be followed by treatment with Ovide® as the products are from two totally different insecticide classes. It has been suggested that ivermectin could be effective in treating resistant head lice.[12,13] Ivermectin is an antiparasitic drug discovered in the early 1970s that is a derivative of avermectin B1, a compound produced by *Streptomyces avemitilis*. It is a systemic parasiticide/acaricide, killing the pests as they feed on their host. Ivermectin already has approval in the U.S. for treatment of onchocerciasis and strongyloidiasis. Treatment with oral ivermectin for resistant head lice seems to be a promising approach, but additional data are needed to demonstrate the safety of this approach.[13]

Nonpesticidal lice control. Nonpesticidal treatments for head lice include shaving the head, coating the hair with a thick layer of Vaseline petroleum jelly, and various combinations of lice combs/conditioners (wet hair combing). Anecdotal reports indicate that petroleum jelly indeed kills/smothers lice, but is extremely difficult to subsequently remove from the hair. One person told the author that she had to wash her hair > 200

PUBIC LICE

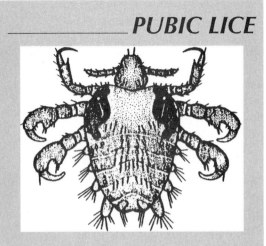

Figure 6
Adult pubic louse.
(From U.S. DHEW, PHS, CDC Pictorial Keys.)

Importance
Itching; irritation; secondary infection

Distribution
Essentially worldwide

Lesion
Variable — sometimes bluish spots

Disease Transmission
None

Key References
Kim, K.C., Pratt, H.D., and Stojanovich,
C.J., *The Sucking Lice of North America*,
Pennsylvania State University Press,
University Park, PA, 1986
Witkowski, J.A. and Parish, L.C., What's
new in the management of lice, *Infect.
Med.*, 14, 287, 1997

Treatment
Pediculicidal shampoos

---

times to remove all of the petroleum jelly. Wet-combing of the hair of lice-infested individuals along with conditioners has been advocated for several years by the U.K. Charity Community Hygiene Concern. "Bug Buster" kits containing fine-toothed lice combs and conditioner have been developed by the Community Hygiene Concern and used in several locations in England, Wales, and Scotland. One evaluation of the Bug Buster kit showed it to deliver a 57% cure rate against head lice.[14]

III. PUBIC OR CRAB LICE

A. General and Medical Importance

Since the 1960s, widespread sexual freedom, particularly among young unmarried individuals, has contributed significantly to the growing incidence of pubic lice infestation.[15] The condition is known as *pediculosis pubis* or *pthiriasis* and is caused by *Pthirus pubis*. The lice occur almost exclusively in the pubic or perianal areas, rarely on eyelashes, eyebrows, or other coarse-haired areas. They feed on human blood through piercing–sucking mouthparts. Often, louse bites produce discrete, round, slaty-gray or bluish spots (maculae caeruleae).[15] The bites may also cause intense itching owing to the host's reaction to proteins in the louse saliva. Secondary infections may occur if the infestation is not treated. Excoriation of the skin from extensive scratching may lead to inflammation of the skin and lymph glands due to bacterial infection. If infestation occurs in the eyelashes, there may be blepharitis, an inflammation of the eyelids. Pubic lice are not known to transmit disease organisms. However, pediculosis pubis frequently coexists with other venereal diseases, particularly gonorrhea and trichomonas. One study indicated that one third of patients with pubic lice may have other sexually transmitted diseases.[16]

B. General Description

Adult pubic lice are dark gray to brown in color. They are called *crab lice* because of their

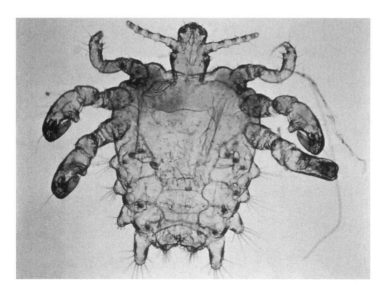

Figure 7
Typical adult pubic louse (CDC photo).

Figure 8
Microscopic view of immature pubic louse holding hair shaft.

crablike shape. They are distinctly flattened, oval, and much wider than body or head lice (Figure 6 and Figure 7). As with head and body lice, the head bears a pair of simple lateral eyes and a pair of short five-segmented antennae. They are 1.5 to 2.0 mm long; their second and third legs are enlarged and contain a modified claw with a thumblike projection, which aids them in grasping hair shafts (Figure 8). The individual egg, or *nit*, is dark brown in color, opalescent, and smaller than that of the body louse.

C. Geographic Distribution

Pubic lice occur on humans virtually worldwide.

D. Biology and Behavior

Pubic lice require human blood to survive. They are only found on humans and do not infest rooms, carpets, beds, pets, etc. If lice happen to be forced off their host, they will die within 24 to 48 h. Fisher and Morton[15] report that this time is less than 20 h. Female pubic lice deposit their eggs (nits) mainly on the coarse hairs of the pubic area and rarely on hairs of the chest, armpits, eyebrows, eyelashes, and mustache. In very rare cases, they have been found in the scalp. They lay approximately 30 eggs during their 3- to 4-week life span. There are three nymphal molts. Nymphs look almost identical to adults, only smaller.

Pubic lice do not fly, jump, or even crawl very much. They often spend their entire life feeding in the same area where the eggs were deposited. Pubic lice are transmitted from person to person most often by sexual contact, although it is possible (though rare) for transmission to occur via toilet seats, clothing, or bedding.

E. Treatment of Infestation

As pubic lice infestations are usually transmitted through sexual contact, it is important to have the sexual contacts of the infested person examined and treated if needed. Likewise, as some family members all sleep in the same bed, if one member of a family has an infestation, all family members should be examined and the infested ones treated. As with head and body lice control products, some are sold over-the-counter and some are by prescription. KWELL® (lindane) shampoo, applied for 5 min in the shower, is quite effective for treatment of infestations in the pubic, perianal, chest, or underarm areas. Also effective are KWELL® lotion (applied for 12 h), synergized pyrethrins (over-the-counter products such as RID®, A200®, and R&C®), and permethrin (NIX®). Treatment is usually repeated in 1 week. Treatments should closely follow directions on the box, bottle, or package insert. For pubic louse infestations of the eyebrows or eyelashes, mechanical removal of nits is used; twice daily applications of petrolatum for 7 to 10 d, or possibly anticholinesterase eye ointments, yellow oxide of mercury, or fluorescein are also used.[17] At the same time of treatment, infested persons should wash all their underclothes and bedding in hot water for 20 min or more, and dry them on the hottest setting. Because of the limited survivability of pubic lice off their hosts, insecticidal sprays, fogs, etc., in the patient's home, work, or school are not necessary.

REFERENCES

1. Foucault, C., Brouqui, P., and Raoult, D., *Bartonella quintana* characteristics and clinical management, *Emerging Infect. Dis.*, 12, 217, 2006.

2. Harwood, R.F. and James, M.T., *Entomology in Human and Animal Health*, Macmillan, New York, 1979.

3. Alexander, J.O., *Arthropods and Human Skin*, Springer-Verlag, Berlin, 1984.

4. Kim, K.C., Pratt, H.D., and Stojanovich, C.J., *The Sucking Lice of North America*, Pennsylvania State University Press, University Park, PA, 1986.

5. Witkowski, J.A. and Parish, L.C., What's new in the management of lice, *Infect. Med.*, 14, 287, 1997.

6. Meinking, T.L., Taplin, D., Kalter, D.C., and Eberle, M.W., Comparative efficacy of treatments for pediculosis capitis infestations, *Arch. Dermatol.*, 122, 267, 1986.

7. Baumgarttner, E.T., Head lice and their management, *MMWR*, 8, 1–2, 1989.

8. Brandenburg, K., Deinard, A.S., DiNapoli, J., Englander, S.J., Orthoefer, J., and Wagner, D., One percent permethrin cream rinse vs. 1% lindane shampoo in treating pediculosis capitis, *Am. J. Dis. Child.*, 140, 894, 1986.

9. Bailey, A.M. and Prociv, P., Persistent head lice following multiple treatment: evidence for insecticide resistance in *Pediculus humanus capitus*, *Aust. J. Dermatol.*, 41, 250, 2000.

10. Downs, A.M.R., Stafford, K.A., and Coles, G.C., Head lice: prevalence in schoolchildren and insecticide resistance, *Parasitol. Today*, 15, 1, 1999.

11. Picollo, M.I., Vassena, C.V., Casadio, A.A., Massimo, J., and Zerba, E.N., Laboratory studies of susceptibility and resistance to insecticides in *Pediculus capitis*, *J. Med. Entomol.*, 35, 814, 1998.

12. Burkhart, K.M., Burkhart, C.N., and Burkhart, C.G., Update on therapy: ivermectin is available for use against lice, *Infect. Med.*, 14, 689, 1997.

13. Estrada, B., Head lice: What about ivermectin? *Infect. Med.*, 15, 823, 1998.

14. Hill, N., Moor, G., Cameron, M.M., Butlin, A., Preston, S., Williamson, M.S., and Bass, C., Single blind, randomised, comparative study of the Bug Buster kit and over the counter pediculicide treatments against head lice in the United Kingdom, *BMJ*, 331, 384, 2005.

15. Fisher, I. and Morton, R.S., *Phthirus pubis* infestation, *Brit. J. Vener. Dis.*, 46, 326, 1970.

16. Chapel, T.A., Katta, T., and Kuszmar, T., *Pthrirus pubis* in clinic for treatment of sexually transmitted diseases, *Sex. Trans. Dis.*, 6, 257, 1979.

17. Buntin, D.M., Rosen, T., Lesher, J.L., Jr., Plotnick, H., Brademas, E., and Berger, T.G., Sexually transmitted diseases: viruses and ectoparasites, *J. Am. Acad. Dermatol.*, 25, 527, 1991.

CHAPTER 23

MILLIPEDES

TABLE OF CONTENTS

I. GENERAL AND MEDICAL IMPORTANCE

Millipedes, sometimes called *thousand leggers*, are elongate, wormlike arthropods that are commonly found in soft, decomposing plant matter.[1] They do not bite or sting but some species secrete (and in some cases forcefully discharge) defensive body fluids containing quinones such as toluquinone and p-benzoquinones and/or hydrogen cyanide that may discolor and burn human skin. Alexander[2] said that affected skin becomes yellowish-brown in color, turning to a dark mahogany brown within 24 h. The mahogany discoloration is attributed to oxidation of quinones on contact with the skin.[3] There may be blistering in a day or two, exfoliating to expose a raw surface.[4] Radford[5] provided an excellent review of millipede burns in people.

MILLIPEDES

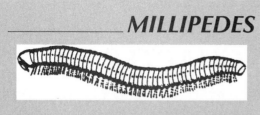

Figure 1
Millipede.
(From U.S. DHEW, PHS, CDC Pictorial Keys.)

Importance
Some species cause burns on skin

Distribution
Numerous species worldwide

Lesion
Yellow to brown discoloration; blisters

Disease Transmission
None

Key Reference
Radford, A. J., *Trop. Geogr. Med.* 27, 279, 1975

Treatment
Skin — wash thoroughly to remove fluids, apply antiseptics; eye — immediate, thorough washing; consult an ophthalmologist for treatment guidelines

II. GENERAL DESCRIPTION

Millipedes are somewhat similar to centipedes except that they have two pairs of legs on most body segments and are generally rounded instead of flattened (Figure 1). In addition, the mouthparts point downward, rather than forward as is the case with centipedes. Millipedes have one pair of antennae. Many species of millipedes are cylindrical (although some are flat) with a hardened, burnished metallic look. Some of the tropical species can attain a length of 30 cm.[2]

III. GEOGRAPHIC DISTRIBUTION

There are numerous species of millipedes worldwide. A genus commonly encountered in leaf litter in North America is *Narceus,* which contains several species. *Rhinochrichus latespargor* found in Haiti, *Spirostreptus* spp. in Indonesia, and *Orthoporus* spp. in Mexico are reported to cause human burns.

IV. BIOLOGY AND BEHAVIOR

Millipedes are commonly found under rocks, in soil, and in leaf litter. They are mostly nocturnal in habit and may be active year round; however, millipedes are more commonly encountered during the wet season. When uncovered, they coil up into a tight spiral (Figure 2). Some of the more slender, agile species are attracted to light and may congregate on front porches, patios, or sidewalks of homes. Female millipedes lay their eggs in the soil. Upon hatching, immature millipedes have few body segments and three pairs of legs; additional legs are added with each molt. They go through 2 to 7 instars before reaching the adult stage. Depending on species, millipedes live from 1 to 7 years. Concerning the poisonous secretions, for the majority of millipede species the secretions ooze out and form droplets around the foramina, but a few species (from genera *Spirobolida, Spirostreptus,* and *Rhinocrichus*) can squirt their secretions for some distance.[5]

Figure 2
Millipede coiled up in tight spiral

V. TREATMENT

Exposed skin should be washed with copious amounts of water as soon as possible. Alexander[2] and Radford[5] recommended using the solvents ether or alcohol to help remove the noxious fluids. Antiseptics may need to be applied. Antibiotics are indicated if secondary infection is suspected.[2] Eye exposure requires thorough irrigation with warm water as soon as possible. An ophthalmologist should be consulted for current treatment recommendations.

REFERENCES

1. Borror, D.J., Triplehorn, C.A., and Johnson, N.F., *An Introduction to the Study of Insects*, Saunders College Publishing, Philadelphia, PA, 1989, p. 138.
2. Alexander, J.O., *Arthropods and Human Skin*, Springer-Verlag, Berlin, 1984, p. 385.
3. Shpall, S. and Freiden, I., Mahogany discoloration of the skin due to the defensive secretion of a millipede, *Pediatr. Dermatol.*, 8, 25, 1991.
4. Harwood, R.F. and James, M.T., *Entomology in Human and Animal Health*, Macmillan, New York, 1979, p. 460.
5. Radford, A.J., Millipede burns in man, *Trop. Geogr. Med.*, 27, 279, 1975.

CHAPTER 24

MITES

TABLE OF CONTENTS

I. CHIGGER MITES

A. General and Medical Importance

Larvae of mites in the family Trombiculidae, sometimes called *chiggers*, *harvest mites*, or *red bugs*, are medically important pests around the world. Over 3000 species of chigger mites occur in the world, but only about 20 species cause dermatitis or transmit diseases such as the agent of scrub typhus. Larval chiggers crawl up on blades of grass or leaves and subsequently get on passing vertebrate hosts. On humans, they generally crawl to and attach where clothing fits snugly or where flesh is tender, such as ankles, groin, or waistline. Chiggers then pierce the skin with their mouthparts, inject saliva into the wound (which dissolves tissue), and then suck up this semidigested material. Feeding generally lasts a few days. Most U.S. pest chiggers produce itching within 3 to 6 h, followed by dermatitis consisting of macules and wheals. At approximately 10 to 16 h, red, dome-shaped papules appear, and itching increases in severity over the next 20 to 30 h.[1] Not all medically important chiggers produce the familiar itch reactions; those serving as vectors of scrub typhus (central, eastern, and southeastern Asia) are not associated with itching or skin reactions.[2]

B. General Description

Adult chigger mites are oval shaped (approximately 1 mm long) with a bright red, velvety appearance. But it is the larval stage that attacks vertebrate hosts; adults do not bite. Accordingly, it is the larval stage or accompanying lesion that is ordinarily collected/seen in a clinical setting. Chigger larvae are very tiny (0.2 mm long), round mites with numerous setae (Figure 1 and Figure 1a). The mites may be red, yellow, or orange in color and have a single dorsal plate (scutum) bearing two sensillae and four to six setae.

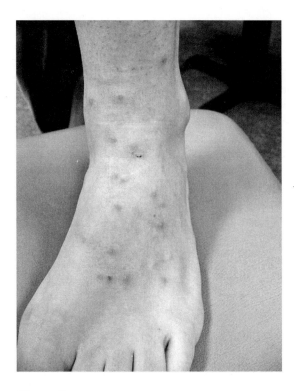

Figure 1a
Chigger bites on ankle.
(Photo courtesy of Wendy C. Varnado.)

C. Geographic Distribution

Chigger species are mostly tropical and sub-tropical, although they occur from Alaska to New Zealand and from sea level to over 16,000 ft in altitude. Some particularly common pest chiggers are *Eutrombicula alfreddugesi* in the U.S. and parts of Central and South America, *Neotrombicula autumnalis* in Europe, and *E. sarcina* in Asia and Australia. Chiggers in the genus *Leptotrombidium* transmit the agent of scrub typhus in Japan, Southeast Asia, and parts of Australia.

D. Biology and Behavior

Chigger mites are found in moist microenvironments within grassy, weedy, or wooded areas, especially forest edges and wild blackberry patches (at least in the southern U.S.). Adult chiggers are predaceous. The female lays eggs singly on soil or litter, and the eggs hatch in about 1 week. After hatching, the life

CHIGGERS

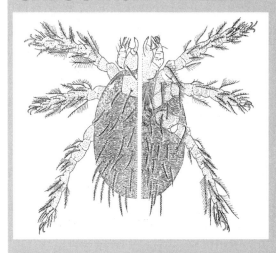

Figure 1
Chigger mite, Leptotrombidium akamushi, *dorsal (left) and ventral (right) views. (From USDA, ARS, Agriculture Handbook No. 655, February 1991.)*

Importance
Irritation; intense itching; disease transmission in the Far East

Distribution
Numerous species worldwide

Lesion
Variable — often a red, dome-shaped papule

Disease Transmission
None in U.S.; scrub typhus in central, eastern, and southeastern Asia

Key Reference
Jenkins, D.W., *Am. J. Hyg*, 48, 22, 1948

Treatment
Antipruritics and/or topical corticosteroids

cycle includes six stages: (1) the inactive prelarva (deutovum or maturing larva), (2) the parasitic larval stage, (3) quiescent protonymph/first nymphal stage (nymphochrysalis), (4) predaceous deutonymph/second nymphal stage, (5) quiescent tritonymph/third nymphal stage (imagochrysalis), and (6) the free-living adult stage. The entire life cycle from egg to adult may be completed in about 60 d. In the northern U.S. the mites are active from about May through September, but in the South they may be active essentially year round. The parasitic larvae normally feed on rodents, insectivores, and ground-frequenting birds, but given the chance some species will avidly feed on people. Larval chigger mites inject saliva that dissolves host cellular tissue. The mites then ingest this mixture of lymph, dissolved body tissues, and stray blood cells. They do not "suck blood" in the sense that other ectoparasites do.

E. Treatment of Bites

After exposure to infested outdoor areas, hot soapy baths or showers will help remove any chiggers, attached or unattached. Antiseptic, hydrocortisone, and/or anesthetic (benzocaine) solutions or ointments are often used as treatments to minimize itching and reduce chances of secondary infection.

II. HOUSE DUST MITES

A. General and Medical Importance

Mites in the genus *Dermatophagoides* are commonly found in houses worldwide and have been associated with house dust allergy. Since the mid-1960s, considerable research on house dust allergy has revealed that both *D. pteronyssinus* and *D. farinae* possess powerful allergens in the mites themselves, as well as in their secretions and excreta. The fecal pellets are especially allergenic. Although dust mites are harmless in that they do not sting or bite, a considerable amount of allergic rhinitis, asthma, and childhood eczema is attributable to their presence in the human environment.[3] Some studies have investigated the link between house dust mite allergens and atopic dermatitis.[4]

B. General Description

Adult house dust mites are white to light tan in color and about 0.5 mm long (Figure 2). Their cuticle has numerous fine striations. The mites have plump bodies (not flattened), well-developed chelicerae, and suckers at the ends of their tarsi.

C. Geographic Distribution

Contrary to its common name, the European house dust mite, *D. pteronyssinus*, actually is widespread throughout the world. It was the first species recognized as being significantly associated with house dust allergy.[5] The American house dust mite is *D. farinae*, and it is also a cosmopolitan inhabitant of houses.

D. Biology and Behavior

House dust mites are associated with furniture (especially mattresses, sofas, and recliner chairs) and debris in household carpets. They are generally more numerous in mattresses and bedrooms than other areas of a house. In a study in Hawaii,[6] dust samples taken from carpets always had more mites than those from noncarpeted floors. Old carpets not cared for properly contained more dust mites than new ones or those often cleaned. The preferred food source of the mites is believed to be shed human skin scales, although they will eat mold, fugal spores, pollen grains, feathers, and animal dander. The mites are most abundant in warm homes with high humidities. Laboratory populations exhibit maximum growth at 25°C and about 75% relative humidity and only survive 10 d or less at 40 or 50% relative humidity (RH).[7,8] High mite levels occur during periods of high RH, and these levels become low when RH drops below the critical level for extended periods. Both *D. pteronyssinus* and *D. farinae* have five developmental stages: egg, larva, protonymph, tritonymph, and adult. The life cycle is completed in about 1 month, and adults may live 2 months or so at optimum temperatures. Cultures of house dust mites are maintained in beds on a year-round basis, and seasonal fluctuations occur.[3]

E.   Treatment of Infestation

House dust mite allergy is managed by immunotherapy using mite extracts and by efforts to minimize the level of dust mites in the patient's home. As long-fibered carpets are difficult to clean, tile, wood, or other "hard" floor covering may be needed in those homes. Vacuuming should be done regularly, especially in the bedroom (although the allergic person probably should not be the one vacuuming). Double-thickness filters or HEPA filters are needed for maximum results. The mattresses should be vacuumed intensely. Studies have demonstrated that after mattresses were vacuum-cleaned, there was as much as an eightfold reduction in the number of mites that became airborne during bedmaking.[9,10] In addition, synthetic pillows and plastic mattress covers should be used and there should be a yearly replacement of mattresses. Sheets and blankets should be cleaned with *hot* water on a regular basis. Efforts should also be made to reduce

HOUSE DUST MITES

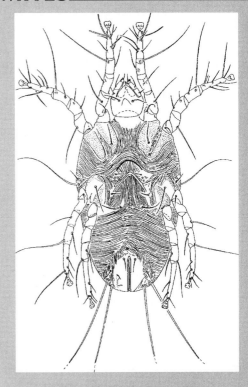

Figure 2
Adult female house dust mite, Dermatophagoides farinae. *(From USDA, ARS, Agriculture Handbook No. 655, February 1991.)*

Importance
Allergies

Distribution
Almost worldwide

Lesion
Generally none — nonbiting

Disease Transmission
None

Key Reference
van Bronswijk, J.E.M.H., *J. Allergy,* 47, 31, 1971

Treatment
Dust and dust mite control; standard treatment for allergies

household humidity levels, as high humidity favors mite growth and reproduction. Although results have been somewhat equivocal, some studies have demonstrated the effectiveness of benzyl benzoate for carpet treatments to reduce mite allergen levels.[11,12] However, reducing the number of mites does not always lower allergen levels right away.

III.   IMAGINARY MITES

A.   Introduction

Patients claiming to be infested with mites (that subsequently cannot be seen, collected, or controlled) may be suffering from delusions of parasitosis (DOP) (see Chapter 7). This emotional disorder is covered more thoroughly in Part I of this book, but basically involves the unwarranted belief that tiny, almost invisible insects or mites are present on the body. Characteristically, the patient presents to the clinic with pieces of tissue paper or small bags or cups containing the presumed mites or insects. However, these usually only contain dust, specks of dirt, dried blood, pieces of skin, and occasionally common (nonharmful) household insects or their body parts. Skin lesions may be present, but self-induced causes are possible. The majority of such patients are elderly females.

B.   Treatment

Cases of imaginary insect or mite infestations must be carefully investigated to rule out actual arthropod causes. Skin scrapings of lesions should be examined for scabies mites. A competent entomologist should examine the patient's workplace and residence. Careful attention should be given to rat- or bird-infested dwellings as they may harbor parasitic mites. If repeated collection attempts and insecticidal treatments are unsuccessful, and if the patient exhibits symptoms consistent with DOP, then DOP should be strongly considered as the cause of the problem. After ruling out actual arthropod causes and any underlying medical conditions, physicians often refer the patient to a psychiatrist.

IV.   HUMAN-BITING MITES

A.   General and Medical Importance

There are actually only two human-associated parasitic mites, the scabies mite and the hair follicle mite (see the following sections on these two mites). However, many other species of mites are important to public health, such as chiggers and house dust mites, which were covered previously. This section includes a hodgepodge of the remaining mite species that may be involved in human bites or cases of dermatitis (clover mites are included — they do not bite, but often invade homes *en masse*). Most mites mentioned here are important only as causes of itch or dermatitis, but the house mouse mite, *Liponyssoides* (formerly *Allodermanyssus*) *sanguineus*, is a vector of rickettsialpox in Massachusetts, Connecticut, New York, Pennsylvania, and Ohio.[13]

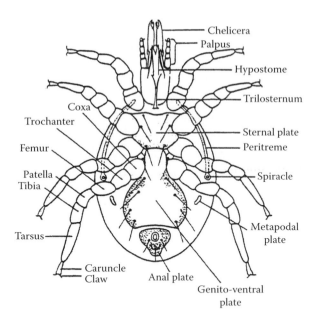

Figure 3
Female rat mite with parts labeled. (From U.S. DHHS, CDC Publication No. 83-8297.)

B. Mite Biology and Discussion of Species

Generalized mite life cycle. It is difficult to describe a "typical" mite life cycle and accompanying life stages as there is tremendous variation among the various orders; however, most mites generally display the following life cycle. Adult females lay eggs that hatch into larvae, pass through one to three nymphal stages, and finally become adults. The larvae have only three pairs of legs, whereas nymphs and adults have four pairs. The first nymphal stage is called a *protonymph*, the second is called a *deutonymph*, and the third nymphal stage, if present, is called a *tritonymph*. In some mite orders, one of the nymphal stages is a nonfeeding phoretic stage for passive transport (in the fur of animals, in bird feathers, on insects, etc.). Figure 3 is provided to familiarize the reader with some of the more prominent morphological characters of mites.

Discussion of Species

Order Mesostigmata

Tropical rat mite. *Ornithonyssus bacoti* is fairly easy to recognize. Females have scissorlike chelicerae, narrow, tapering dorsal and genitoventral plates, and an egg-shaped anal plate (Figure 4). In addition, the protonymphal stage and adult females suck blood, and become tremendously distended after feeding. They look like tiny engorged ticks.   The tropical rat mite is an ectoparasite of rats. The adult mites produce a nonfeeding larval stage, a bloodsucking protonymphal stage, and a nonfeeding deutonymphal stage. A complete generation of this species usually takes about 2 weeks. Females can live unfed 10 d or more after rats have been trapped out of a

BITING MITES

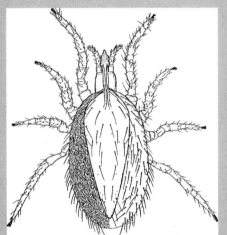

Figure 4
Adult tropical rat mite, Ornithonyssus bacoti.
(From USDA, ARS, Agriculture Handbook #655, February 1991.)

Importance
Painful bites; itching; irritation

Distribution
Numerous species worldwide

Lesion
Variable — erythema, bright red papules of varying sizes

Disease Transmission
Generally none; possibly rickettsialpox

Key Reference
Sections on "Biting Mites" in Alexander, *Arthropods and Skin*, Springer-Verlag, Berlin, 1984

Treatment
Generally palliative treatment only; eliminate source of exposure

building. Tropical rat mites bite people, readily producing a papulovesicular dermatitis with accompanying urticaria.

The tropical rat mite is widely distributed on all continents in association with rats.

Tropical fowl mite. *Ornithonyssus bursa* is similar in appearance to the tropical rat mite, but has a wider dorsal plate (Figure 5).

The biology of the tropical fowl mite is similar to that of the tropical rat mite, except it is found on domestic and wild birds. It may be found on rodents, but rarely. This species infests poultry, but it is also a significant parasite of the English sparrow. Tropical fowl mites often infest wild birds roosting on roofs and around the eaves of homes and office buildings. Nestling birds are especially infested with the mites.[14] These mites may subsequently enter human habitations and bite people when the birds abandon their nests.

Tropical fowl mites are distributed over most of the world but are more commonly seen in tropical and subtropical areas.

Northern fowl mite. *Ornithonyssus sylviarum* also is similar in appearance to the tropical rat mite but has a much shorter sternal plate (Figure 6). This plate has only four setae; the setae on the dorsal plate are quite short. The northern fowl mite is a pest of domestic fowl, pigeons, sparrows, and starlings. The species overwinters in bird nests or cracks and crevices of buildings. Unlike the chicken mite, *D. gallinae*, the northern fowl mite spends its entire life on the host. In poultry houses the mites are usually only found on the birds, but they have been found on eggs and cage litter. Northern fowl mites cannot survive more than a month or so in the absence of their poultry hosts. The northern fowl mite occurs in temperate regions worldwide.

Spiny rat mite. *Laelaps echidnina* is easily recognized by its large genitoventral plate, with a concaved posterior margin into which the anal plate fits (Figure 7). Spiny rat mites are ectoparasites of the Norway rat and roof rat. This species is probably the most prevalent mite species occurring on rats in the U.S.,

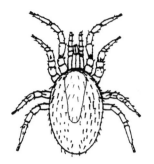

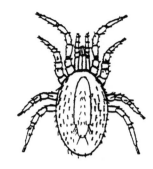

Figure 5
Adult tropical fowl mite. (Redrawn in part from U.S. DHEW, PHS, CDC Pictorial Keys.)

Figure 6
Adult northern fowl mite.

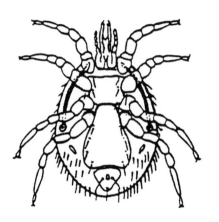

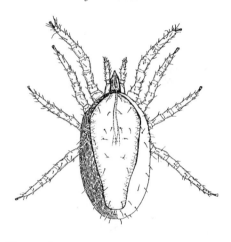

Figure 7
Adult spiny rat mite. (From U.S. DHHS, CDC Publication No. 83-8297.)

Figure 8
Adult chicken mite, Dermanyssus gallinae. (From USDA, ARS, Agriculture Handbook No. 655, February 1991.)

particularly in the central and northern regions of the country. The spiny rat mite is found worldwide.

Chicken mite. *Dermanyssus gallinae* has large dorsal and anal plates, a short sternal plate, and needlelike chelicerae (Figure 8).

The chicken mite, also known as the *red mite of poultry,* is commonly found on domestic fowl, pigeons, English sparrows, starlings, and other birds. This mite is one of the most common species causing human dermatitis in poultry houses, farms, ranches, and markets where chickens are traded or sold. Poultry workers are often bitten on the backs of the hands and on the forearms. *D. gallinae* is nocturnal; during the day the mites hide in cracks and crevices in chicken houses or buildings where infested birds nest. Eggs are deposited in these hiding places. The chicken mite occurs worldwide.

House mouse mite. *Liponyssoides sanguineus* looks similar to the chicken mite (*D. gallinae*). Females of the house mouse mite can be distinguished from most other mites by the presence of two dorsal shields, a large anterior plate, and a small posterior plate bearing one pair of setae (Figure 9).

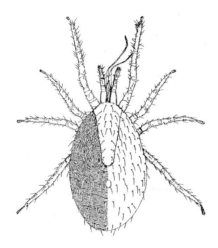

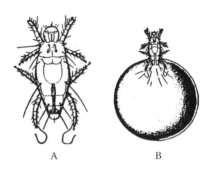

A   B

Figure 9
Adult house mouse mite, Liponyssoides san-
guineus. *(From USDA, ARS, Agricultural
Handbook No. 655, February 1991.)*

Figure 10
*(A) Adult straw itch mite and (B) gravid adult
female. (From U.S. Navy Laboratory Guide to
Medical Entomology.)*

The house mouse mite is primarily an ectoparasite of mice, but has been collected
from rats and other rodents. The protonymphs, deutonymphs, and adults all suck
blood. The life cycle from egg to adult takes about 18 to 23 d.

The house mouse mite occurs in northern Africa, Asia, Europe, and the U.S. (primarily the Northeast).

Order Prostigmata

Straw itch mite. *Pyemotes tritici* is an elongate species that has the first and second
pairs of legs widely separated from the third and fourth pairs (Figure 10A). Also, this
mite has a club-shaped hair between the base of the first and second pairs of legs. The
male is so small that it is almost invisible to the naked eye. Females with eggs may
become enormously swollen (about 1 mm) with their abdomen resembling a tiny pearl
(Figure 10B). Straw itch mites are parasites of several insect species, but will bite
people readily. Bites occur when humans come into contact with infested straw, hay,
grasses, beans, peas, grains, and other materials. Accordingly, agricultural workers,
persons who process or handle grain, or those who sleep on straw mattresses are most
prone to infestation. Alexander[1] said that severe human infestations may occur, with
upward of 10,000 lesions on an individual. Attacks are more frequent during the hot
weather months. This mite has a curious life cycle. As the female feeds, the opisthosoma becomes enormously distended (Figure 10B). Also, all the nymphal stages occur
within the egg so that after hatching, active young occur in the body of the gravid
female. Each adult female may produce 200 to 300 young. Males mate with emerging
young females at or near the genital opening of the mother.

Straw itch mites occur in most areas of the world.

In the last few years, there have been reports of another *Pyemotes* itch mite affecting
thousands of people in Kansas, Missouri, Nebraska, Oklahoma, and Texas.[15] These mites,
identified by entomologists as *Pyemotes herfsi*, cause a rash in exposed individuals, primarily on the limbs, face, and neck. Approximately 77% of the rashes were erythematous, well-demarcated, and papular; less than 22% were pustular, macular, or confluent.[15]

Rashes were said to be extremely itchy. Researchers from Kansas State University, Pittsburg State University, and the University of Nebraska determined that the mites were infesting gall-making midge larvae on oak trees. After itch mites feed on the midge larvae inside the galls, an estimated 16,000 mites can emerge from a single leaf! Such a high mite population is a ubiquitous source of exposure to persons living nearby.

Clover mite. Clover mites in the *Bryobia praetiosa* complex are tiny reddish-brown mites about 0.8 mm long. They have extremely long first pairs of legs and wedge-shaped body hairs (Figure 11).

Clover mites are commonly found on herbaceous plants such as ivy, grass, and clover, as well as some deciduous trees (especially fruit trees). They tend to develop most where there is abundant and actively growing vegetation. This may present a problem when homeowners plant shrubs or ivy immediately next to their houses. The mites will invade homes in great numbers, especially in the fall, as the weather starts to get cold. Quite often they swarm by the thousands over outer walls of buildings and make their way indoors around doors, windows, or cracks. They do not bite people, but often cause alarm.

Clover mites are cosmopolitan in distribution.

Cheyletiellid mite. *Cheyletiella yasguri, C. blakei,* and *C. parasitivorax* have fused cheliceral bases further fused with the subcapitulum forming a capsular gnathosoma (Figure 12). This makes it look as though they are wearing a helmet. They have free and highly developed palpi with strong curved claws that look like an extra pair of legs near the mouthparts.

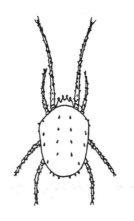

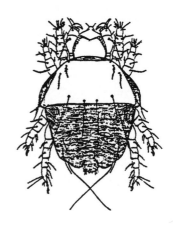

Figure 11
*Adult clover mite. (From U.S. DHHS, CDC
Publication No. 83-8297.)*

Figure 12
Adult cheyletiellid mite.

The *Cheyletiellidae* are parasites of birds and various species of small mammals. The *Cheyletiella* spp. mentioned in this section are obligate parasites of small- or medium-sized mammals (including pet dogs, cats, and pet rabbits), living on the keratin layer of the epidermis; they do not burrow. These mites may cause a mangelike condition on pets and a transient itching dermatitis on humans who handle these pets. In most cases the patient complains of itching and has papules or papulovesicles (2- to 6-mm diameter) on the flexor side of the arms, on the breasts, or on the abdomen. Mites are rarely found in scrapings of the lesions, as they have usually left the person by the time medical advice is sought. *Cheyletiella* eggs are attached to hairs on the host (pet dog, cat, etc.) about 2 to 3 mm above the host's skin. These mites cannot survive more than 48 h off their hosts.

In a review of the *Cheyletiella*,[16] specimens were reported from small pets in the U.S., South America, Western Europe, Australia, New Zealand, and few widely scattered spots in Africa, India, and Japan.

Order Astigmata

Grain and flour mites.   There are several species of grain and flour mites that can cause grocer's itch, copra itch, and other lay-named itches. They are tiny mites (0.5 mm or less) that are pale gray or yellowish white, and they have the first and second pairs of legs widely separated from the third and fourth pairs (Figure 13 to Figure 15). Also, they have conspicuous long hairs (Figure 16). Mites causing grocer's itch, copra itch, vanillism, wheat pollard itch, and dried fruit dermatitis are basically scavengers on a wide variety of organic matter including flour, meal, grains, dried fruits, vanilla pods, meats, and other similar products. These tiny mites have brief developmental times and can multiply into the billions in a stored product in a very short period of time. They do not suck blood, but they will penetrate the superficial epidermis, producing a temporary pruritus.

Grain and flour mites occur worldwide wherever people have shipped food products.

C.   Treatment of Infestation and Bites

Human infestations with the previously mentioned mite species are mostly transitory and the reaction is variable. Sensitive people may develop dermatitis. These mite species will not take up permanent residence on human skin and thus perpetuate the

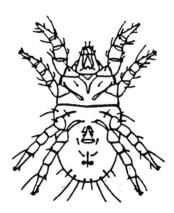

Figure 13
*Adult grain mite. (From U.S. DHHS, CDC
Publication No. 83-8297.)*

Figure 14
*Adult ham mite. (From U.S. DHHS, CDC Pub-
lication No. 83-8297.)*

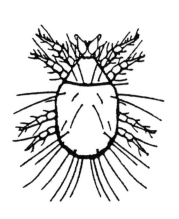

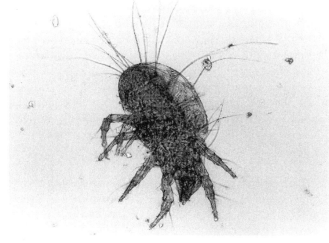

Figure 15
*Adult cheese mite. (From U.S.
DHHS, CDC Publication No.
83-8297.)*

Figure 16
Microscopic view of grain mite, Acarus siro.

infestation. Accordingly, treatment primarily involves alleviation of the symptoms,
and avoidance and/or eradication of the mites from the pet, home, or workplace.
Acute urticarial lesions may respond satisfactorily to topical corticosteroid lotions and
creams, which reduce the intensity of the inflammatory reaction. Oral antihistamines
may relieve itching and burning sensations.

V.   SCABIES MITES (HUMAN ITCH OR MANGE MITES)

A.   General and Medical Importance

Scabies, caused by *Sarcoptes scabei*, is probably the most important disease caused
by mites, with at least 300 million cases annually.[17] It occurs worldwide, affecting all
races and socioeconomic classes in all climates. The tiny mites burrow under the skin,

leaving small open sores and linear burrows that contain the mites and their eggs. When a person is infested with scabies mites for the first time, there is little pathology for about a month, until sensitization develops. When that happens, there is severe itching, especially at night and frequently over much of the body. Large patches of erythema or rash may occur on the body. The patient's tissues apparently become sensitized to various proteins liberated by the mites. Interestingly, the generalized rash may not correspond to the sites where the mites are burrowing. The burrows are usually located on the hands, wrists, and elbows, especially in the webbing between the fingers and the folds of the wrists.[18,19] Alexander[1] said that burrows are sometimes few in number and difficult to find. Genital lesions are common in scabies. The presence of crusted, excoriated, pruritic papules on the penis or buttocks is virtually pathognomonic.[20] In many cases severe itching causes the patient to scratch himself or herself vigorously, leading to secondary infections such as impetigo, eczema, pustules, and boils.

The author has encountered scabies problems frequently in nursing homes. In the elderly, reactions to the mite are often not inflammatory — as seen in younger people — and are muted. Accordingly, scabies is often missed by the attending health care provider. Bedridden patients with scabies complain of intense itching. Mites may be found on the back of such patients (unusual).

People who cannot scratch themselves and the immunocompromised (e.g., AIDS patients) may develop more serious scabies infestations in which millions of mites may inhabit thick crusts over the skin — a condition called *Norwegian scabies* or *crusted scabies*. Alexander[1] says there are two principal components of the eruption, localized horny plaques and a more diffuse erythematosquamous appearance. He also says that patients with Norwegian scabies are frequently inmates of long-stay institutions, and when presented to the dermatologist appear to be suffering from chronic exfoliative eczema.

It should be noted here that animal forms of scabies (such as canine or equine) are also caused by "races" of *S. scabei*, but these mites cannot propagate in human skin. Canine scabies can be temporarily transferred to humans from dogs, causing itching and papular or vesicular lesions primarily on the waist, chest, or forearms. However, treatment or removal of the infested dog will result in a gradual resolution of this type of scabies.

B. General Description

Sarcoptes scabei are very tiny (0.2 to 0.4 mm long), oval, saclike, eyeless mites (Figure 17). Their legs are rudimentary; the anterior two pairs have bell-shaped suckers on their tips. The body is covered with striations and has several stout blunt spines and a few long setae (Figure 18). Scabies mite mouthparts are composed of toothed chelicerae and one-segmented palps fused to the central hypostome. Nymphs look almost identical to the adults, except that they are smaller.

C. Geographic Distribution

Human scabies mites occur worldwide. There are several scabies mites that occur on domestic animals (dogs, cats, horses, and camels) worldwide also. These "races" of *S. scabei* are virtually indistinguishable from the human form, but do not produce sustained infestations on people.

CASE HISTORY

SCABIES IN AN INSTITUTION FOR THE MENTALLY HANDICAPPED

On August 10, 1999, the medical director of a local institution for the mildly retarded called saying that several of his patients were obviously scratching themselves, even to the point of bleeding. He was wondering what different types of arthropod pests might be responsible. On August 13, I visited the institution and found approximately 12 male patients complaining of intense itching. Most had obvious rashes on the arms and trunks. One man had numerous, self-inflicted, deep fingernail scratch marks on his body that were unbandaged and bleeding. I told the medical director that scabies was likely the cause of the itching, but skin scrapings, performed by a dermatologist, would be needed for confirmation. Later I interviewed maintenance staff asking questions about the layout of the building, history of pest (such as rat or bird) problems, and bird nesting (or lack thereof) in the attic. Nothing out of the ordinary was noted. I then took dust samples — alcohol swipes of furniture, bedding, etc. — from affected patient's rooms. Back at the lab, dust samples were examined microscopically for the presence of human biting mites. A sample from one patient's bed contained dust, debris, pieces of excoriated skin, and a scabies mite, *Sarcoptes scabei*, thus confirming scabies infestation. Patients were subsequently treated with scabicidal lotions, and the infestation was eliminated.

Comment: Investigation of this event revealed several interesting features. The patients were itching so severely that they were scratching themselves to the point of bleeding. Perhaps this was due to their altered mental status or lowered immune status. Second, the infestation was so great that scabies mites could be found in the bed. Most scabies infestations involve only a few mites; finding them is a difficult task. Except in cases of Norwegian (also called *crusted*) scabies, it is extremely rare to find the mites on furniture, bedding, and the like.

Figure 17
Microscopic view of scabies mite removed from human skin.

SCABIES MITES

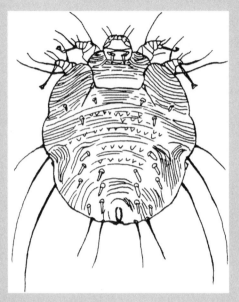

Figure 18
Human scabies mite.

Importance
Intense itch; secondary infection

Distribution
Worldwide

Lesion
Burrows and papules where mites are located; generalized rash may occur in other areas

Disease Transmission
None

Key References
Scabies section in Alexander, *Arthropods and Skin*, Springer-Verlag, Berlin, 1984
Mellenby, K., Classey Ltd. Publishers, Hampton, Middlesex, U.K., 1972

Treatment
Scabicides (cream or lotion) per label instructions

D. Biology and Behavior

Scabies is transmitted by close, human-to-human contact with infested individuals. There is some evidence that fomites may be important sources of infestation or reinfestation.[21] Touching or shaking the hands of infested persons is a major mode of transmission. The practice of several family members sleeping in one bed contributes to its spread, as does sexual activity. In addition, institutionalized children (day care) and elderly (nursing homes) seem to be contributing to an increase in the incidence of scabies.

A female mite infests a new host by burrowing beneath the outer layer of skin and laying her eggs in the tunnels that she excavates. A 6-legged larval stage emerges from each egg and molts to the first nymphal stage in 2 or 3 d. Nymphal and adult stages have eight legs. Larvae and nymphs are often found in short burrows or in hair follicles. After a few days the nymphs molt to the next nymphal stage. After the second nymphal stage adults are formed. The entire life cycle takes 10 to 17 d. The mites apparently eat human skin, although the immatures may feed on hair follicle secretions.

Survival of scabies mites off-host is probably a few days, more likely hours. Studies have shown that the mites may survive 1 to 5 d at room conditions, but may have a difficult time infesting a host after being off the host (presumably due to the mite's weakened condition).[21]

E. Treatment of Infestation

First of all, scabies should be confirmed by isolating the mites in a skin scraping, as other forms of dermatitis may resemble scabies infestation. Scrapings should be made at the burrows, especially on the hands between the fingers and the folds of the wrists. Some dermatologists may choose to scrape burrows located on the feet.[1] To do the scraping, mineral oil is placed on a sterile scalpel blade and allowed to flow onto suspected lesions. By gentle scraping with the blade, the tops of burrows or papules are removed. The oil and scraped material is then transferred to a glass slide and a coverslip applied. Diagnosis can be made either by finding mites, ova, or fecal pellets. Alternatively, mites can be extracted from a burrow by gently pricking open the burrow with a needle and working toward the end where the tiny mites usually are. A hand lens may be useful for this task.

Once a scabies infestation is confirmed, treatment can be initiated. As the mites cannot live very long off a human host, insecticide treatments of bedding, furniture, rooms, etc., are unnecessary. It is recommended, however, that upon initiation of treatment, the patient's bedcovers, pillowcases, and undergarments be removed and washed on the hot wash cycle. If the patient is a child, toys, stuffed animals, etc., should be removed from human contact for a week or so.

There have been several products used for scabies treatment in the past, such as sulfur ointment, benzyl benzoate, lindane, crotamiton, and thiabendazole. The most widely used in the U.S. today are lindane (Kwell®), permethrin (Elimite®), and crotamiton (Eurax®). There have been reports of lindane-resistant scabies, especially in cases of immigrants or recent travelers to Central and South America or Asia.[22] Regardless of the product used, package instructions should be followed carefully. For most scabicides, the product is applied to the entire body, except the head (see package insert; sometimes the head is treated), and left on for 14 to 48 h depending on instructions. After that, a cleansing bath may be taken. A second treatment may be called for in the instructions. Itching may persist for weeks or more after treatment and does not necessarily indicate treatment failure. The posttreatment itch/rash is often treated with cortisone cream. Some physicians prescribe oral antipruritic agents (antihistamines or hydroxyzine) simultaneously with the application of scabicides.

F. Management of Scabies in Nursing Homes

As scabies control in institutions can be difficult and frustrating, a brief outline of a control strategy is offered here. First of all, the diagnosis should be confirmed. Frequently, scabies control measures are implemented with only weak evidence of infestation, e.g., "Looks sorta' like a scabies rash to me." However, implementing a large-scale control effort on the suspicion that an outbreak is occurring in the nursing home is medically and administratively unsound. The active ingredients in scabicidal creams or lotions are pesticides. A dermatologist should be consulted to perform skin scrapings on affected patients. Once the infestation is confirmed, treatment with one of the scabicidal products can be initiated. Both patients and employees with direct, close contact with affected patients should be treated. If cases reappear, aggressive treatment strategies may have to be used such as simultaneously treating all patients and employees in the nursing home (and possibly even family members of employees).[23,24] Consultation with state or local health department epidemiology personnel is often helpful as well. There have been studies showing the effectiveness of a single, oral dose of ivermectin for treatment of human scabies;[25] however, this antihelminthic drug is not yet approved for scabies and additional controlled studies are needed.

VI. FOLLICLE MITES

A. Introduction

Although there are numerous species of *Demodex* (family Demodecidae) infesting wild and domestic animals, only two species of the mites are specific human-associated mites and are called *follicle mites*. The minute, wormlike mites live exclusively in hair follicles

FOLLICLE MITES

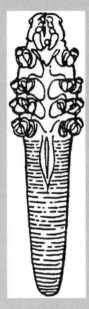

Figure 19
Hair follicle mite.
(From U.S. DHHS, CDC
Publication No. 83-8297.)

Importance
Lives in human skin but mostly harmless

Distribution
Worldwide

Lesion
Generally none

Disease Transmission
None

Key Reference
Desch, C. and Nutting, W. B., *J. Parasitol.,*
58, 169–177, 1972

Treatment
Generally none needed

or sebaceous glands. They have no proven detrimental effect on humans, although some authors have attributed various pathological conditions of the skin to *Demodex*. Alexander[1] provides a good review of this issue. He concludes, "It should be emphasized that, in general, *Demodex* is a harmless saprophyte. It is only exceptionally that it appears to exercise a pathogenic influence, as, for example, when excessive amounts of cosmetics prepare the ground for its proliferation or when it escapes into the dermis." Various estimates of the incidence of human *Demodex* infestation range from about 25 to 100%, and clinicians should be aware of mite appearance, as they may be seen during skin-scraping examination.

B. Discussion of Species

Demodex folliculorum lives in the hair follicles and *D. brevis* in the sebaceous glands. Both species are similar in appearance (with the exception that *D. brevis* is a shortened form) and are elongated, wormlike mites with only rudimentary legs (Figure 19). They are approximately 0.1 to 0.4 mm long and have transverse striations over much of the body. These mites most commonly occur on the forehead, malar areas of the cheeks, nose and nasolabial fold, but they can occur anywhere on the face, around the ears, and occasionally elsewhere.[1] Most people acquire *Demodex* mites early in life from household contacts — primarily maternal.

REFERENCES

1. Alexander, J.O., *Arthropods and Human Skin*, Springer-Verlag, Berlin, 1984, p. 357.

2. Traub, R. and Wisseman, C.L., The ecology of chigger-borne rickettsiosis (scrub typhus), *J. Med. Entomol.*, 11, 237, 1974.

3. Wharton, G.W., House dust mites, *J. Med. Entomol.*, 12, 577, 1976.

4. Cameron, M.M., Can house dust mite-triggered atopic dermatitis be alleviated using acaricides?, *Br. J. Dermatol.*, 137, 1, 1997.

5. Voorhorst, R., Spieksma-Boezeman, M.I.A., and Spieksma, F.T.M., Is a mite (*Dermatophagoides* sp.) the producer of the house dust allergen?, *Allerg. Asthma*, 10, 329, 1964.

6. Sharp, J.L. and Haramoto, F.H., *Dermatophagoides pteronyssinus* and other acarines in house dust in Hawaii, *Proc. Hawaii. Entomol. Soc.*, 20, 583, 1970.

7. Arlian, L.G., Dehydration and survival of the European house dust mite, *Dermatophagoides pteronyssinus*, *J. Med. Entomol.*, 12, 437, 1975.

8. Murton, J.J. and Madden, J.L., Observations on the biology, behavior, and ecology of the house dust mite in Tasmania, *J. Aust. Entomol. Soc.*, 16, 281, 1977.

9. van Bronswijk, J.E.M.H., Schoonen, J.M.C.P., and Berlie, M.A.F., On the abundance of *Dermatophagoides pteronyssinus* in house dust, *Res. Popul. Ecol.*, 13, 67, 1971.

10. van Bronswijk, J.E.M.H. and Sinha, R.N., Pyroglyphid mites and house dust allergy, *Res. Popul. Ecol.*, 13, 67, 1971.

11. Chang, J.H., Becker, A., Ferguson, A., Manfreda, J., Simons, E., Chan, H., Noertjojo, K., and Chan-Yeung, M., Effect of application of benzyl benzoate on house dust mite allergen levels, *Ann. Allerg. Asthma Immunol.*, 77, 187, 1996.

12. Huss, R.W., Huss, K., Squire, E.N., Jr., Carpenter, G.B., Smith, L.J., Salata, K., and Hershey, J., Mite allergen control fails, *J. Allerg. Clin. Immunol.*, 94, 27, 1994.

13. Huebner, R.J., Jellison, W.L., and Pomerantz, C., Rickettsialpox — a newly recognized rickettsial disease. IV. Isolation of a rickettsia apparently identical with the causative agent of rickettsialpox, from *Allodermanyssus sanguineus,* a rodent mite, *Public Health Rep.*, 61, 1677, 1946.

14. Denmark, H.A. and Cromroy, H.L., Tropical fowl mite, *Ornithonyssus bursa* (Berlese), in Florida Dep. Agric. and Consumer Serv., Tallahassee, FL, Entomology Circular No. 299, 1987, pp. 1–4.

15. CDC, Outbreak of pruritic rashes associated with mites — Kansas, 2004, *MMWR*, 54, 952–955, 2005.

16. van Bronswijk, J.E.M.H. and de Kreek, E.J., *Cheyletiella* of dog, cat, and domesticated rabbit, *J. Med. Entomol.*, 13, 315, 1976.

17. Service, M.W., *Medical Entomology for Students*, Chapman and Hall, New York, 1996, p. 248.

18. Bartley, W.C. and Mellanby, K., The parasitology of human scabies (women and children), *Parasitology*, 35, 207, 1944.

19. Johnson, C.G. and Mellanby, K., The parasitology of human scabies, *Parasitology*, 34, 285, 1942.

20. Buntin, D.M., Rosen, T., Lesher, J.L., Jr., Plotnick, H., Brademas, E., and Berger, T.G., Sexually transmitted diseases: viruses and ectoparasites, *J. Am. Acad. Dermatol.*, 25, 527, 1991.

21. Arlian, L.G., Biology, host relations, and epidemiology of *Sarcoptes scabei*, *Annu. Rev. Entomol.*, 34, 139, 1989.

22. Purvis, R.S. and Tyring, S.K., An outbreak of lindane-resistant scabies treated successfully with permethrin 5% cream, *J. Am. Acad. Dermatol.*, 25, 1015, 1991.

23. Juranek, D.D., Currier, R., and Millikan, L.E., Scabies control in institutions, in *Cutaneous Infestations and Insect Bites*, Vol. 4, Orkin, M. and Maibach, H.I., Eds., Marcel Dekker, New York, 1985, p. 139.

24. Paules, S.J., Levisohn, D., and Heffron, W., Persistent scabies in nursing home patients, *J. Fam. Pract.*, 37, 82, 1993.

25. Meinking, T.L., Taplin, D., Hermida, J.L., Pardo, R., and Kerdel, F.A., The treatment of scabies with ivermectin, *New Engl. J. Med.*, 333, 26, 1995.

CHAPTER 25

MOSQUITOES

TABLE OF CONTENTS

MOSQUITOES

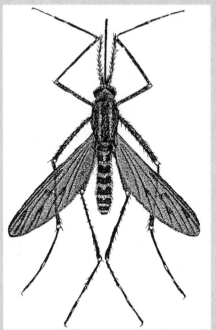

**Adult mosquito
(USDA, ARS, Agri. Hndbk. No. 182, 1961)**

Importance
Annoyance and disease transmission

Distribution
Numerous species worldwide

Lesion
Punctate hemorrhages, papular lesions, or large wheals with edema

Disease Transmission
Number one arthropod vector of disease agents — malaria, dengue, encephalitis, etc.

Key References
Carpenter, S.J. and Lacasse, W.J., University of California Press, Berkeley, 1955

Darsie, R.F. Jr. and Ward, R.A., University Press of Florida, Gainesville, 2005

Treatment
Generally, palliative creams or lotions; topical corticosteroids and antibiotic creams or ointments may be needed; oral antihistamines may be effective in reducing symptoms of mosquito bites

I. GENERAL AND MEDICAL SIGNIFICANCE

Mosquitoes are by far the most important of the bloodsucking arthropods worldwide, giving annoyance to and causing disease in humans, other mammals, and birds. Swarms of salt marsh mosquitoes torment people living on seacoasts throughout the world. In those areas, it is not unusual to experience a mosquito landing rate of over 200 mosquitoes per minute. Inland, rice fields and wetlands produce tremendous mosquito problems each year. Discarded automobile tires, ditches, bird baths, and paint cans are excellent mosquito breeding grounds in urban areas. In many cases swarms of mosquitoes often interfere with normal outdoor work and other activities.

About 3500 species of mosquitoes have been described worldwide. Relatively few of them are significant vectors of human diseases; however, the mosquito-transmitted disease problem worldwide is quite severe. Table 1 details biological data on medically important mosquitoes in the U.S. Below is a brief description of some of the major mosquito-transmitted diseases.

A. Malaria

Malaria remains prevalent in several large areas of the world (especially so in Africa), resulting in staggering case numbers and millions of deaths (Figure 1). In 1986 there were an estimated 489 million cases of the disease worldwide and 2.3 million deaths.[1] To make matters worse, mosquito vectors of malaria are becoming resistant to many of the pesticides being used to control them, and the malaria parasites themselves are becoming resistant to the prophylactic drugs used to prevent the disease.

Malaria in humans is caused by any one of four species of protozoans in the genus *Plasmodium* (*P. vivax, P. malariae, P. ovale*, and *P. falciparum*). The life cycle is quite complicated, and its description is fraught with technical terms (Figure 2). Malaria parasites are

OFTEN-ASKED QUESTION

WHY CAN'T MOSQUITOES TRANSMIT THE AIDS VIRUS?

Because human immunodeficiency virus (HIV) is a bloodborne pathogen, concerns have been raised about the possible transmission of HIV by blood-feeding arthropods. Laboratory studies and epidemiologic surveys indicate that this possibility is extremely remote. For biological transmission, the virus must avoid digestion in the gut of the insect, recognize receptors on and penetrate the gut, replicate in insect tissue, recognize and penetrate the insect salivary glands, and escape into the lumen of the salivary duct. In one study by Webb and colleagues, the virus persisted for 8 d in bed bugs.[1] Another study by Humphrey-Smith and colleagues[2] showed the virus to persist for 10 d in ticks artificially fed meals with high levels of virus ($\geq 10^5$ tissue culture infective doses per milliliter (TCID/ml), but there was no evidence of viral replication. Intra-abdominal inoculation of bed bugs and intrathoracic inoculation of mosquitoes was used to bypass any gut barriers, but again the virus failed to multiply.[1] Likewise, *in vitro* culture of HIV with a number of arthropod cell lines indicated that HIV was incapable of replicating in these systems. Thus, biological transmission of HIV seems extremely improbable.

Mechanical transmission would most likely occur if the arthropod were interrupted while feeding, and then quickly resumed feeding on a susceptible host. Transmission of HIV would be a function of the viremia in the infected host and the virus remaining on the mouthparts or regurgitated into the feeding wound. The blood meal residue on bed bug mouthparts was estimated to be 7×10^{-5} ml, but 50 bed bugs, interrupted while feeding on blood containing 1.3×10^5 TCID/ml HIV, failed to contaminate the uninfected blood on which they finished feeding or the mouse skin membrane through which they refed.[1]

Within minutes of being fed blood with 5×10^4 TCID of HIV, stable flies regurgitated 0.2 μl of fluid containing an estimated 10 TCID.[3] The minimum infective dose for humans contaminated in this manner is unknown, but under conditions such as those in some tropical countries where there are large populations of biting insects and a high prevalence of HIV infection, transfer might be theoretically possible, if highly unlikely. In these countries, however, other modes of transmission are overwhelmingly important, and, although of fatal importance to the extremely rare individual who might contract HIV through an arthropod bite, arthropods are of no significance to the ecology of the virus.

An epidemiologic survey of Belle Glade, a south Florida community believed to have a number of HIV infections in individuals with no risk factors, provided no evidence of HIV transmission by insects.[4] Interviews with surviving patients with the infections revealed that all but a few had engaged in the traditional risk behavior (e.g., drug use and unprotected sex). A serosurvey for exposure to mosquito-borne viruses demonstrated no significant association between mosquito contact and HIV status. Nor were repellent use, time outdoors, or other factors associated with exposure to mosquitoes related to risk of HIV infection. A serosurvey for HIV antibodies detected no positive individuals between 2 and 10 years of age or 60 and older. No clusters of cases occurred in houses without other risk factors. There was thus no evidence of insect-borne HIV transmission.

Source: Adapted from McHugh, C.P., *Lab. Med.*, 25, 436, 1994. Copyright 1994, by the American Society of Clinical Pathologists. With permission.

REFERENCES

1. Webb, P.A., Happ, C.M., Maupin, G.O., Johnson, B.J.B., Ou, C.-H., and Monath, T.P., Potential for insect transmission of HIV: experimental exposure of *Cimex hemipterous* and *Toxorhynchites amboinensis* to human immunodeficiency virus, *J. Infect. Dis.*, 60, 970, 1989.

2. Humphrey-Smith, I., Donker, G., Turzo, A., Chastel, C., and Schmidt-Mayerova, H., Evaluation of mechanical transmission of HIV by the African soft tick, *Ornithodoros moubata*, AIDS, 7, 341, 1993.

3. Brandner, G., Kloft, W.I., Schlager-Vollmer, Platten, E., and Neumann-Opitz, P., Preservation of HIV infectivity during uptake and regurgitation by the stable fly, *Stomoxys calcitrans, L.*, AIDS-Forschung., 5, 253, 1992.

4. Castro, K.G., Lieb, S., Jaffe, H.W. et al., Transmission of HIV in Belle Glade, Florida: lessons for other communities in the United States, *Science*, 239, 1988, p. 193.

Table 1
Biological Data on Medically Important Mosquitoes in the U.S.[a]

Species	Larval Habitat(s)[b]	Biting Time[c]	Flight Range[d] (mi.)	Disease Agent[e]
Aedes aegypti	AC	C, D	<0.5	DG, YF
Ae. albopictus	AC, TH	D	<0.5	DG, YF, (CE)
Ochlerotatus dorsalis	SM, LM	D	10-20	(WEE)
Oc. melanimon	IP, FW	D	1-2	WEE, CE
Oc. nigromaculis	IP, FW	D	1-2	(WEE), (CE)
Oc. sollicitans	SM	C	5-10	EEE
Oc. taeniorhynchus	SM	C, N	5-10	VEE, (CE), (WNV)
Oc. triseriatus	TH, AC	D	0.5-1	CE, (WNV)
Oc. trivittatus	GP, WP, FW	C, N	0.5-1	CE
Ae. vexans	FW, GP, IP	C, N	1-5	CE, (EEE), (WNV)
Anopheles crucians complex	SM, FS, LM	C	1-2	(VEE), (EEE)
An. freeborni	RF, DD	C	1-2	M, (WEE), (SLE)
An. quadrimaculatus complex	FW, GP, LM, RF	C	0.5-1	M, (WNV)
Coquillettidia perturbans	FS, GP, LM	C, N	1-2	EEE, (VEE)
Culex nigripalpus	GP, FW, DD	C	0.5-1	SLE, (WNV)
Cx. pipiens/ quinquefasciatus	AC, SCB, GRP	C, N	<0.5	WNV, SLE, (WEE), (VEE)
Cx. restuans	WP, GRP, DD	C, N	1-2	WNV, (EEE), (WEE)
Cx. salinarius	GP, LM, FS	C, N	1-5	WNV, (EEE)
Cx. tarsalis	IP, RF, GRP	C, N	1-2	WEE, WNV, SLE
Culiseta inornata	GRP, DD	C,N	1-2	(WEE), (CE)
Mansonia titillans	FS, GP, LM	C, N	1-5	(VEE)
Psorophora columbiae	IP, RF, GRP	C, N	1-5	VEE, EEE, (WNV)

[a] *Adapted from U.S. Air Force Mosquito Surveillance Data.*

[b] *AC = artificial containers; DD = drainage ditches; FS = freshwater swamps; FW = floodwaters; GP = grassland pools; GRP = ground pools; IP = irrigated pastures; LM = lake margins; RF = rice fields; SCB = sewer catch basins; SM = salt marshes; TH = tree holes; WP = woodland pools.*

[c] *C = crepuscular (dusk and dawn); D = day; N = night.*

[d] *Values given are estimates of normal flight ranges. For some species, seasonal migratory flights may be 10 times these values.*

[e] *Parentheses indicate secondary or suspected vectors, otherwise primary vectors. CE = California group encephalitis; DG = dengue; EEE = Eastern equine encephalitis; M = malaria; SLE = St. Louis encephalitis; VEE = Venezuelan equine encephalitis; WEE = Western equine encephalitis; WNV = West Nile virus; YF = yellow fever.*

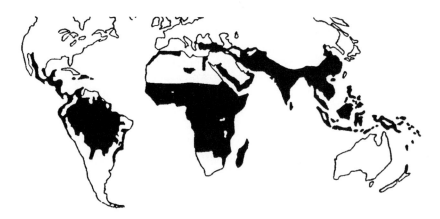

Figure 1
Approximate worldwide distribution of malaria.
(From Florida Department Health and Rehabilitative Services.)

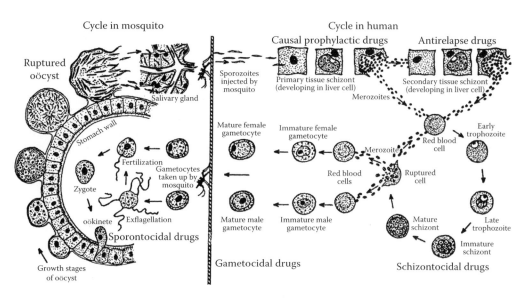

Figure 2
Life cycle of malaria. (From U.S. Navy Guide to Malaria Prevention and Control.)

acquired and transmitted to humans by bites from *Anopheles* mosquitoes only. However, not every species of *Anopheles* is a vector; less than half of the 470 or so known species are considered vectors. Falciparum malaria is the worst of the four species and is often fatal in infants and young children. Most malaria occurring in the U.S. each year is a result of people having relapses from former cases or from cases recently acquired in foreign countries where malaria is endemic (introduced malaria). During 1988, there were 1023 malaria cases reported in the U.S., 991 of which were introduced cases.[2] In recent years small foci of autochthonous malaria have become more frequent in the U.S.[3–5]

Malaria treatment is complicated and ever-changing because of development of resistance to choroquine and other commonly used antimalarial drugs. There have even been reports of resistance to the new artemisinin products. Treatment of resistant strains of malaria involves combination drug therapy. For the most current treatment recommendations, physicians should call the Centers for Disease Control and Prevention, Parasitic Diseases Division, telephone (770) 488-7776.

B.   Yellow Fever

The causative agent of yellow fever (YF) is probably the most lethal of all the arboviruses and has had a devastating effect on human social development. The causative agent, a flavivirus, occurs in Africa and South and Central America, and is maintained in a sylvatic form among monkeys by forest or scrub mosquitoes. An urban form of the disease occurs when humans become infected and transmission occurs from person to person by *Aedes aegypti*. Mild cases may be characterized by fever, headache, generalized aches and pains, and nausea. Persons with severe YF may exhibit high fever, headache, dizziness, muscular pain, jaundice, hemorrhagic symptoms, and profuse vomiting of brown or black material. There is often collapse and death. Even though there is an effective vaccine, the WHO estimates that 200,000 people are infected with YF every year, resulting in 30,000 deaths.[6,7] Many large epidemics have occurred in the past, especially during the time of exploration of the New World. For example, 20,000 of 27,000 British troops were killed by YF during an expedition to conquer Mexico in 1741.

C.   Dengue Fever

Dengue fever, caused by a virus in the family Togaviridae, is responsible for widespread morbidity (breakbone fever) and some mortality (dengue hemorrhagic fever, DHF) in much of the tropics and subtropics each year (Figure 3). There are four closely

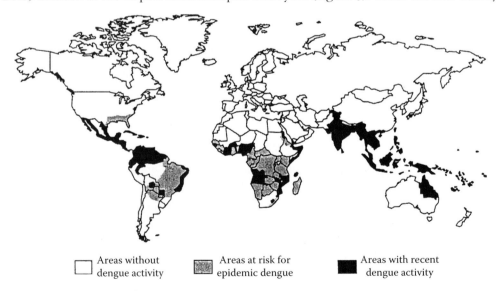

☐ Areas without dengue activity   ▨ Areas at risk for epidemic dengue   ■ Areas with recent dengue activity

Figure 3
Worldwide distribution of dengue, 1993.
(From U.S. DHHS, PHS, CDC, MMWR, 43, No. SS-2, July 22, 1994.)

related, but antigenically distinct, virus serotypes (DEN-1, DEN-2, DEN-3, and DEN-4). DHF (sometimes called *dengue shock syndrome* in its most severe form) is a hemorrhagic complication occurring mostly in children, and is thought to be a result of sequential infection by more than one dengue serotype or variations in viral virulence. Most dengue infections result in relatively mild illness characterized by fever, headache, myalgia, rash, nausea, and vomiting. However, DHF is often severe and is characterized by petechiae, purpura, mild gum bleeding, nosebleeds, gastrointestinal bleeding, and dengue shock syndrome. The case fatality rate of DHF in most countries is about 5%, with most deaths occurring among children. In one recent study of DHF in the Cook Islands, deaths occurred in patients 14 to 22 years old and were due to acute upper gastrointestinal bleeding.[8] The virus is transmitted among humans by *Ae. aegypti* and *Ae. albopictus*. Dengue fever seems to be increasing in prevalence and geographic distribution, especially in southeast Asia, India, the Caribbean, and Central and South America.[9] After an almost 60-year absence, there were more than 100 cases of locally-acquired dengue in Hawaii in 2001.[10] Worldwide, as many as 100 million cases of dengue occur annually, and several hundred thousand cases of DHF.[11]

D. Lymphatic Filariasis

Lymphatic filariasis (Figure 4 and Figure 4a) is an important human disease occurring in much of the world. Malayan filariasis, caused by *Brugia malayi*, is mostly confined to southeast Asia, and the Bancroftian form, *Wucheria bancrofti*, is prevalent over much of the tropical world. The World Health Organization estimated that 250 million cases of Bancroftian or Brugian filariasis occurred worldwide in 1974.[12] There was at one time a small endemic center of human filariasis near Charleston, SC, that has apparently disappeared. However, we still have an efficient vector in the U.S., *Culex quinquefasciatus*, that could transmit the filarial worms should a reintroduction occur. Bancroftian filariasis is an interesting disease in that there is no other known vertebrate host of the worms. It is transmitted solely by mosquitoes, and there is no multiplication of the parasite, only development, in the mosquito vector.[8] In addition, the adult worms may live up to 10 years in humans.[13]

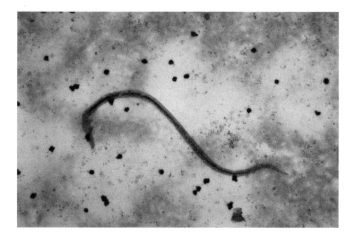

Figure 4
Filarial worm, Wuchereria bancrofti *(CDC photo).*

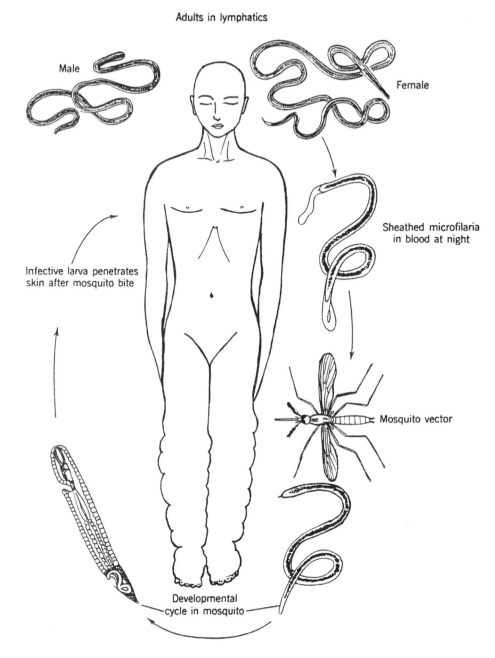

Adults in lymphatics

Male

Female

Sheathed microfilaria
in blood at night

Infective larva penetrates
skin after mosquito bite

Mosquito vector

Developmental
cycle in mosquito

Figure 4a
Life cycle of Wuchereria bancrofti. *(Reprinted from Essentials of Medical Parasitology by Thomas J. Brooks, Jr., Macmillan Publishing Company. Copyright 1963, Thomas J. Brooks, Jr. With permission.)*

E.   *Other Human-Infesting Filarial Worms*

Numerous filarial worms are transmitted to humans and other mammals by mosquitoes and black flies. Examples include the causative agents of Bancroftian and Malayan filariasis (discussed earlier), loiasis, onchocerciasis, and dirofilariasis (dog heartworm).

CASE HISTORY

FATAL CASE OF
MOSQUITO-TRANSMITTED ENCEPHALITIS

An 11-year-old Native American from an Indian community developed a fever of 103°F and diarrhea on the 31st of July. Gastroenteritis was reportedly "going around" in the community at the time. He was taken to a local emergency room and given symptomatic treatment. He was somewhat better until the night prior to admission when he developed a headache, stomachache, and decreased appetite. He went to bed early, which was unusual for him. The next morning he went with his family to a scheduled ophthalmologic exam and slept most of the 1-h drive. He was drowsy and nauseated on arrival at the clinic, then "turned pale" and began grand mal seizure activity. He was taken to the emergency room, loaded with Dilantin and transferred to the admitting hospital. On admission, he was responsive but lethargic. His admission temperature was 102.2°F. His admission laboratory showed a white blood cell count of 19,500 with 59% neutrophils, 19% band forms, and 16% lymphocytes. The hematocrit was 34.4% and the serum glucose with 184 mg per 100 ml. Spinal fluid showed a white blood cell count of 980, with 91% neutrophils, no organisms seen on Gram stain, negative latex agglutination, a protein of 68, and a glucose of 105 mg per 100 ml. Additional blood, spinal fluid, and stool cultures were obtained. The patient was placed on Claforan and Dilantin. He remained febrile up to 105°F, but became more responsive and was ambulatory by the second day after admission. At approximately 2 P.M., the 5th of August, the patient experienced another seizure with eye deviation to the right and head turning to the right. CT showed enhancement of the cisterna, but only mild increased intracranial pressure. Respirations became irregular and the patient was electively intubated and hyperventilated. He was started on Streptomycin, PZA, and INH for possible TB, and Acyclovir for possible CNS herpes. His condition deteriorated over the next 24 hours until he showed no evidence of brain stem function. A lumbar puncture was performed for viral studies as none of his previous cultures were growing. He was taken off the ventilator the evening of August 6th. At autopsy the patient's meninges were relatively clear, but cerebral edema was present. Confirmation of infection with eastern equine encephalitis (EEE) virus was made by the CDC (Ft. Collins); two separate serum samples indicated a fourfold rise in HI antibody to EEE virus and ELISA tests indicated presence of specific IgM.

Comment: Investigation of this EEE case revealed several interesting features. The patient lived in a house without window screens. This likely led to increased exposure to mosquitoes (and thus, biting) — a risk factor for any mosquito-borne disease. An environmental survey of the Indian community revealed numerous prime *Coquillettidia perturbans* (the suspected mosquito vector in this case) breeding sites. In addition, *Cq. perturbans* were collected by CDC light traps in the community at the time of the survey. No other known vectors of EEE virus were collected. Finally, mosquito trapping a year later (same month as patient's infection) at the Indian community revealed that this species was the predominant species in the area.

Source: Adapted from *J. Agromed.* 2, 53, 1995, Copyright 1995, the Hayworth Press, Binghamton, New York. With permission.

Other filarial worms may or may not cause symptomatic disease and are less well known (and thus have no common name), such as *Mansonella ozzardi, M. strepto-cerca, M. perstans, Dirofilaria tenuis, D. ursi, D. repens,* and others. Beaver and Ori-hel[14] reported 39 such cases that were caused by *Dirofilaria immitis* (dog heartworm), other *Dirofilaria* sp., *Dipetalonema* sp., and *Brugia* sp.

The dog heartworm, *Dirofilaria immitis,* occurs mainly in the tropics and subtropics but also extends into southern Europe and North America. This worm infects several canid species, sometimes cats, and, rarely, humans. Numerous mosquito species are capable of transmitting dog heartworm, especially those in the genera *Aedes, Och-lerotatus, Anopheles,* and *Culex.* Mosquitoes pick up the microfilariae with their blood meal when feeding on infected dogs. In endemic areas, a fairly high infection rate may occur in local mosquitoes.

Undoubtedly, thousands of people in the U.S. are bitten each year by mosquitoes infected with *D. immitis.* Fortunately, humans are accidental hosts, and the larvae usually die. However, they may occasionally be found as a subadult worm in the lung (seen as a coin lesion on x-ray exam).[15] The incidence of dog heartworm in humans may well be decreasing in the U.S. because of widespread — and fairly consistent — treatment of domestic dogs for heartworm. The closely related *D. tenuis* is commonly found in the subcutaneous tissues of raccoons (again, mosquito-transmitted) and may accidentally infest humans as nodules in subcutaneous tissues.

F.   Encephalitides

In temperate North America, the worst mosquito-borne diseases are probably the encephalitides. Certainly not all cases of encephalitis are mosquito-caused (enteroviruses and other agents are often involved), but mosquito-borne encephalitis has the potential to become a serious cause of morbidity and mortality covering widespread geographic areas each year.

Eastern equine encephalomyelitis.   Eastern equine encephalomyelitis (EEE) is generally the most virulent, being severe and frequently fatal (mortality rate 30 to 60%). Fortunately, large and widespread outbreaks are not common; between 1961 and 1985 only 99 human cases were reported.[16] EEE occurs in late summer and early fall in the central and north central U.S., parts of Canada, southward along the coastal margins of the eastern U.S. and the Gulf of Mexico, and sparsely throughout Central and South America. The ecology of EEE is complex. The virus circulates in wild bird populations, and the exact mechanism of spread to humans is not well known. The mosquitoes most likely involved include *Ochlerotatus sollicitans, Coquillettidia perturbans, Culex salinarius,* and *Ae. vexans.*[17–19]

St. Louis encephalitis.   St. Louis encephalitis (SLE) produces lower mortality rates than EEE (3 to 20%), but it occurs occasionally in large epidemics in much of the U.S. As with EEE, most cases occur in late summer (Figure 5). In 1933 there were 1095 cases in St. Louis with more than 200 deaths.[20] From 1975 to 1976 there were over 2000 cases reported from 30 states, primarily in the Mississippi Valley.[20] SLE is a bird virus that is transmitted by *Cx. tarsalis* (western and southwestern U.S.), *Cx. quinquefasciatus* (central U.S.), and *Cx. nigripalpus* (southeastern U.S.) (Figure 6).

West Nile encephalitis.   The West Nile virus (WNV) was identified for the first time in the Western Hemisphere in New York in 1999. By the end of the year, it had caused encephalitis in 62 people and numerous horses in and around New York City, resulting in 7 human and 10 equine deaths.[21] The virus continued to spread in subsequent years,

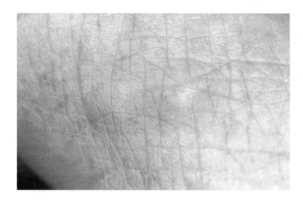

COLOR FIGURE 3.1 Fire ant stings on thumb 1 h after sting. (Photo courtesy of Dr. James Jarratt, Mississippi Cooperative Extension Service, Mississippi State University.)

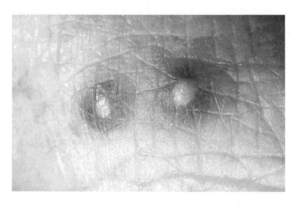

COLOR FIGURE 3.2 Fire ant stings on thumb 72 h after sting. (Photo courtesy of Dr. James Jarratt, Mississippi Cooperative Extension Service, Mississippi State University.)

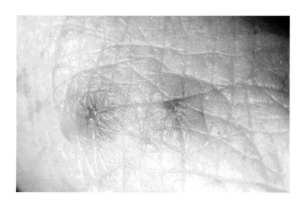

COLOR FIGURE 3.3 Fire ant stings on thumb 1 week after sting. (Photo courtesy of Dr. James Jarratt, Mississippi Cooperative Extension Service, Mississippi State University.)

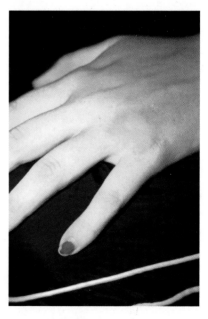

COLOR FIGURE 4.4 Wheal and flare reaction 30 min after mosquito bite. (Photo courtesy of Julianne Goddard.)

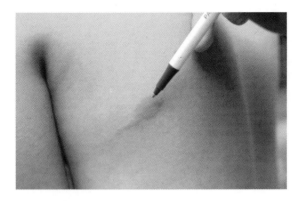

COLOR FIGURE 4.5 Lesion from tick bite on back.

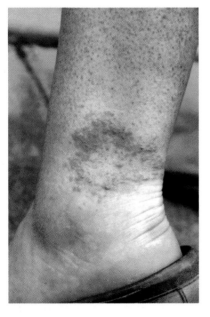

COLOR FIGURE 4.6 Possible hypersensitivity reaction to tick bite — 48 h postbite.

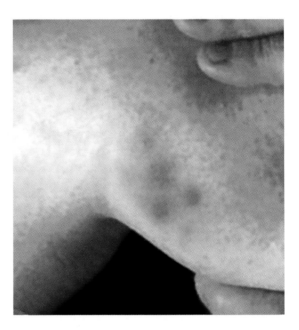

COLOR FIGURE 4.7 Chigger bites on leg.

COLOR FIGURE 6.8 Human bot fly larva removed from a patient. (Photo courtesy of Dr. Mark Mehrany, Department of Dermatology, Mayo Clinic, Rochester, MN.)

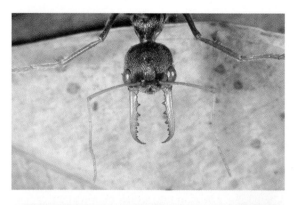

COLOR FIGURE 10.9 Head of bull ant. (Photo courtesy of Dr. Richard Russell and Stephen Doggett, Department of Medical Entomology, Westmead Hospital, NSW, Australia.)

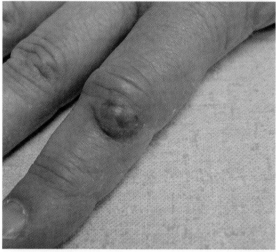

COLOR FIGURE 10.10 Fire ant sting on knuckle. (Photo copyright 2004 by Jerome Goddard.)

COLOR FIGURE 10.11 Hand in fire ant nest after approximately 10 sec.

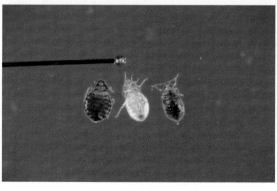

COLOR FIGURE 13.12 Bed bugs and cast skin from molt.

COLOR FIGURE 13.13 Immature reduviid bug which bit a patient causing intense pain. Note stout beak.

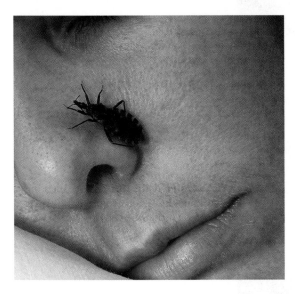

COLOR FIGURE 13.14 Kissing bug on face. (Photo courtesy of Wendy C. Varnado. Copyright 2005 by Jerome Goddard. With permission.)

COLOR FIGURE 16.15 Cockroach on human ear. (Photo courtesy of Joseph Goddard.)

COLOR FIGURE 16.16 Cockroach feces surrounding electrical outlet.

COLOR FIGURE 19.17 Adult biting midge, *Culicoides* spp. (USDA photo.)

COLOR FIGURE 21.18 Fly larvae (*Hermetia illucens*), supposedly causing intestinal myiasis, sent in to the Mississippi Department of Health.

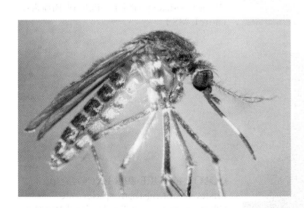

COLOR FIGURE 25.19 Adult *Culex annulirostris*, one of the most medically important mosquitoes in Australia. (Photo courtesy of Dr. Richard Russell and Stephen Doggett, Department of Medical Entomology, Westmead Hospital, NSW, Australia.)

COLOR FIGURE 28.20 Typical scorpion.

COLOR FIGURE 29.21 Garden spider. (Photo courtesy of Gretchen Waggy. Copyright 2005 by Jerome Goddard. With permission.)

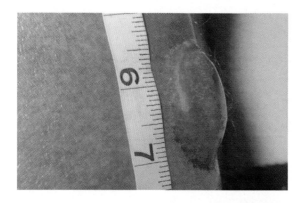

COLOR FIGURE 29.22 Huge blister resulting from spider bite. (Photo courtesy of Dr. Michael Brooks, Laurel, MS.)

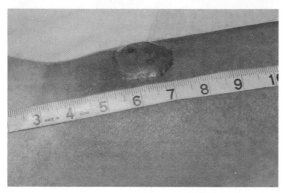

COLOR FIGURE 29.23 Close-up of blister resulting from spider bite. (Photo courtesy of Dr. Michael Brooks, Laurel, MS.)

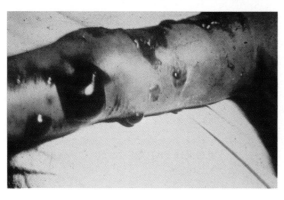

COLOR FIGURE 29.24 Extensive hemolytic condition of the arm due to a brown recluse spider. (Photo courtesy of U.S. Armed Forces Institute of Pathology, Negative Number 75-5876-1.)

COLOR FIGURE 29.25 Brown recluse spider. (Photo courtesy of Dr. Barry Engber, North Carolina Department of Environment and Natural Resources.)

COLOR FIGURE 29.26 Black widow spider. (Photo courtesy of Dr. James Jarratt, Mississippi Cooperative Extension Service, Mississippi State University.)

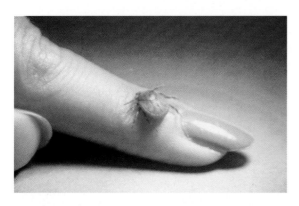

COLOR FIGURE 30.27 Hard tick on finger.

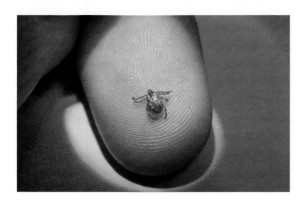

COLOR FIGURE 30.28 Adult American dog tick, *Dermacentor variabilis.*

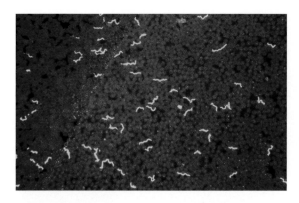

COLOR FIGURE 30.29 *Borrelia hermsi*, causative agent of tick-borne relapsing fever in several western U.S. states. (Photo courtesy of Dr. Tom Schwan, National Institutes of Health, Rocky Mountain Laboratories.)

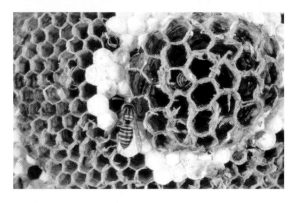

COLOR FIGURE 31.30 Yellowjacket worker on nest. (Photo courtesy of Dr. James Jarratt, Mississippi Cooperative Extension Service, Mississippi State University.)

COLOR FIGURE 31.31 Huge yellowjacket nest found in a ditch bank near Starkville, Mississippi. The nest was approximately 4 ft tall. (Photo courtesy of Dr. James Jarratt, Mississippi Cooperative Extension Service, Mississippi State University.)

COLOR FIGURE 31.32 European hornets building nest. (Photo courtesy of Dr. James Jarratt, Mississippi Cooperative Extension Service, Mississippi State University.)

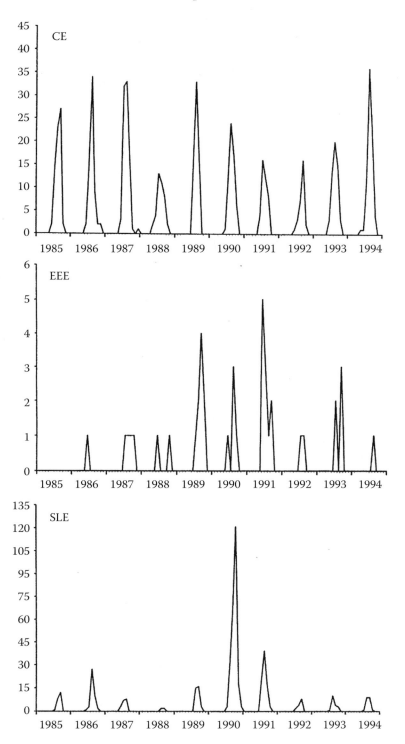

Figure 5
Case numbers of California serogroup (CE) encephalitis, eastern equine encephalitis (EEE), and St. Louis encephalitis (SLE), showing appearance of cases in mid-to-late summer. (From U.S. DHHS, Publ. No. CDC 93–8017.)

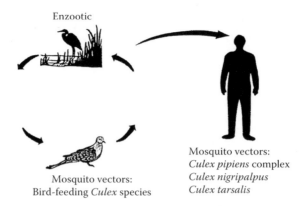

Figure 6
Life cycle of St. Louis encephalitis.

and now evidence of WNV has been found in virtually all the continental states and the District of Columbia. WNV will likely eventually spread all the way into Central and South America. As far as severity of the disease, WNV is no more dangerous than SLE (one of our "native" encephalitis viruses). Approximately 80% of all WNV infections are asymptomatic, approximately 20% cause West Nile fever, and less than 1% cause West Nile neuroinvasive disease.[22] As with SLE, WNV is more dangerous to older patients. Interestingly, of the first five patients in New York City admitted to hospitals, four had severe muscle weakness and respiratory difficulty, a finding atypical for encephalitis.[23] Also, GI complaints such as nausea, vomiting, or diarrhea occurred in four of five patients.[23] Much remains to be learned about the ecology of WNV in the U.S., but we do know the virus causes a bird disease, and is transmitted by mosquitoes. The house sparrow has been found to be one of the best amplifying hosts in nature, producing highest viremias for the longest period of time. Although the virus has been isolated from many mosquito species, the main vectors are believed to be *Culex pipiens, Cx. quinquefasciatus, Cx. salinarius, Cx. restuans,* and *Cx. tarsalis.*[24–26]

Western equine encephalitis. Western equine encephalitis (WEE), occurring in the western and central U.S., parts of Canada, and parts of South America, has occurred in several large outbreaks. There were large epidemics in the northcentral U.S. in 1941 and in the central valley of California in 1952. The 1941 outbreak involved 3000 cases. During 1964–1997, there were 639 human WEE cases reported to the CDC, for a national average of 19 cases per year.[27] WEE is generally less severe than EEE and SLE, with a mortality rate of only 2 to 5%. Cases appear in early to midsummer, and are primarily due to bites by infected *Culex tarsalis* mosquitoes.

LaCrosse encephalitis. LaCrosse encephalitis (LAC) has historically affected children in the midwestern states of Ohio, Indiana, Minnesota, and Wisconsin. However, it is increasingly being diagnosed in the southern states. The mortality rate of LAC is generally less than 1%, but seizures (even status epilepticus) and cerebral herniation may result from LAC infection. In fact, most LAC patients present with seizures. The national average for LAC is 73 per year.[27] Cases occur in July, August, and September. LAC is a bunyavirus that is transmitted to humans by *Ochlerotatus triseriatus* and *Ae. albopictus.* Interestingly, the virus may be transferred vertically from adult female *Oc. triseriatus* to her offspring through ovarial contamination. Some amplification of the virus takes place in nature through an *Oc. triseriatus,* small mammal cycle.

CASE HISTORY

SEIZURES ASSOCIATED WITH LACROSSE ENCEPHALITIS

In late June of 1999, a six-month-old, previously healthy infant was brought to an emergency department (ED) with a several-hour history of fever up to 101.6°F. He was experiencing a focal seizure characterized by uncontrollable blinking of the left eye, twitching of the left side of the mouth, and random tongue movement. In the ED, seizures continued intermittently in spite of the administration of diazepam and lorazepam. The infant was started on phenytoin and admitted to the hospital. Examination of CSF upon admission showed 294 white blood cells (47% polymorphonuclear leukocytes, 41% histiocytes, 12% lymphocytes), and 3 red blood cells. Protein and glucose were within normal limits. Admission CT scan was read as normal. The infant was started on acyclovir because of the possibility of herpes encephalitis, and cefotaxime and vancomycin were also initiated to cover possible bacterial infection. The patient then became seizure-free. He was taken off intravenous phenytoin and started on oral phenobarbital. Focal seizures, which progressed to generalized tonic clonic seizures, recurred on the fourth hospital day, at which time he had a therapeutic level of phenobarbital. After a repeat CT scan, which was read as within normal limits, the child was transferred to another hospital. On admission, the infant was noted to have continuous seizure-like movements of the chin and face. A repeat lumbar puncture revealed 307 white blood cells (33% polymorphonuclear leukocytes, 29% lymphocytes, and 38% histiocytes), 812 red blood cells, a protein of 104 mg/dl and a glucose of 74 mg/dl. He was continued on acyclovir, cefotaxime, and vancomycin. The patient was intubated secondary to excessive secretions and to avoid respiratory compromise. He was again treated with lorazepam and restarted on phenytoin. Seizure activity ended. The patient continued to be intermittently febrile. He was extubated on the sixth continuous hospital day. Herpes PCR results from admission samples were negative. In addition, admission blood, urine, and CSF cultures were negative, and on the ninth hospital day, cefotaxime and vancomycin were discontinued. Patient's maximum temperature on that day was 100.1°F, and he was becoming more alert and playful. The patient subsequently became and remained afebrile and without seizure activity, so he was transferred back to the original hospital on the twelfth hospital day for completion of 21 d of intravenous acyclovir. On day 21 he was discharged home on oral phenytoin. He had some residual left-sided weakness requiring several weeks of physical therapy. A sample of CSF was sent to a reference laboratory for testing for antibodies to various encephalitis agents including herpes and arboviruses. Results showed an indirect fluorescent antibody titer of 1:8 to LaCrosse (LAC) virus. Serum was sent to CDC for confirmation, which showed the presence of IgM antibody to LaCrosse virus.

Comment: Differentiation of LAC infection must be made from other encephalitides (postvaccinal or postinfection), tick-borne encephalitis (not common in the U.S.), rabies, nonparalytic polio, mumps meningoencephalitis, aseptic meningitis from enteroviruses, herpes encephalitis, various bacterial, mycoplasmal, protozoal, leptospiral, and mycotic meningitides or encephalitides, and others. Any cases of encephalitis in late summer should be suspect. Specific identification is usually made (with the help of the Centers for Disease Control in Ft. Collins) by finding specific IgM antibody in acute serum or CSF, or antibody rises (usually HI test) between early and late serum samples. Serological identification of the particular virus is complicated because of cross-reactivity with heterologous viruses of the same group. For example, IgM antibody from patients with LAC virus infection has the highest titers to LAC virus itself, but also reacts to a lesser extent with Snowshoe Hare virus and to a still lesser extent with Jamestown Canyon virus (two other viruses within the California serogroup).

Other California group encephalitis. Although LAC (described earlier) encephalitis is probably the most notorious, several other California group encephalitis viruses exist. North American forms include California encephalitis (CE), Jamestown Canyon (JC), Jerry Slough (JS), Keystone (KEY), San Angelo (SA), Trivittatus (TVT), and others. Viruses in the California serogroup are primarily pathogens of rodents and lagomorphs. They are transmitted to people by several species of mosquitoes, but especially the tree-hole, floodwater, and snow pool mosquitoes in the genera *Ochlerotatus* and *Aedes* spp. California group encephalitis viruses generally produce only mild illness in humans (mortality rates 1% or so).

Venezuelan equine encephalitis. Venezuelan equine encephalitis (VEE) is relatively mild in humans and rarely affects the central nervous system, but it will be included here as an encephalitid. VEE is endemic in Mexico and Central and South America; epidemics occasionally reach the southern U.S. Cases generally appear during the rainy season. Although the mortality rate is generally less than 1%, significant morbidity is produced by this virus. In an outbreak in Venezuela from 1962 to 1964, there were more than 23,000 reported human cases with 156 deaths.[28] In 1971, an outbreak of VEE in Mexico extended into Texas, resulting in 84 human cases.[29] There was a more recent outbreak in Colombia and Venezuela during the summer of 1995 with 75,000 to 100,000 human cases.[30]

Japanese encephalitis. Japanese encephalitis (JE) does not occur in the U.S., but it is the principal cause of epidemic viral encephalitis in the world, with approximately 50,000 clinical cases each year and 10,000 deaths.[31,32] JE epidemics have, at times, been widespread and severe. In 1924 there were 6125 cases with 3797 deaths.[33] Historically, JE has been focused in the northern areas of countries in southeast Asia, East Asia, and midsouthern Asia, especially China and Vietnam. Recently there has been a steady westward extension of reported epidemic activity into northern India, Nepal, and Sri Lanka, and a southern movement into Australia. JE is highly virulent. Approximately 25% of the cases are rapidly fatal, 50% lead to neuropsychiatric sequelae, and only 25% fully recover.[32] In temperate zones JE has a summer–fall distribution, but in the tropics no seasonal peak is apparent. There are several mosquito vectors of JE, but probably the most important is *Culex tritaeniorhynchus*, a rice-field-breeding species. Hogs may serve as amplifying reservoirs for JE.

G. *Other Arboviral Diseases*

Rift Valley fever. Rift Valley fever (RVF), occurring throughout sub-Saharan Africa and Egypt, is a *Phlebovirus* in the family Bunyaviridae. It causes abortions in sheep, cows, and goats, and heavy mortality of lambs and calves. In humans, it may result in fever, myalgias, encephalitis, hemorrhage, or retinitis. Permanent visual impairment has been reported.[28] Huge epidemics occasionally occur; in 1950 to 1951, there were an estimated 100,000 sheep and cattle cases and at least 20,000 human cases.[28] Outbreaks continue: in 1998 there were 27,500 human cases and 170 deaths from RVF in Kenya.[34] Although RVF may be acquired by handling infective material of animal origin during necropsy or butchering, it is also transmitted by mosquitoes, especially *Ochlerotatus/ Aedes* mosquitoes.

Ross River virus disease. Ross River (RR) disease, also called *epidemic polyarthritis*, occurs throughout most of Australia and occasionally New Guinea and some Pacific Islands such as Fiji, Tonga, and the Cook Islands. It causes fever, headache, fatigue, rash, and — most notably — arthritis in the wrist, knees, ankles, and small joints of the

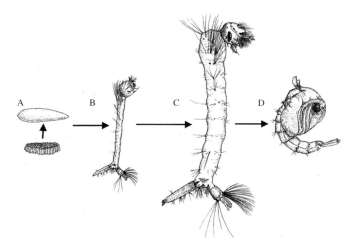

Figure 7
Immature stages of a mosquito.
(From Joe McGown, Mississippi Cooperative Extension Service, Mississippi State University.)

extremities. The disease is not fatal, but may be debilitating with symptoms occurring for weeks or months. RR disease is the most common arboviral disease in Australia with more than 5000 cases annually. During 1997, there were 6683 cases reported.[35] Peak incidence of RR occurs from January through March, when the mosquito vectors are most abundant. Serological studies and laboratory research have indicated that most likely kangaroos and wallabies are natural hosts for RR virus. There are several mosquito vectors of RR virus in Australia, but particularly *Culex annulirostris* inland and *Ochlerotatus vigilax* and *O. camptorhynchus* in northern and southern coastal areas, respectively.

II.   BASIC BIOLOGY AND ECOLOGY

Mosquitoes undergo complete metamorphosis, having egg, larval, pupal, and adult stages (Figure 7). Larvae are commonly referred to as *wigglers* and pupae as *tumblers*. Larvae and pupae of mosquitoes are always found in water. The water source may be anything from water in discarded automobile tires to water in the axils of plants, to pools, puddles, swamps, and lakes. Mosquito species differ in their breeding habits, biting behavior, flight range, etc. However, a generalized description of their life cycle is presented here and will serve as a useful basis for understanding mosquito biology and ecology. Most larvae in the subfamily Culicinae hang down just under the water surface by a breathing tube (siphon), whereas anopheline larvae lie horizontally just beneath the water surface supported by small notched organs of the thorax and clusters of float hairs along the abdomen. They have no prominent siphon. Mosquito larvae feed on suspended particles in the water as well as microorganisms. They undergo four molts (each successively larger), the last of which results in the pupal stage. With optimum food and temperature conditions, the time required for larval development can be as short as 7 d.

Mosquito pupae are quite active and will "tumble" toward the bottom of their water source upon disturbance. Pupae do not feed. They give rise to adult mosquitoes in 2

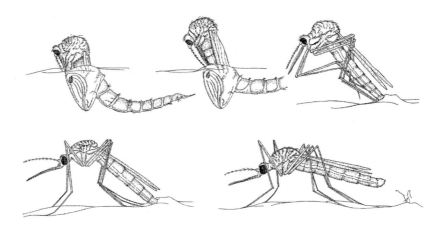

Figure 8
Adult mosquito emerging from pupal stage.
(From Bowles, E., Mississippi Department of Health Publication, The Mosquito Book.)

to 4 d. This process begins with the splitting of the pupal skin along the back. The emerging adult must dry its wings and separate and groom its head appendages before it can fly away (Figure 8). This is a critical stage in the survival of mosquitoes. If there is too much wind or wave action, the emerging adult will fall over while leaving its skin, becoming trapped on the water surface to soon die.

Adult mosquitoes of both sexes obtain nourishment for basic metabolism and flight by feeding on nectar. In addition, females of most species obtain a blood meal from birds or mammals for egg development. An excellent review of host-finding behavior is provided by Bowen.[36] Breeding sites selected for egg laying differ by species, but generally mosquitoes can be divided into three major breeding groups: permanent water breeders, floodwater breeders, and artificial container/tree hole breeders. *Anopheles* and many *Culex* mosquitoes select permanent water bodies, such as swamps, ponds, lakes, and ditches that do not usually dry up. Floodwater mosquitoes lay eggs on the ground in low areas subject to flooding. During heavy rains, water collecting in these low areas covers the eggs, which hatch from within minutes up to a few hours. Salt marsh mosquitoes (*Aedes sollicitans*), inland floodwater mosquitoes (*Ae. vexans*), and dark rice field mosquitoes (*Psorophora columbiae*) are included in this group. Artificial container/tree hole breeders are represented by yellow fever mosquitoes (*Ae. aegypti*), Asian tiger mosquitoes (*Aedes albopictus*), tree hole mosquitoes (*Oc. triseriatus*), and others. However, several species of *Anopheles* and *Culex* may also occasionally oviposit in these areas. Some of these container-breeding species lay eggs on the walls of a container just above the water line. The eggs are flooded when rains raise the water level. Other species oviposit directly on the water surface.

Female *Anopheles* mosquitoes generally lay eggs on the surface of the water at night. Each batch usually contains 100 to 150 eggs. The *Anopheles* egg is cigar shaped, about 1 mm long, and bears a pair of air-filled floats on the sides. Under favorable conditions, hatching occurs within 1 or 2 d.

Aedes and *Ochlerotatus* mosquitoes lay their eggs on moist ground around the edge of the water or, as previously mentioned, on the inside walls of artificial containers just above the waterline. *Aedes* and *Ochlerotatus* eggs will die if they become too dry when first laid. However, after the embryo in the egg develops, the eggs can withstand dry conditions for long periods of time. This trait has allowed *Aedes* and *Ochlerotatus* mosquitoes to use temporary water bodies for breeding, such as artificial containers, periodically flooded salt marshes or fields, tree holes, and storm water pools. Also, *Aedes* mosquitoes have inadvertently been carried to many parts of the world as dry eggs in tires, water cans, or other suitable containers. The Asian tiger mosquito (*An. albopictus*) was introduced into the U.S. in the 1980s in shipments of used truck tire casings imported from Taiwan and Japan. After these tires were stacked outside and began to collect rainwater, the eggs hatched. A more recent introduction into the U.S. has been *Oc. japonicus*.[37]

Psorophora mosquitoes also lay dry-resistant eggs. These mosquitoes are often a major problem species in rice fields. Eggs are laid on the soil and hatch once the field is irrigated.

Culex mosquitoes lay batches of eggs that are attached together to form little floating rafts. Upon close inspection of a suitable breeding site, these egg rafts can often be seen floating on the surface of the water.

In tropical areas, mosquito breeding may continue year round, but in temperate climates many species undergo a diapause in which the adults enter a dormant state similar to hibernation. In preparation for this, females become reluctant to feed, cease ovarian development, and develop fat body. They may seek a protected place to pass the approaching winter. Some species, instead of passing the winter as hibernating adults, produce dormant eggs that can survive the harsh effects of winter.

Mosquitoes vary in their biting patterns as well. Most species are diurnal in activity, biting mainly in the early evening. However, some species, especially *Ae. aegypti* and *Ae. albopictus*, bite in broad daylight (although there may be a peak of biting very early and late in the day). Others, such as the salt marsh species and many members of the genus *Psorophora* do not ordinarily bite during the day but will attack if disturbed (such as walking through high grass harboring resting adults).

III. GENERAL DESCRIPTION OF MOSQUITOES

Adult mosquitoes can be distinguished from other flies by several characteristics (Figure 9 is provided to familiarize the reader with these characteristics) (see also Color Figure 25.19). They have long, 15-segmented antennae, a long proboscis for bloodsucking, and scales on the wing fringes and wing veins. Scale patterns on the "back" (scutum) of mosquitoes can sometimes be used to identify the species (Figure 10). The mosquito head is rounded, bearing large compound eyes that almost meet. Males can usually be distinguished from females by their bushy antennae (Figure 11C, Figure 11D). Female *Anopheles* mosquitoes have palpi as long as the proboscis (Figure 11B); other groups do not have this characteristic. Larval mosquitoes, commonly called *wiggle tails* or *wigglers*, are white to dark gray in color and possess no legs or wings. They bear simple or branched tufted setae (hairs) along the body, anal gills, antennae, and chewing mouthparts. Culicine larvae have a prominent siphon tube for respiration at the water surface. Anopheline species do not, and thus must lie horizontally just beneath the water surface. Pupal mosquitoes, often called *tumblers*, also occur in water. They are comma-shaped, with two prominent respiratory trumpets on the thorax and a set of paddles on the last abdominal segment.

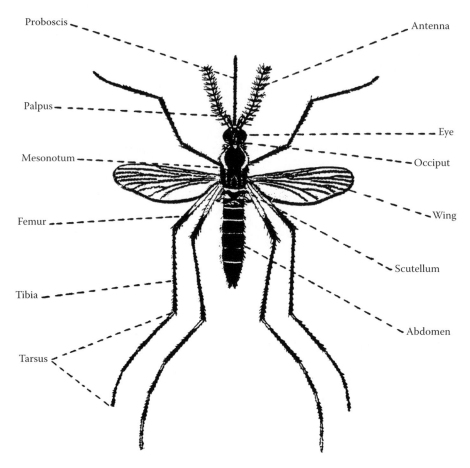

Figure 9
Adult mosquito with parts labeled. (From U.S. DHEW, PHS, CDC Pictorial Keys.)

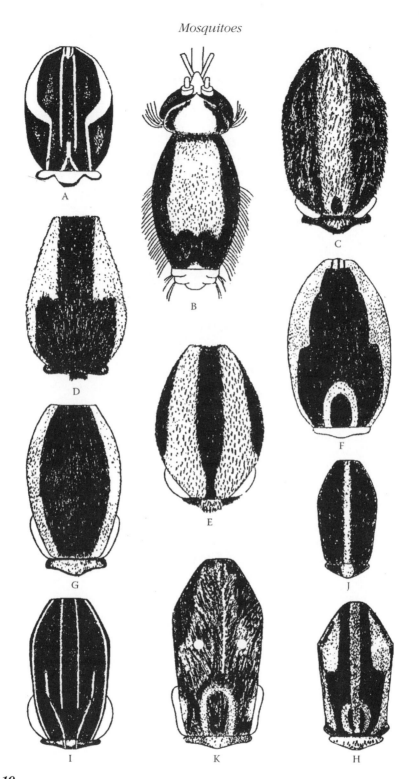

Figure 10

Examples of thoracic markings of mosquitoes (not necessarily drawn to scale): (A) Aedes aegypti, *(B)* Ochlerotatus infirmatus, *(C)* O. atlanticus, *(D)* O. thibaulti, *(E)* O. trivittatus, *(F)* O. triseriatus, *(G)* Psorophora varipes, *(H)* P. ciliata, *(I)* Orthopodomyia signifera, *(J)* Uranotaenia sapphirina, *(K)* Culex restuans. *(From USDA, ARS, Agriculture Handbook. No. 173, 1960.)*

275

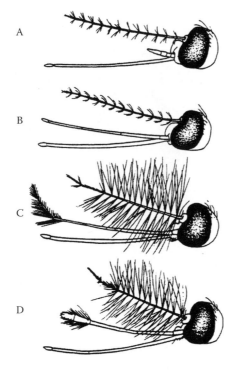

Figure 11
Mosquito head and appendages: (A) Aedes *female, (B)* Anopheles *female, (C)* Aedes *male, and (D)* Anopheles *male. (From USDA, ARS, Tech. Bull. No. 1447.)*

IV.   DISCUSSION OF SOME COMMON U.S. SPECIES*

Yellow Fever Mosquito
Aedes aegypti (Linnaeus)

Medical Importance: Known vector of yellow fever and dengue fever

Distribution: Cosmotropical in distribution, occurring worldwide within the 20°C isotherms (Figure 12)

---

* The descriptions for each mosquito species are only general comments about their macroscopic appearance. Mosquito identification to the species level is quite difficult, utilizing a number of microscopic characteristics. Specific identifications should be performed by specialists at institutions that routinely handle such requests (universities, extension services, state health departments, the military, etc.).

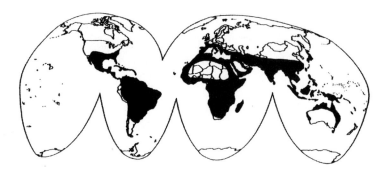

Figure 12
Approximate geographic distribution of Aedes aegypti.

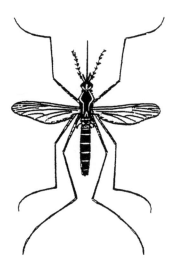

Figure 13
Adult Aedes aegypti.

Description: Typical *Aedes* mosquito with a pointed abdomen (at tip) and black and white rings on the legs. Small black species with a silver-white lyre-shaped figure on the upper sides of its thorax (Figure 13); also silver-white bands on the hind tarsi and abdomen

Remarks: Breeds in shaded artificial containers around buildings such as tires, cans, jars, flowerpots, and gutters; usually bites during the morning or late afternoon; readily enters houses and prefers human blood meals, biting principally around the ankles or back of the neck; flight range from 100 ft to 100 yd; in many places in the U.S. where *Ae. albopictus* has been introduced, this species has virtually disappeared, apparently being displaced

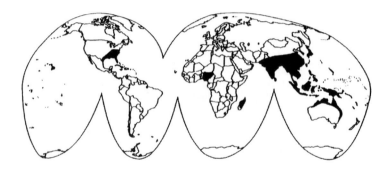

Figure 14
Approximate geographic distribution of Aedes albopictus.

Asian Tiger Mosquito
Aedes albopictus (Skuse)

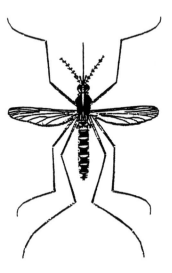

Medical Importance: Aggressive, daytime biting mosquito often found associated with piles of used automobile tires; vector of the agents of yellow fever and dengue fever; has been found naturally infected with LaCrosse encephalitis virus

Distribution: Widely distributed in the Asian region, the Hawaiian islands, and parts of the eastern U.S., where it was accidentally introduced in 1986 (Figure 14); recently reported from the African continent (Nigeria)

Description: Very similar in appearance to *Ae. aegypti* with a black body and silver-white markings (Figure 15); major difference between the two: *Ae. albopictus* has a single, silver-white stripe down the center of the dorsum of the thorax (instead of the lyre-shaped marking)

Figure 15
Adult Aedes albopictus.

Remarks: Breeds in artificial containers such as cans, gutters, jars, tires, flowerpots, etc., and seems especially fond of discarded tires; aggressive mosquito, often landing and biting immediately; flight range is generally less than a quarter mile

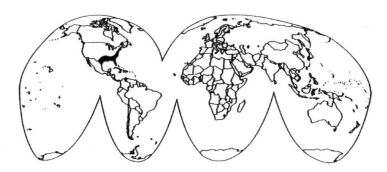

Figure 16
Approximate geographic distribution of Ochlerotatus sollicitans.

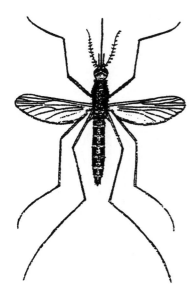

Figure 17
Adult Ochlerotatus sollicitans.

The Salt Marsh Mosquito
Ochlerotatus sollicitans (Walker)

(Note: Reinert[38] recently reclassified many Aedes *species as genus* Ochlerotatus*)*

Medical Importance: Fierce biter; can often discourage coastal development (in Louisiana, one of the most important pest species in coastal parishes[39]); known vector of the agent of EEE

Distribution: Nearctic region along the eastern coastal and inland saline areas (Figure 16)

Description: Bronze-brown species with golden yellow markings (Figure 17); whitish-yellow bands on upper side of abdomen with yellowish stripe down center; short palps with small white tips, banded legs, and dark proboscis with a single band in the middle

Remarks: Breeds primarily in salt marshes flooded by tides and rain; breeds throughout the year, although developmental time is longer in winter; aggressive biter that feeds at night (however, adults rest on vegetation during the day and will readily bite when disturbed); flight range is between 5 and 10 miles

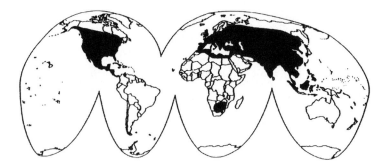

Figure 18
Approximate geographic distribution of Aedes vexans.

The Inland Floodwater Mosquito
Aedes vexans (Meigen)

Medical Importance: Bothersome pest and possible vector of EEE and WN virus

Distribution: Holarctic and Oriental regions, Pacific Islands, South Africa, Mexico, and parts of Central America (Figure 18)

Description: Typical Aedes mosquito with a pointed abdomen (at tip) and black and white rings on the legs. Medium-sized brown to golden brown mosquito with light gray or white markings shaped like wide "B's" on the dorsum of the middle segments of abdomen (Figure 19); hind tarsi with narrow white rings on all segments and short and dark palps with a few white scales at the tip

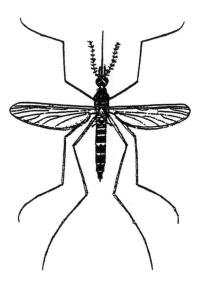

Figure 19
Adult Aedes vexans.

Remarks: Breeds in any temporary body of freshwater in both wooded and open areas; many broods produced each year from May through September when breeding areas are flooded by rains; vicious biter, active mainly at dusk and just after dark; flight range is 5 to 10 miles

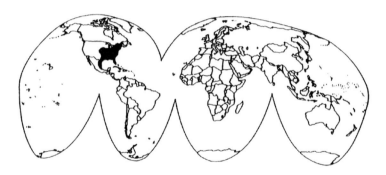

Figure 20
Approximate geographic distribution of Anopheles quadrimaculatus.

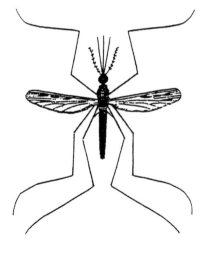

Figure 21
Adult Anopheles quadrimaculatus.

Malaria Mosquito
Anopheles quadrimaculatus Say

Medical Importance: One of many *Anopheles* mosquitoes that can transmit the agent of malaria to humans; principal vector in the eastern half of the U.S. until 1951

Distribution: Central and eastern U.S., southern Canada, and Mexico (Figure 20)

Description: Actually, five sibling species (named *An. quadrimaculatus, An. smaragdinus, An. diluvialis, An. maverlius,* and *An. inundatus*) that all look alike but differ in behavior and ecology; medium to large dark brown mosquitoes with palps as long as the proboscis, dark unbanded legs, and wings marked by four dark spots (Figure 21); long maxillary palps in both sexes (typical characteristics of all *Anopheles* mosquitoes); feed with their proboscis, head, and body in an almost straight line (many species feed almost perpendicular to skin surface of host); it looks as though they are standing on their heads when they bite (Figure 22) (also characteristic of all *Anopheles* mosquitoes)

Remarks: Breed in permanent freshwater sites such as ponds, pools, swamps, and rice fields containing emergent or floating vegetation (as most *Anopheles* mosquitoes); breeding may be continuous throughout year, especially if winters are mild; however, peak populations usually during July or August (decline rapidly during September and October); adults rest during the day in cool, dark, damp areas and come out to feed at night; adults usually fly no more than 4 miles from their breeding area

281

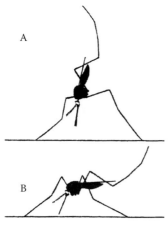

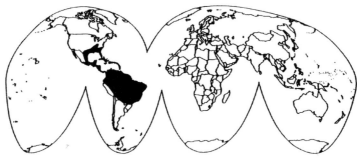

Figure 22
Feeding positions of mosquitoes: (A) Anopheles, *(B)* Culex.
(From USDA, ARS, Agriculture Handbook. No. 173).

Figure 23
Approximate geographic distribution of Culex nigripalpus.

Culex nigripalpus Theobald

Medical Importance: Major vector of SLE and WN viruses in the southeastern U.S. (in 1990, central and southern counties of Florida experienced outbreak of 212 SLE cases with 10 deaths[40])

Distribution: Southern U.S., Antilles, Mexico, Central America, Trinidad, Ecuador, Colombia, Venezuela, the Guianas, Brazil, and Paraguay (Figure 23)

Description: Typical *Culex* mosquito with a blunt abdomen (at tip) and hardly any striking features; adults medium-sized dark mosquitoes with rounded abdomen, unbanded dorsally (Figure 24); white patches may be present laterally; difficult to distinguish from other small *Culex* mosquitoes.

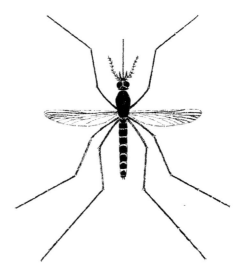

Figure 24
Adult Culex nigripalpus.

Remarks: Active in spring, summer, and autumn, but generally most abundant in autumn; breeds in shallow rainwater pools and semipermanent ponds

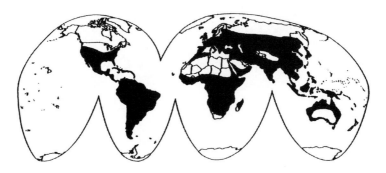

Figure 25
Approximate geographic distribution of Culex quinquefasciatus *(including* C. pipiens*).*

Southern House Mosquito
Culex quinquefasciatus Say

Medical Importance: One of the major vectors of the SLE virus (numerous outbreaks of SLE associated with this species); primary vector of WNV in southern states; highly anthropophilic species in many parts of the tropical and subtropical world, where it is also a major vector of lymphatic filariasis

Distribution: *Culex pipiens* complex, which includes *Cx. quinquefasciatus*, cosmotropical in distribution (Figure 25); in North America this group is represented by three main members: the northern house mosquito, *Cx. pipiens*, occurs in areas above 39° N latitude; the southern house mosquito, *Cx. quinquefasciatus*, occurs at latitudes less than 36° N latitude; between 36° and 39° N latitudes, *Cx. pipiens* and *Cx. quinquefasciatus*, as well as hybrids between the two, are encountered

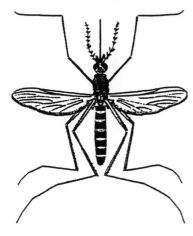

Description: Typical *Culex* mosquito with a blunt abdomen (at tip) and hardly any striking features; medium-sized brown mosquito with a few white bands on upper side of the abdominal segments (Figure 26); bands widest in middle of dorsum and narrow laterally; legs and proboscis dark and unbanded; difficult to distinguish from other *Culex* mosquitoes

Remarks: Breeds in ditches, storm sewer catch basins, cesspools, and polluted water (also breeds in artificial containers around homes such as cans, jars, and tires); readily enters homes, producing the familiar "singing" sound around persons at night; flight range variable, but may be as far as 1100 meters in a single night

Figure 26
Adult Culex quinquefasciatus.

283

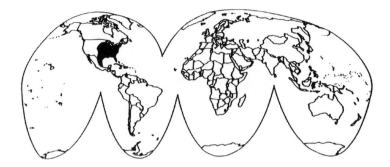

Figure 27
Approximate geographic distribution of Culex salinarius.

Culex salinarius Coquillett

Medical Importance: Often occurs in extremely high numbers and thus a nuisance pest; possibly a major bridge vector of WNV; secondary vector of SLE and EEE viruses

Distribution: U.S. and Canada east of Rocky Mountains, as well as Bermuda and parts of Mexico (Figure 27); especially common along the Atlantic and Gulf coasts

Description: Typical *Culex* mosquito with a blunt abdomen (at tip) and hardly any striking features; brown species has both proboscis and hind tarsi entirely dark (no white bands) (Figure 28); dorsal abdominal segments have drab and often inconspicuous basal bands of yellowish or brownish scales; bands usually irregular and narrow; last two abdominal segments often coppery colored

Remarks: Breeds in ditches, pools, artificial containers, and often salty water, such as that in salt marshes; found breeding year round in Gulf coastal states; feeds readily on people, mostly outdoors, but occasionally inside buildings; feeding heaviest at dusk and first hours of darkness; flight range 1 mile or more

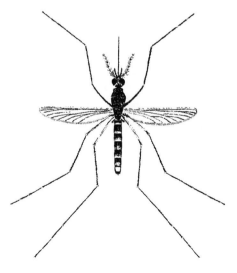

Figure 28
Adult Culex salinarius.

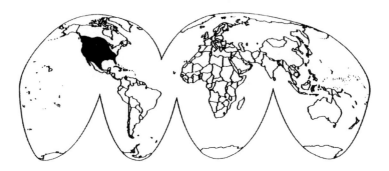

Figure 29
Approximate geographic distribution of Culex tarsalis.

Encephalitis Mosquito
Culex tarsalis Coquillett

Medical Importance: Abundant and widespread species; main vector of WEE, WN, and SLE viruses in central, western, and southwestern U.S.; probably the most ubiquitous mosquitoes in California

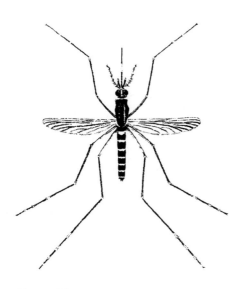

Figure 30
Adult Culex tarsalis.

Distribution: Most of U.S., parts of southwestern Canada, and Mexico (Figure 29)

Description: Medium-sized mosquito, dark brown to black in color, having white bands on the legs and abdomen (Figure 30); broad white band in middle of proboscis; ventor of abdomen is white with a black chevron on each segment

Remarks: Primarily a rural species that breeds in both polluted and clear water in ground pools, grassy ditches, and artificial containers; in arid and semiarid regions, frequently found in canals and ditches associated with irrigation; domestic and wild birds are the preferred food source, but people are also readily attacked; most active soon after dusk; active primarily in summer, but may remain active all winter in southern parts of range

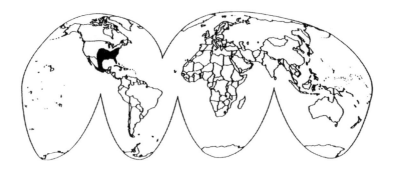

Figure 31
Approximate geographic distribution of Psorophora columbiae.

Dark Rice Field Mosquito
Psorophora columbiae (Dyar and Knab)

Medical Importance: Although occasionally involved as secondary vector of several encephalitides, primary adverse effect on people is nuisance biting; vicious biter and, thus, severe nuisance pest

Distribution: Much of the southeastern U.S. as far north as New York state and west to southern South Dakota, south to Texas and northern Mexico (Figure 31); records from Arizona and New Mexico may be closely related species in the *Ps. confinnis* complex

Description: Large mosquito, dark brown to black (Figure 32); white bands on the legs and yellow band on otherwise dark proboscis; first segment of hind tarsus is brown with a white ring in the middle; light and dark scales give much of the body and wings a "salt-and-pepper" effect

Remarks: Breeds in open, temporary pools of freshwater such as ditches, rice fields, and low, flooded areas; eggs deposited on moist soil subject to flooding either by rain or irrigation; several broods produced in active season (April to October); bites aggressively during day or night and can fly 6 to 8 miles

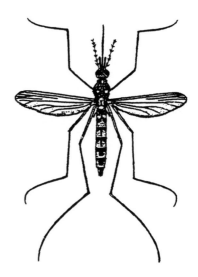

Figure 32
Adult Psorophora columbiae.

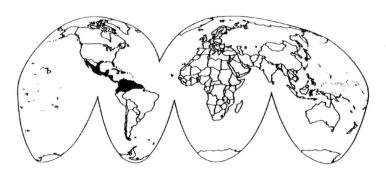

Figure 33
Approximate geographic distribution of Anopheles albimanus.

V. DISCUSSION OF SOME MAJOR PEST SPECIES IN OTHER AREAS OF THE WORLD

Anopheles albimanus Weidemann

Medical Importance: One of the most important vectors of malaria in Mexico, Central and South America, and Caribbean

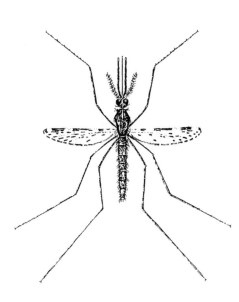

Figure 34
Adult Anopheles albimanus.

Distribution: Extreme southern U.S. (Texas and Florida), Mexico, Central America, South America (probably only Colombia, Venezuela, and Ecuador), Antilles, and Caribbean (Figure 33)

Description: All *Anopheles* females have long maxillary palps and wings often spotted; *An. albimanus* is a spotted species (Figure 34) with pale hind legs similar to many other related *Anopheles* species; black proboscis and white terminal segment of palpi

Remarks: Prefer hot, humid climates; most abundant during the wet season, where they breed in sunny pools, pits, puddles, ponds, marshes, lagoons, and artificial containers, especially those water sources containing floating or grassy vegetation; females feed on people and domestic animals, both indoors and outdoors; usually rest outdoors after feeding

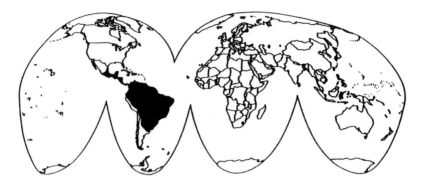

Figure 35
Approximate geographic distribution of Anopheles darlingi.

Anopheles darlingi Root

Medical Importance: A major contributor to endemic malaria in extreme southern Mexico and Central and South America; adults feed readily on people and often collect in large numbers inside houses

Distribution: Mexico through Central America southward into Argentina and Chile (Figure 35)

Description: Last three segments of the hind tarsi entirely white (Figure 36); long, slender, black proboscis; white terminal segment of palpi; light spot at the cross vein on the front wing margin smaller than the preceding dark spot

Remarks: Breeds in shaded areas of freshwater marshes, swamps, lagoons, rice fields, lakes and ponds, and the edges of streams, especially those with vegetation; feeds on humans indoors and rests indoors after feeding

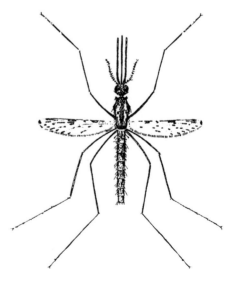

Figure 36
Adult Anopheles darlingi.

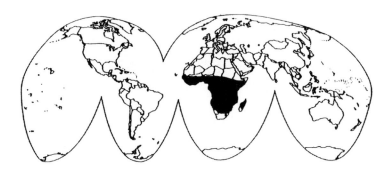

Figure 37
Approximate geographic distribution of the Anopheles gambiae *complex.*

Anopheles gambiae Giles (consists of several species in a complex)

Medical Importance: Most efficient malaria transmitter in Africa (including all members of the complex); highly anthropophilic (blood meal analyses have indicated that more than 50% of fed females contain human blood)

Distribution: At least six members of the complex and their distributions in Africa are presented below (Figure 37):

1. *An. arabiensis:* Tropical Africa, southwest Arabia, the Cape Verde Islands, Zanzibar, Pemba, Madagascar, and Mauritius
2. *An. gambiae:* Tropical Africa, Bioko, Zanzibar, Pemba, and Madagascar
3. *An. melas:* Coastal West Africa
4. *An. merus:* Brackish water habitats in East and South Africa
5. *An. quadriannulatus:* Ethiopia, South Africa, Swaziland, Zambia, Zanzibar, and Zimbabwe
6. *An. bwambae:* Uganda

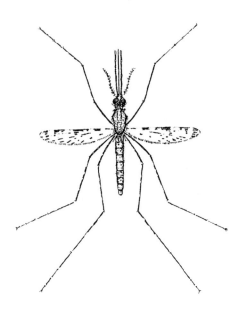

Figure 38
Adult Anopheles gambiae.

Description: Members of complex generally look like the one depicted in Figure 38; precise identification of members of species complex done by genetic methods, comparing enzymes, chromosome characteristics, and PCR

Remarks: Each member of complex may exhibit different behaviors, but larvae of *An. gambiae* and *An. arabiensis* (formerly species A and B of the complex) occur in all types of water-containing depressions close to human habitations including pools, puddles, hoofprints, borrow pits, and even rice fields; especially abundant during rainy season; females have average flight range of about 1 mile

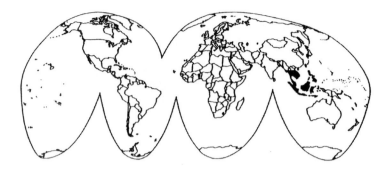

Figure 39
Approximate geographic distribution of the Anopheles leucosphyrus *group.*

The *Anopheles leucosphyrus* Group

Medical Importance: Group of *Anopheles* mosquitoes (containing about 20 closely related species); contains several main vectors of malaria in Southeast Asia

Distribution: Distributions and descriptions are based on two recent papers.[41,42] Distributions of the main malaria vectors in the group are given below (Figure 39):

1. *An. balabacensis*: Philippines (Balabac, Culion, and Palawan Islands), Malaysia (Sabah and Sarawak), Indonesia (Kalimantan, Java, Lombok), and Brunei
2. *An. dirus*: Thailand, Kampuchea, Vietnam, China (Hainan Island)
3. *An. leucosphyrus*: Sumatra, Indonesia
4. *An. lateens:* Malaysia (Sabah and Sarawak, and states in Peninsular Malaysia), Indonesia (Kalimantan), Thailand (southern provinces)
5. *An. baimaii:* Thailand (western and southern provinces), Myanmar, India (eastern states), Bangladesh

Description: Members of group have numerous spots within the wings and on the legs (Figure 40); conspicuously white joints where the tibia and tarsus meet on each hind leg

Remarks: Breeds in shaded freshwater pools in and among rocks, in hoofprints, vehicle ruts, and the like; many are forest species that bite humans and animals outdoors; fed females generally rest outdoors (in the case of *A. balabacensis*, the females rest on forest vegetation during the day, congregate around human dwellings at sundown, remain quiescent on vegetation for some time after dark, then enter the dwellings to feed on sleeping persons)

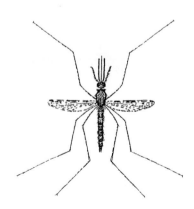

Figure 40
Anopheles balabacensis, *a member of the* A. leucosphyrus *group.*

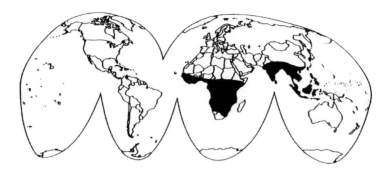

Figure 41
Approximate geographic distribution of the Anopheles minimus *complex.*

The *Anopheles funestus* Group

Medical Importance: Important vectors of malaria in tropical Africa, the Indian subcontinent, and Southeast Asia[43,44]

Distribution: The distributions of three members of the *A. minimus* complex are given below (Figure 41):

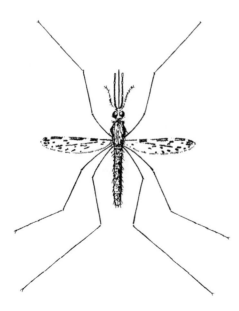

Figure 42
Adult Anopheles minimus.

1. *An. flavirostris*: The Philippines
2. *An. funestus*: Most of tropical Africa
3. *An. minimus*: Burma, Bangladesh, India, Cambodia, Laos, Malaysia, Thailand, Vietnam, and Taiwan

Description: Small, dark specimens having narrow white bands on the female palps (Figure 42). *An. minimus* has a totally dark proboscis

Remarks: *An. flavirostris* bites indoors but rests outdoors afterward; both *An. flavirostris* and *An. minimus* breed in grassy edges of small foothill streams, springs, irrigation ditches, or seepages; they both prefer shaded areas; *An. funestus* is an extremely important vector of malaria in tropical Africa (second only to *An. gambiae*); larvae of *An. funestus* prefer marshes, swamps, and edges of streams; they feed indoors and outdoors, but rest outdoors after feeding

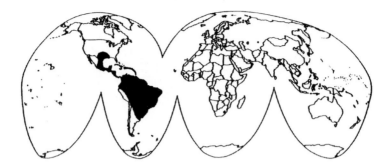

Figure 43
Approximate distribution of Anopheles pseudopunctipennis.

Anopheles pseudopunctipennis Theobald

Medical Importance: Important vector of malaria in Mexico, Central America, and parts of South America; primary vector of a large malaria outbreak in Oaxaca, Mexico during 1998

Distribution: Southern U.S. southward to Argentina, and Antilles (Figure 43)

Description: Females are large gray mosquitoes with black legs (Figure 44); two narrow white bands and a white tip on the palps; totally dark proboscis; entirely black hind tarsus; two light spots on the front wing margin

Remarks: Highland species occurring most abundantly during the dry season; heavy rains during the wet season flush the larvae out of their breeding sites, which include shallow, quiet, or slow-moving water such as drying streambeds; larvae often associated with green algae, like *Spirogyra,* in stream pools; adults enter houses and bite people readily

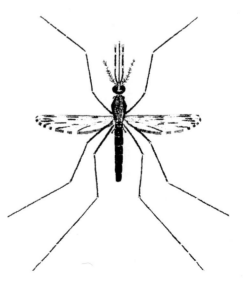

Figure 44
Adult Anopheles pseudopunctipennis.

292

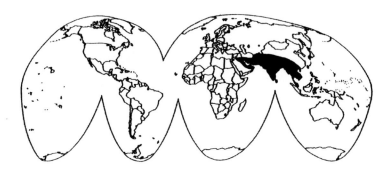

Figure 45
Approximate geographic distribution of Anopheles stephensi.

Anopheles stephensi Liston

Medical Importance: Widely distributed urban vector of malaria

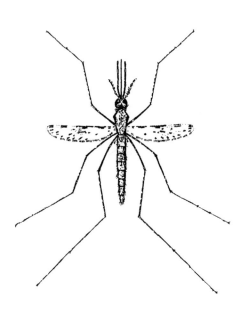

Figure 46
Adult Anopheles stephensi.

Distribution: Afghanistan, Bangladesh, Burma, China, India, Indochina, Iraq, Nepal, Oman, Pakistan, Saudi Arabia, and United Arab Emirates (Figure 45); one of the most abundant species in the Persian Gulf region

Description: Females have numerous white spots on both the legs and wings (Figure 46); hind tarsal segments have narrow white band apically (last segment is dark); females have white-tipped palps and one dark spot on the front margin of the wing proximal to the outermost spot; abdomen is almost totally covered with scales

Remarks: Breeds in artificial habitats such as wells, cisterns, gutters, pools, and many other peridomestic water sources; accordingly, malaria transmitted usually localized and limited to urban centers; adults bite people both indoors and outdoors and rest mainly indoors after feeding; flight range usually does not exceed ½ mile

293

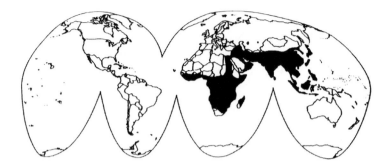

Figure 47
Approximate geographic distribution of Culex tritaeniorhynchus.

Culex tritaeniorhynchus Giles

Medical Importance: Vector of JE virus in Far East; also most important nuisance mosquito during July and August in Japan

Distribution: Saudi Arabia, Borneo, Celebes, China, India, Iran, Iraq, Japan, Okinawa, Israel, Lebanon, Syria, Turkey, Java, Sumatra, Philippines, Sri Lanka, Madagascar, Egypt, Gold Coast, Nigeria, Cameroon, Sudan, Zaire, Kenya, Tanzania, Angola, Namibia, Uganda, Zanzibar, Mozambique, and Ivory Coast (Figure 47); especially abundant in Korea, Japan, and eastward through central China

Description: Very small mosquito (average size is 3 mm, the Japanese forms may be larger), dark reddish-brown in color with a banded proboscis (Figure 48)

Remarks: Breeds in freshwater collections such as natural and artificial water impoundments, ground pools, and drainage and irrigation ditches; readily bites people; also feeds on horses, cows, pigs, and other animals

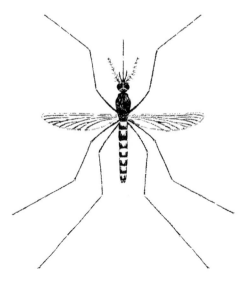

Figure 48
Adult Culex tritaeniorhynchus.

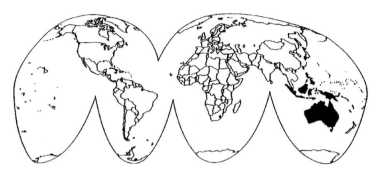

Figure 49
Approximate geographic distribution of Culex annulirostris.

Culex annulitostris Skuse

Medical Importance: Most medically important mosquito in Australia; main vector of the flaviviruses Murray Valley encephalitis and Kunjin (a subtype of West Nile virus), and the main inland vector of the alphaviruses Ross River and Barmah Forest

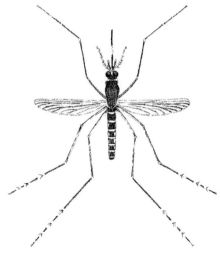

Distribution: Southern and western Australasian region (including the entire Australian mainland), Indonesia, Philippines (Figure 49)

Description: Medium-sized, brown mosquito; similar to U.S. species, *Culex tarsalis,* tarsi banded, scutum dark, proboscis with complete white ring in middle (Figure 50); may be difficult to distinguish from other, closely related, Australian species

Remarks: Active mid-spring to late fall; breeds in freshwater areas such as natural and artificial water impoundments, ground pools, and drainage and irrigation ditches; does not breed in artificial containers; readily bites people; also feeds on other mammals and birds

Figure 50
Adult Culex annulirostris.

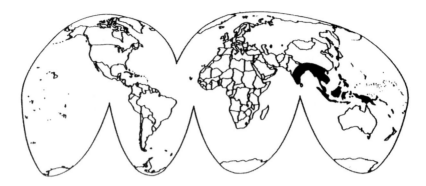

Figure 51
Approximate geographic distribution of Mansonia annulifera.

Mansonia annulifera (Theobald)

Medical Importance: An important vector of *Wuchereria malayi* (lymphatic filariasis) in Asia, Philippines, and New Guinea; significant nuisance, being strongly anthropophilic

Distribution: Borneo, Myanmar, Celebes, India, Indochina, Java, New Guinea, Philippines, Sumatra, and Thailand (Figure 51)

Description: Similar to other *Mansonia* mosquitoes (except lighter) with legs, palps, wings, and body covered with a mixture of brown and pale (white or creamy) scales, giving them a dusty appearance; pale species ranging in color from yellow to light brown; mottled wings with very broad and asymmetrical scales and banded legs (Figure 52); very prominent silvery-white scales on midlobe of scutellum

Remarks: Most commonly found in ponds, pools, backwaters, swamps, and marshes that contain the aquatic plants *Pistia* spp. and *Eichhornia* spp; larvae obtain oxygen by puncturing the underwater stems of aquatic plants (similar to other members of genus)

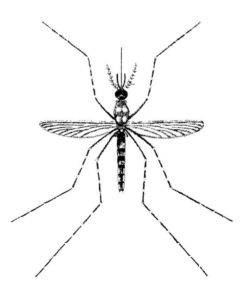

Figure 52
Adult Mansonia annulifera.

296

REFERENCES

1. Sturcher, D., How much malaria is there worldwide?, *Parasitol. Today*, 5, 39, 1989.

2. Morris, C.D., Baker, R.H., and Nayer, J.K., Human Malaria, Florida Medical Entomology Laboratory (IFAS), Fact Sheet, July 1990.

3. CDC, Multifocal autochthonous transmission of malaria — Florida, 2003, *MMWR*, 53, 412–413, 2004.

4. Robert, L.L., Santos-Ciminera, P.D., Andre, R.G., Schultz, G.W., Lawyer, P.G., Nigro, J., Masuoka, P., Wirtz, R.A., Neely, J., Gaines, D., Cannon, C.E., Pettit, D., Garvey, C.W., Goodfriend, D., and Roberts, D.R., Plasmodium-infected Anopheles mosquitoes collected in Virginia and Maryland following local transmission of Plasmodium vivax malaria in Loudoun County, Virginia, *J. Am. Mosq. Control Assoc.*, 21, 187, 2005.

5. Strickman, D., Gaffigan, T.G., Wirtz, R.A., Benedict, M.Q., Rafferty, C.S., Barwick, R.S., and Williams, H.A., Mosquito collections following local transmission of *Plasmodium falciparum* malaria in Westmoreland County, Virginia, *J. Am. Mosq. Control Assoc.*, 16, 219, 2000.

6. WHO, Yellow fever, in WHO/EPI/GEN/98.11, Geneva, 1998, p. 6.

7. WHO, Yellow fever technical consensus meeting, in WHO/EPI/GEN/98.08, Geneva, 1998, p. 3.

8. CDC, Dengue surveillance summary, in San Juan Laboratories Rep (CDC), 62, 1–3, 1991.

9. Calisher, C.H., Persistent emergence of dengue, *Emerging Infect. Dis.*, 11, 738, 2005.

10. Effler, P.V., Pang, L., Kitsutani, P., Vorndam, V., Nakata, M., Ayers, T., Elm, J., Tom, T., Reiter, P., Rigau-Perez, J.G., Hayes, J.M., Mills, K., Napier, M., Clark, G.G., and Gubler, D.J., Dengue fever, Hawaii, 2001–2002, *Emerging Infect. Dis.*, 11, 742, 2005.

11. Gubler, D.J., Epidemic dengue and dengue hemorrhagic fever: a global public health problem in the 21st century, in *Emerging Infections*, Vol. 1, Scheld, W.M., Armstrong, D., and Hughes, J.M., Eds., ASM Press, Washington, D.C., 1998, p. 1.

12. WHO, Expert Committee on Filariasis, in Third report, WHO Tech. Rep. Ser. No. 542, Geneva, 1974, p. 1.

13. Laurence, B.R., The global distribution of bancroftian filariasis, *Parasitol. Today*, 5, 260, 1989.

14. Beaver, P.C. and Orihel, T.C., Human infection with filariae of animals in the United States, *Am. J. Trop. Med. Hyg.*, 14, 1010, 1965.

15. Thomas, J.G., Sundman, D., Greene, J.N., Coppola, D., Lu, L., Robinson, L.A., and Sandlin, R.L., A lung nodule: malignancy or the dog heartworm?, *Infect. Med.*, 15, 105, 1998.

16. Morris, C.D., Eastern equine encephalomyelitis, in *The Arboviruses: Epidemiology and Ecology*, Vol. 3, Monath, T.P., Ed., CRC Press, Boca Raton, FL, 1988, p. 1.

17. Cupp, E.W., Klingler, K., Hassan, H.K., Viguers, L.M., and Unnasch, T.R., Transmission of eastern equine encephalomyelitis virus in central Alabama, *Am. J. Trop. Med. Hyg.*, 68, 495, 2003.

18. Cupp, E.W., Tennessen, K.J., Oldland, W.K., Hassan, H.K., Hill, G.E., Katholi, C.R., and Unnasch, T.R., Mosquito and arbovirus activity during 1997–2002 in a wetland in northeastern Mississippi, *J. Med. Entomol.*, 41, 495, 2004.

19. Wozniak, A., Dowda, H.E., Tolson, M.W., Karabatsos, N., Vaughan, D.R., Turner, P.E., Ortiz, D.I., and Wills, W., Arbovirus surveillance in South Carolina, 1996–1998, *J. Am. Mosq. Control Assoc.*, 17, 73, 2001.

20. Chamberlain, R.W., History of St. Louis encephalitis, in *St. Louis Encephalitis*, Monath, T.P., Ed., American Public Health Association, Washington, D.C., 1980, p. 3.

21. CDC, Outbreak of West Nile-like viral encephalitis in New York, *MMWR*, 48, 845–848, 1999.

22. Mostashari, F., Bunning, M.L., and Kitsutani, P., Epidemic West Nile encephalitis, New York, 1999: results of a household-based seroepidemiological survey, *Lancet*, 358, 261, 2001.

23. Asnis, D.W., Conetta, R., and Teixeira, A., The West Nile virus outbreak of 1999 in New York: the Flushing Hospital experience, *Clin. Infect. Dis.*, 30, 2000.

24. Molaei, G., Andreadis, T.G., Armstrong, P.M., Anderson, J.F., and Vossbrinck, C.R., Host feeding patterns of *Culex* mosquitoes and West Nile virus transmission, northeastern United States, *Emerging Infect. Dis.*, 12, 468, 2006.

25. Kilpatrick, A.M., Kramer, L.D., Campbell, S.R., Alleyne, E.O., Dobson, A.P., and Daszak, P., West Nile virus risk assessment and the bridge vector paradigm, *Emerging Infect. Dis.*, 11, 425, 2005.

26. Hayes, E.B., Komar, N., Nasci, R.S., Montgomery, S.P., O'Leary, D.R., and Campbell, G.L., Epidemiology and transmission dynamics of West Nile virus disease, *Emerging Infect. Dis.*, 11, 1167, 2005.

27. CDC, Western equine encephalitis case numbers, Arbovirus Diseases Branch, Division of Vector-borne Infectious Diseases, Ft. Collins, Co., 1998.

28. Harwood, R.F. and James, M.T., *Entomology in Human and Animal Health*, Macmillan, New York, 1979.

29. Anonymous, Venezuelan equine encephalitis, a national emergency, in USDA, APHIS-81-1, U.S. Department of Agriculture, Washington, D.C., 1972.

30. Weaver, S.C., Salas, R., Rico-Hesse, R., Ludwig, G.V., Oberste, M.S., Boshell, J., and Tesh, R.B., Re-emergence of epidemic Venezuelan equine encephalomyelitis in South America, *Lancet*, 348, 436, 1996.

31. Anonymous, Facts about Japanese encephalitis, http://www.jepn.org/jepn/RegionResources.aspx, 2005.

32. Burke, D.S. and Leake, C.J., Japanese encephalitis, in *The Arboviruses: Epidemiology and Ecology*, Vol. 3, Monath, T.P., Ed., CRC Press, Boca Raton, FL, 1988, p. 63.

33. Rappleye, W.C., Epidemic Encephalitis, Etiology, Epidemiology, Treatment, in Third Report of the Matheson Commission, Columbia University Press, New York, 1939.

34. Woods, C.W., Karpati, A.M., Grein, T., McCarthy, N., Gaturuku, P., Muchiri, E., Dunster, L., Henerson, A., Khan, A.S., Swanepol, R., Bonmarin, I., Martin, L., Mann, P., Smoak, B.L., Ryan, M., Ksizek, T.G., Arthur, R.R., Ndikuyeze, A., Agata, N.N., and Peters, C.J., An outbreak of Rift Valley fever in northwestern Kenya, *Emerging Infect. Dis.*, 8, 138, 2002.

35. Anonymous, Ross River Virus, in New South Wales Dep. Health, Fact Sheet, Sydney, Australia, 2002, p.1.

36. Bowen, G.S. and Francy, D.B., Surveillance, in *St. Louis Encephalitis*, Monath, T.P., Ed., American Public Health Association, Washington, D.C., 1980, p. 473.

37. Peyton, E.L., Campbell, S.R., Candeletti, T.M., Romanowski, M., and Crans, W.J., *Aedes (Finlaya) japonicus japonicus* (Theobald), a new introduction into the United States, *J. Am. Mosq. Control Assoc.*, 15, 238, 1999.

38. Reinert, J.F., New classification for the composite genus *Aedes*, elevation of subgenus *Ochlerotatus* to generic rank, reclassification of the other sungenera, and notes on certain subgenera and species, *J. Am. Mosq. Control Assoc.*, 16, 175, 2000.

39. Chapman, H.C. and Johnson, E.B., The Mosquitoes of Louisiana, in Louisiana Mosquito Control Assoc. Tech. Bull. No. 1, 1986.

40. Anonymous, St. Louis encephalitis outbreak in central Florida, in Florida Epigram, 1, 1, 1991.

41. Sallum, M.A.M., Peyton, E.L., Harrison, B.A., and Wilkerson, R.C., Revision of the *Leucosphyrus* Group of *Anopheles* (*Cellia*), *Rev. Bras. Entomol.*, 49(Suppl. 1), 1, 2005.

42. Sallum, M.A.M., Peyton, E.L., and Wilkerson, R.C., Six new species of the *Anopheles leucosphyrus* group, reinterpretation of *An. elegans*, and vector implications, *Med. Vet. Entomol.*, 19, 158, 2005.

43. Garros, C., Harbach, R.E., and Manguin, S., Systematics and biogeographical implications of the phylogenetic relationships between members of the funestus and minimus groups of Anopheles (Diptera: Culicidae), *J. Med. Entomol.*, 42, 7, 2005.

44. Garros, C., Harbach, R.E., and Manguin, S., Morphological assessment and molecular phylogenetics of the Funestus and Minimus groups of Anopheles (Cellia), *J. Med. Entomol.*, 42, 522, 2005.

MOTHS (SPECIES WHOSE SCALES OR HAIRS CAUSE IRRITATION)

TABLE OF CONTENTS

I. INTRODUCTION AND MEDICAL IMPORTANCE

As discussed in Chapter 14, several species of Lepidoptera have larvae, commonly called *urticating caterpillars*, that possess stinging hairs or spines and can inflict a painful sting upon exposure to human skin (erucism).[1] Some moth species as adults bear scales or hairs, which may detach, become airborne, and cause urticaria and irritation in humans (see Chapter 5). Irritation from adult moths is called *lepidopterism*. Several moth species in the families Notodontidae, Saturniidae, and Lymantriidae have been reported as causes of lepidopterism.[2] The dermatitis may present as localized,

LEPIDOPTERISM

Douglas fir tussock moth,
Orgyia pseudotsugata

Importance
Urticaria; irritation; respiratory problems from airborne moth hairs or scales

Distribution
Several species worldwide, especially those in genus *Hylesia* in South America

Lesion
Variable — often itchy papules or maculopapules

Disease Transmission
None

Key References
Delgado, A., Venoms of Lepidoptera, in *Arthropod Venoms*, Bettini, S., Ed., Springer-Verlag, Berlin, 1978, p. 555
Rosen, T., Caterpillar dermatitis, *Dermatol. Clin.*, 8, 245, 1990

Treatment
Antihistamines, topical corticosteroids, or calamine products; occasional oral steroids may be indicated

widespread, or generalized erythematous macules that rapidly evolve into urticarial wheals.[3] Wheals are often, but not always, replaced by small infiltrated papules or papulovesicles. There was a recent outbreak of 40 cases of dermatitis, conjunctivitis, and pulmonary affection in Europe due to the oak processionary caterpillar, *Thaumetopoea processionea*.[4] Lepidopterism is especially severe in South America during certain times of the year owing to moths in the genus *Hylesia*.[2] Severe problems have occurred elsewhere. DeLong[5] reported an epidemic of 500,000 cases of caterpillar dermatitis in and around Shanghai, China, which was associated with an unexpected population explosion of yellow-tail moths in surrounding rural areas and appropriate climatic conditions conducive to windblown spread of the disease.

In the U.S., the tussock moths and their relatives in the family Lymantriidae may cause irritation. The female Douglas fir tussock moth, *Orgyia pseudotsugata*, covers her egg masses with froth and body hairs. These hairs, along with other airborne hairs from the tips of the female abdomens, cause rashes, upper respiratory irritation, and eye irritation to forest workers in the Pacific Northwest.[6] A related species, the white-marked tussock moth, *O. leucostigma*, occurs throughout most of North America and may also be involved in cases of lepidopterism, especially in the East. In addition, airborne gypsy moth hairs, silken threads, and shed skins have been reported to cause a pruritic cutaneous reaction in sensitive individuals.

II. GENERAL DESCRIPTION OF SOME SPECIES INVOLVED

Moths in the genus *Hylesia* are about 20 mm long and have a wingspan of about 40 mm. Alexander[7] states that the northern species (Mexico and Central America) are a golden color, whereas those from Peru and Argentina have dark brown or black stripes on the body. The wings and body are covered with numerous scales and long hairs.

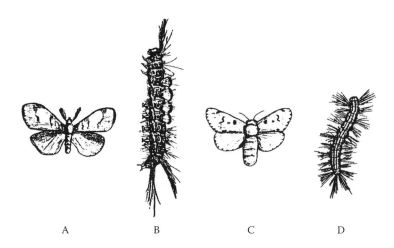

Figure 1
Some irritating moths: (A) Douglas fir tussock moth male (females are wingless), (B) Douglas fir tussock moth larva, (C) gypsy moth female, and (D) gypsy moth larva. (Taken in part from USDA Agriculture Handbook No. 536.)

Tussock moths and their relatives are medium-sized moths whose larvae are quite hairy. Female Douglas fir tussock moths are wingless. Males are brownish gray with a wingspan of about 25 mm (Figure 1A). The mature caterpillars are about 30 mm long and are gray to brown in color with a black head (Figure 1B). They bear two long, dark tufts or pencils of hair, similar to horns, right behind the head; a similar but longer pencil is on the posterior. Four dense, buff-colored tussocks are located forward along the middle of the back.

The white-marked tussock moth is larger than the Douglas fir tussock moth. Females are wingless; males are gray and have lighter hind wings with a wingspan of about 30 mm. The mature caterpillars are about 35 mm long and are light brown with yellow and black stripes. In contrast to the Douglas fir tussock moth, *O. leucostigma* caterpillars have a bright red head. Again, there are tufts of white hair on the dorsal side of the first four abdominal segments.

The female gypsy moth is winged (in contrast to the tussock moths mentioned above), and has a wingspan of approximately 50 mm, with white and black markings (Figure 1C). The males are slightly smaller than females and are gray. Mature gypsy moth caterpillars, like other members of the Lymantriidae, have long tufts of hair along the body (Figure 1D). They are approximately 30 to 50 mm long, gray in color, and have yellow stripes running lengthwise down the body. They may also appear to have red or blue spots along the sides and top of the body.

III. GEOGRAPHIC DISTRIBUTION

Hylesia moths occur from Mexico to Argentina. The Douglas fir tussock moth is a serious pest in western North America. Some of the worst outbreaks have occurred in British Columbia, Idaho, Washington, Oregon, Nevada, California, Arizona, and New Mexico. The white-marked tussock moth occurs throughout most of North America. Gypsy moths occur in Europe and in the eastern U.S.

IV.   BIOLOGY AND BEHAVIOR

Female *Hylesia* moths have barbed or spiny urticating setae that are called *fleshettes*. Fleshettes may break off and become airborne as large numbers of the moths emerge and gather around lights in towns or cities. *Hylesia* moths lay their eggs in clusters on the branches of trees, and the female coats the eggs with urticating spines. The life cycle is about 3 months in duration.

Tussock moths and their relatives are serious defoliators of forest, shade, and occasionally fruit trees. The Douglas fir tussock moth feeds primarily on fir trees. Usually the first indication of attack appears in late spring. Larvae from newly hatched eggs feed on the foliage of the current year, causing it to turn brown. The small larvae are inconspicuous, but by mid-July they are larger and more colorful. To feed in different areas of the tree, they lower themselves through the tree canopies by silken threads. By wind action they often spread to nearby uninfested trees. By August the larvae stop feeding and drop to the ground, crawling everywhere looking for a place to pupate. Female moths do not fly. Eggs are laid on tree trunks or branches, usually near the cocoon from which the female emerged.[8] Tussock moths overwinter in the egg stage.

Gypsy moths lay eggs on tree trunks during July and August in masses covered with froth and body hairs from the female. Eggs overwinter and the tiny first-stage larvae emerge the following spring, usually in late April or early May.[9] The females can fly, but only weakly. Dispersal of gypsy moth infestation is primarily by young larvae on silken threads traveling tree to tree or limb to limb by "ballooning" in the wind.

V.   TREATMENT

Treatment mainly involves personal protection and avoidance of the offending species. During season of *Hylesia* (South America) moth emergence, it may help to turn off outdoor lights. In affected areas of the U.S., air-conditioning and frequent changes of filters may help reduce airborne levels of scales or hairs. Sensitive persons should avoid walking in forests heavily infested with tussock moths.

Acute urticarial lesions may respond satisfactorily to topical corticosteroid lotions and creams (desoximetasone gel has been frequently used), which reduce the intensity of the inflammatory reaction. Oral antihistamines may relieve itching and burning sensations. In more serious cases, oral steroids may be indicated. Rosen[3] said systemic administration of corticosteroids in the form of intramuscular triamcinolone acetonide has been remarkably effective in relieving severe itching due to gypsy moth dermatitis.

REFERENCES

1. Delgado, A., Venoms of Lepidoptera, in *Arthropod Venoms*, Bettini, S., Ed., Springer-Verlag, Berlin, 1978, p. 555.
2. Harwood, R.F. and James, M.T., *Entomology in Human and Animal Health*, Macmillan, New York, 1979.
3. Rosen, T., Caterpillar dermatitis, *Dermatol. Clin.*, 8, 245, 1990.
4. Gottschling, S. and Meyer, S., An epidemic airborne disease caused by the oak processionary caterpillar, *Pediatr. Dermatol.*, 23, 64, 2006.

5. DeLong, S., Mulberry tussock moth dermatitis, *J. Epidemiol. Comm. Health*, 35, 1, 1981.

6. Perlman, F., Press, E., Googins, G.A., Malley, A., and Poarea, H., Tussockosis: reactions to Douglas fir tussock moth, *Ann. Allerg.*, 36, 302, 1976.

7. Alexander, J.O., *Arthropods and Human Skin*, Springer-Verlag, Berlin, 1984.

8. Borror, D.J., Triplehorn, C.A., and Johnson, N.F., *An Introduction to the Study of Insects*, Saunders College Publishing, Philadelphia, PA, 1989.

9. Nichols, J.O., The Gypsy Moth in Pennsylvania — Its History and Eradication, Pennsylvania Dep. Agric. Misc. Bull. No. 4404, 1962, p. 13.

CHAPTER 27

PENTASTOMES (TONGUE WORMS)

TABLE OF CONTENTS

I. GENERAL AND MEDICAL IMPORTANCE

Pentastomes (Phylum Pentastomida), sometimes also classified as members of Arthropoda because of their chitinous exoskeleton, are wormlike parasites found mainly in the respiratory tracts of carnivorous mammals, reptiles, and birds.[1] Pentastomiasis (human parasitism by pentastomes) may occur in the viscera, where nymphs develop in the liver, spleen, lungs, and other organs, or in the nasopharyngeal area, including the nasal passages, larynx, or eustachian tubes. Fortunately, human pentastomiasis is mostly asymptomatic and only found by radiography, surgery, or autopsy. Occasionally, human infestations may cause discomfort in the throat, paroxysmal coughing, and sneezing.

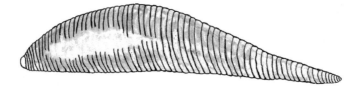

Figure 1
The tongue worm, Linguatula serrata.

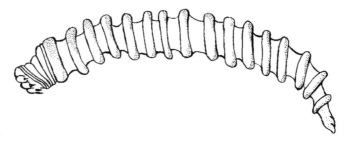

Figure 2
Armillifer armillatus, *a common pentastomed found in snakes.*

II.   GENERAL DESCRIPTION OF SOME SPECIES INVOLVED

Pentastome worms vary in color and shape, depending on the species, but are generally colorless to yellow, ringed organisms with no apparent legs or body regions. The body may be cylindrical or flattened. Females may be up to 130 mm long and 10 mm wide; males are smaller, being up to 30 mm long. On either side of the mouth are two pairs of hollow, fanglike hooks which can be retracted into grooves like the claws of a cat.

III.   GEOGRAPHIC DISTRIBUTION

Two species of pentastomes account for 99% of the infections in humans.[2] The tongue worm, *Linguatula serrata* (Figure 1), which occurs worldwide, is often found in the nasal passages and frontal sinuses of several host animals, especially dogs. *Armillifer armillatus* (Figure 2) ordinarily inhabits the respiratory tract of certain snakes in central Africa.

IV.   BIOLOGY AND BEHAVIOR

The life cycle of *L. serrata* is provided here as a fairly typical example of pentastomid biology. Adults are often found in the nasal passages and frontal sinuses of canines and felines. Eggs produced pass out of their host in nasal discharges and are deposited in water or on vegetation. An intermediate host (e.g., rabbit or sheep) swallows contaminated water or vegetation and the eggs hatch into primary larvae, which penetrate the intestinal wall and become lodged in the liver, lungs, or mesenteric nodes. The larvae then pass through two molts resulting in a pupalike stage. Up to seven more molts may occur before the nymphal stage develops (sometimes called an *infective larva*). The nymphs then migrate to the abdominal or pleural cavity of the intermediate host where they become encysted. When a definitive host (dog or cat) eats the intermediate host, the nymphs escape rapidly and migrate anteriorly, subsequently clinging to the lining within the host's mouth. From this point they migrate to the nasal cavities where they develop into adults. The adult stage may survive for up to 2 years.

V. TREATMENT

Treatment of pentostomiasis is usually not necessary, but in symptomatic cases surgical removal of the parasites may be needed.[3]

REFERENCES

1. Markell, E., Voge, M., and John, D., *Medical Parasitology*, W.B. Saunders, Philadelphia, PA, 1992, p. 359.
2. Binford, C.H. and Conner, D.H., *Pathology of Tropical and Extraordinary Diseases*, Armed Forces Institute of Pathology, Washington, D.C., 1976, p. 38.
3. Hobmaier, A. and Hobmaier, M., On the life cycle of *Linguatula rhinaria*, *Am. J. Trop. Med.*, 20, 199, 1940.

CHAPTER 28

SCORPIONS

TABLE OF CONTENTS

I.   GENERAL AND MEDICAL IMPORTANCE

Scorpions are eight-legged arthropods that can inflict a painful sting (Color Figure 28.20). There are about 1250 species occurring on all major land masses except Antarctica.[1] Some species are more dangerous than others, depending on the type of venom; size and appearance do not determine the medical importance of a scorpion. For example, scorpions in the genera *Pandinus* and *Heterometrus* (Old World) and *Hadrurus* (New World) are huge and menacing, but they constitute no serious health hazard. Most species (*Centruroides vittatus* is a common offender in the southwestern U.S.; Figure 1A) can elicit local effects by their stings in a manner similar to that of bees and wasps. There is immediate sharp pain at the site of venom injection and often moderate local edema (which may be discolored). Dr. Scott Stockwell at the Walter Reed Biosystematics Unit compares the sting to hitting your thumb with a hammer. Regional lymph node enlargement, local itching, paresthesia, fever, and occasionally nausea and vomiting may also occur.[2] Signs and symptoms in a person stung by a scorpion with this type of venom usually subside in a few hours; however, it must be noted that a person

SCORPIONS

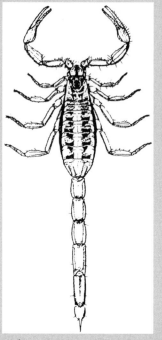

Centruroides vittatus, **a common scorpion species.**

Importance
Painful sting; some species deadly

Distribution
Only one dangerous species in the U.S.; numerous species in other areas

Lesion
Variable — sometimes local swelling and discoloration

Disease Transmission
None

Key Reference
Polis, G.A., *The Biology of Scorpions*, Stanford University Press, Stanford, CA, 1990, chap. 10

Treatment
Ice packs; anticonvulsants; vasodilators; sometimes antivenin serum for deadly species

with insect sting allergy could have a systemic reaction from this type of venom.

The other, more deadly, scorpion species are in the family Buthidae and have a venom that is more neurologic and hemolytic in activity. Systemic effects from such stings include drowsiness, abdominal cramps, blurred vision, spreading partial paralysis, muscle twitching, abnormal eye movements, profuse salivation, perspiration, priapism, hypertension, tachycardia, and convulsions.[3–5] Extreme restlessness, resembling seizure, is a frequent presenting sign in children.[6] In fact, very small children may flail, writhe, and display roving eye movements. Death is probably due to respiratory paralysis, peripheral vascular failure, and/or myocarditis, and it may occur at any interval between 1.5 to 42 h.[2] Four cases of scorpion sting in the southwestern U.S. by our only dangerous species, *C. exilicauda* (formerly called *C. sculpturatus*), caused severe clinical manifestations including respiratory failure, metabolic acidosis, and severe multiorgan system disease.[7] Other than the sting of *C. exilicauda* in Arizona and surrounding areas, scorpion stings in the U.S. are not usually life-threatening (barring allergic reaction), but scorpions constitute a severe problem in other areas of the world, such as northern Africa and Central and South America. For example, the number of scorpion stings in Morocco is estimated to be 40,000 per year.[4]

II. GENERAL DESCRIPTION

Scorpions are crablike in appearance, with pincers attached to their two front appendages (Figure 2). Their five-segmented tail terminates in a bulbous structure with a prominent curved stinger (Figure 3). Adult scorpions vary in size from 2 to 10 cm, depending on the species. Most American species are yellowish brown or brown in color. An excellent identification key to the North American families and genera is provided by Stockwell.[8] *C. exilicauda*, the dangerous U.S. species, is yellow to yellowish brown in color and is relatively small (usually 6 cm maximum; Figure 1B). It

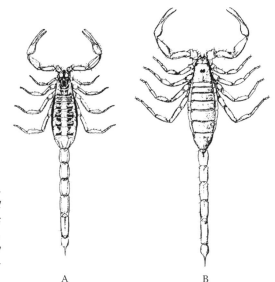

Figure 1
(A) Centruroides vittatus — *a very common species in the southwestern U.S. and (B)* Centruroides exilicauda — *a dangerous U.S. species. (Reprinted from Keegan, H.L., Scorpions of Medical Importance, University of Mississippi Press. With permission. Copyright 1980 University of Mississippi Press.)*

A   B

Figure 2
Typical scorpion in a stinging position.
(U.S. Army Photo. Courtesy of LTC Harlan, H.J., 74th Medical Detachment.)

is a slender species with very narrow and elongate tail segments. Immatures have a diagnostic small spine or tubercle at the base of the stinger (Figure 4). A variant form of *C. exilicauda* may also occur with two irregular black stripes down its back. At one time people thought this striped version of *C. exilicauda* was another species named *C. gertschi*.

III. GEOGRAPHIC DISTRIBUTION

Table 1 lists some of the dangerous scorpions worldwide and their distributions. About 100 species occur in the U.S. and all but 4 of these naturally occur west of the Mississippi River. The most commonly encountered species in the southwestern U.S. is the

Figure 3
Close-up of stinger of Pandinus imperator.

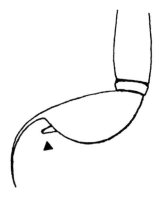

Figure 4
Side view of Centruroides exilicauda *stinger showing diagnostic small spine or tubercle. This structure is less obvious in adults.*

common striped scorpion, *C. vittatus*, whose sting is as painful as a wasp or bee sting. The species is especially common in Texas and is responsible for numerous nonfatal stings each year. The only dangerous scorpion (other than the allergic reaction) in the U.S., *C. exilicauda*, occurs primarily in Arizona, but also in southeastern California, Nevada, southern Utah, and southwestern New Mexico. In Mexico, this species has been found in Baja California at Puerto Punta Penasco, Sonora.

Androctonus australis, the fat-tailed scorpion (Figure 5A), is a highly venomous, very aggressive scorpion found in North Africa in Egypt, Algeria, Tunisia, and Libya. Subspecies of *A. australis* may occur in the Middle East. The fat-tailed scorpion is responsible for 80% of all reported scorpion stings in Algeria; about one third of them are fatal.[9] *Buthus occitanus* (Figure 5B) is a widely distributed, dangerous scorpion composed of several subspecies; it is found in southern France, Spain, Italy, Greece, several Mediterranean islands, Israel, Jordan, and northern Africa. The dangerously venomous scorpion, *Leiurus quinquestriatus*, is found in Turkey southward through Syria, Lebanon, Jordan, Israel, and down into northern Africa. In the Middle East, *Prionurus* (= *Androctonus*) *crassicauda* is responsible for many serious stinging incidents annually. *Tityus serrulatus* and *T. bahiensis* are probably the most dangerous

Table 1
Geographic Distribution of Some
Dangerously Venomous Scorpions

Region	Country	Species
Old World	Algeria	*Buthus occitanus*
		Androctonus australis
	Egypt	*A. australis*
		A. amoreuxi
		Leiurus quinquestriatus
	Iraq	*Hemiscorpion lepturus*
		A. crassicauda
	Israel	*L. quinquestriatus*
		A. crassicauda
		A. bicolor
	Jordan	*B. occitanus*
	Morocco	*A. mauritanicus*
		A. australis
		A. amoreuxi
		B. occitanus
	South Africa	*Parabuthus triradulatus*
		P. transvaalensis
		P. villosus
	Sudan	*B. minax*
	Turkey	*A. crassicauda*
		L. quinquestriatus
		Mesobuthus gibbosus
New World	Argentina	*Tityus bahiensis*
	Brazil	*T. bahiensis*
		T. serrulatus
	Guyana	*T. cambridgei*
	Mexico	*Centruroides noxius*
		C. suffussus
		C. infamatus
		C. elegans
		C. limpidus
	Trinidad	*T. trinitatis*
	U.S.	*C. exilicauda*
	Venezuela	*T. trinitatis*
		C. gracilis

Source: From Keegan, H.L., Scorpions of Medical Importance, University of Mississippi Press, Jackson, 1980, chap. 3. With permission.

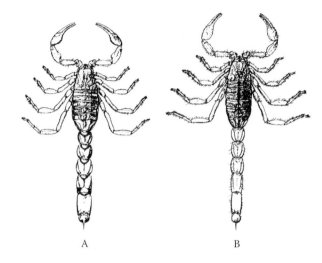

Figure 5
(A) Androctonus australis *and (B)* Buthus occitanus.
(Reprinted from Keegan, H.L., Scorpions of Medical Importance, University of Mississippi Press. Copyright 1980 University of Mississippi Press. With permission.)

species in Brazil. In South Africa, *Parabuthus triradulatus, P. transvaalensis,* and *P. villosus* are dangerous.

IV. BIOLOGY AND BEHAVIOR

Female scorpions have pouches in which the young develop for a certain period, before being born alive. Subsequently, they climb onto the mother's back and remain there until molting about 2 weeks later. After this molt, the young scorpions scatter to live solitary lives, molting six or seven more times before reaching maturity.

Scorpions are nocturnal feeders, feeding mostly on insects and spiders, although there are examples of the larger species being seen feeding on lizards, snakes, and other small vertebrates. They seize prey with their claws and paralyze it by their sting. During the day scorpions remain concealed and are only encountered when people disturb them. Some of the most dangerous species are known as *bark scorpions* because they are often found under loose bark and in the crevices between the bases of palm tree leaves. They do not burrow but hide under boards, bricks, and other rubbish. *Ground scorpions*, on the other hand, burrow into loose sand or gravel and are capable of hibernating without food or water for up to 7 months.

Around homes, scorpions are usually found under the houses and in attics. They enter attics by climbing between wall partitions. However, they will abandon attics as soon as the temperature gets too high, migrating downward into the walls, and if an opening is available, into the house itself. They favor attics with air-conditioning ducts that provide cool temperatures and a place to hide.

Because they will seek water, they are often found in bathrooms, kitchens, or laundry rooms at night. It is not uncommon to find them in bathtubs or sinks, which they entered seeking water and were unable to climb back out.

The amount of venom injected with a sting varies with the size of scorpion. Whittemore and Keegan[10] found that electrically stimulated specimens of a large species from the Middle East yielded an average of 0.48 mg, whereas the average yield from a small, but highly dangerous species of *Centruroides* from Mexico was only 0.075 mg.

V. TREATMENT OF STINGS

Many species of scorpions are innocuous and produce stings that are followed by sharp pain or a burning sensation and a wheal, which usually disappears with no complications. The venom of dangerous species is quite neurotoxic. Death is due to cardiovascular failure or paralysis of the respiratory muscles within a few hours.

After a sting by a dangerous species, it is important for patients to stay calm in order to minimize absorption of venom into the body. A pressure dressing over the sting site may be helpful, as well as application of ice packs. In adults, symptoms may be self-limited with resolution occurring in several hours. Children are at greatest risk of severe reactions.[6,11] In many countries administration of a potent antivenin within a couple of hours of the sting may be needed for treatment of severe envenomization.[2,11] Antivenins are commercially available for many (but not all) of the dangerous scorpions worldwide. There is some controversy concerning antivenin use; the best course may be to evaluate and administer antivenin only as needed. In the U.S. an antivenin for *Centruroides exilicauda* envenomization is produced from goat serum at Arizona State University and is approved by the Arizona Board of Pharmacy for use within the state of Arizona.[12] But there was a 58% rate of serum sickness in one study.[11] Because the antivenin is not an FDA-approved product and cannot be used across state lines; its application is limited.[12]

Scorpion sting therapy may include mechanical ventilation to improve oxygenation and administration of CNS depressants such as midazolam (a benzodiazepine) and vasodilators such as selective α_1 blockers and calcium channel blockers.[4,12] Others have used intravenous propranolol HCL.[13] Patients treated with midazolam should be carefully monitored for underventilation or apnea, which can lead to hypoxia or cardiac arrest.

REFERENCES

1. Fet, V., Sissom, W.D., Lowe, G., and Braunwalder, M.E., *Catalog of the Scorpions of the World (1758–1998)*, New York Entomological Society, New York, 2000, pp. 1–346.

2. Keegan, H.L., *Scorpions of Medical Importance*, University of Mississippi Press, Jackson, MS, 1980, chap. 3.

3. Bucherl, W. and Buckley, E., *Venomous Animals and their Venoms*, Academic Press, New York, 1971.

4. Ghalim, N., El-Hafney, B., Sebti, F., Heikel, J., Lazar, N., Moustanir, R., and Benslimane, A., Scorpion envenomation and serotherapy in Morocco, *Am. J. Trop. Med. Hyg.*, 62, 277, 2000.

5. Harwood, R.F. and James, M.T., *Entomology in Human and Animal Health*, Macmillan, New York, 1979, chap. 17.

6. Polis, G.A., *The Biology of Scorpions*, Stanford University Press, Stanford, CA, 1990, chap. 10.

7. Berg, R.A., Envenomation by the scorpion *Centruroides exilicauda*: severe and unusual manifestations, *Pediatrics*, 87, 930, 1991.

8. Stockwell, S., Systematic observations on North American Scorpionida with a key and checklist of the families and genera, *J. Med. Entomol.*, 29, 407, 1992.

9. Nichol, J., *Bites and Stings — The World of Venomous Animals*, Facts on File, New York, 1989, p. 58.

10. Whittemore, F.W. and Keegan, H.L., Medically important scorpions in the Pacific area, in *Venomous and Poisonous Animals and Noxious Plants of the Pacific Region*, Keegan, H.L. and MacFarlane, W.V., Eds., Pergamon Press, Oxford, 1963, p. 107.

11. Bond, G.R., Antivenin administration for *Centruroides* scorpion sting: risks and benefits, *Ann. Emergency Med.*, 21, 788, 1992.

12. Gilby, R., Williams, M., Walter, F.G., McNally, J., and Conroy, C., Continuous intravenous midazolam infusion for *Centruroides exilicauda* envenomation, in North American Congress of Clinical Toxicology (oral presentation), Orlando, FL, September 1998.

13. Rachesky, I.J., Banner, W., Jr., Dansky, J., and Tong, T., Treatments for *Centruroides exilicauda* envenomation, *Am. J. Dis. Child.*, 138, 1136, 1984.

CHAPTER 29

SPIDERS

TABLE OF CONTENTS

I. SPIDERS IN GENERAL

A. Medical Importance

There are more than 36,000 named species of spiders ranging in size (including legs) from a few millimeters to more than 17 cm (7 in) (Color Figure 29.21). Even some spiders that are not tarantulas are very large (Figure 1). All spiders, with the exception of the family Uloboridae, are venomous and use their venom to immobilize or kill prey; however, the chelicerae of many species are too short to penetrate human skin. Of the species that can inflict human bites, the widow spiders (*Latrodectus* spp.), the violin spiders (*Loxosceles* spp.), and hobo spiders (*Tegenaria agrestis*) are significant causes of local or systemic effects from envenomization in the U.S. In Australia, the Sydney funnel web spider (*Atrax robustus*) and related species can cause serious illness and death. In South America, the Brazilian Huntsman (*Phoneutria fera*) is a threat to humans, and Aranha Armadeira (*Phoneutria nigriventer*) has extremely potent venom.

It is important to note that at least 50 other species have been implicated in bites on humans. Bites from these species, although painful, are not likely to be dangerous. An account given by Carpenter et al.[1] illustrates this point. A 15-year-old boy was bitten by a plectreurid spider, (*Plectreurys tristis*) in Kern County, CA. Initially the bite produced pain, edema, and slight pallor at the site of the wound, which persisted for 15 to 30 min, and subsequently the patient reported vague, diffuse numbness near the site. All these symptoms resolved within 2 h. Because not all spider bites are life-threatening or lead to necrotic lesions, it is important for the patient to bring in to the clinic the offending spider for identification. This might help physicians avoid expensive treatments used for spider bites from dangerous species. In addition to the direct effects of spider venom, spider bites may produce secondary infections and allergic reactions (apparently rare in the U.S.).

B. Treatment of Bites

The treatment for most spider bites involves washing the bite site, applying cold compresses, avoiding exercise, treating with tetanus prophylaxis (if indicated), analgesics for pain, and antibiotics if secondary infection is suspected.[2] Additional measures are needed for bites by dangerous species, such as hobo, widow, brown recluse, and funnel web spiders (see the following text).

II. BROWN RECLUSE AND OTHER VIOLIN SPIDERS

A. General and Medical Importance

Violin spiders have venom that produces cutaneous lesions (Color Figure 29.22 through Color Figure 29.24). The brown recluse spider (*Loxosceles reclusa*) is probably the most

Figure 1
Huge fish-eating spider collected in Mississippi — not a tarantula.
(Photo courtesy of Sheryl Hand and Dr. Sally Slavinski.)

important of the violin spiders in the U.S. However, several other species occur in the Southwest (*L. rufescens, L. deserta,* and *L. arizonica*) and Northeast that can produce necrotic lesions. *L. laeta* is a notorious violin spider in South America.

Every year, the brown recluse spider (also called the *fiddleback spider*) is responsible for numerous cases of envenomization. The bite is usually localized but may produce considerable necrosis resulting in an unsightly scar.[3,4] Research on brown recluse venom has indicated that it produces necrotic and hemolytic effects, but not neurologic. Brown recluse venom contains several fractions, but the primary dermonecrotic component appears to contain the phospholipase enzyme, sphingomyelinase D, which acts on tissues by activation of platelet aggregation, thrombosis, and massive neutrophil infiltration.[2,5] The bite is usually painless until 3 to 8 h later, when it may become red, swollen, and tender. One study reported that the clinical signs of brown recluse bite observed when the patients were first seen in a clinical setting included erythema, cellulitis, rash, and blister.[6] Later, a black scab may develop, and eventually, an area around the site may decay and slough away, producing a large ulcer from 1 to 25 cm in diameter. Finally, the edges of the wound thicken and become raised whereas the central area is filled by scar tissue. Healing may require months, and the end result is often a sunken scar (Figure 2). A helpful distinguishing sign of a brown recluse bite is the tendency of the lesion to extend downward in a gravitationally dependent manner (Figure 2a). The direction of the lesion is dependent on the victim's position during the time the vessel damage was produced. Asymmetrical lesions (especially the "flowing-downhill" type) are rare from other arthropod bites.

Systemic reactions — called *systemic loxoscelism* — characterized by hematuria, anemia, fever, rash, nausea, vomiting, coma, and cyanosis have sometimes been reported from brown recluse bites. Systemic signs and symptoms are rare and, if they occur, usually begin 2 to 3 d following envenomization in contrast to the local cutaneous lesion that is evident by 12 to 24 h.[7] Hemolysis, always heralded by rash and fever, has been reported occurring as early as 24 h and as late as 3 d after the bite.[7] It usually continues for 72 h. There was a recent report of shock in a 13-year-old girl that began 9 h after being bitten by a violin spider.[8]

OFTEN-ASKED QUESTION

HOW DO YOU KNOW IT REALLY IS A BROWN RECLUSE BITE?

More than 2000 brown recluse (BR) bites are reported to poison control centers each year, but if epidemiology and confirmed cases are any indication, most of them are due to something else.[1] Vetter[2] says the problem is that hundreds of BR bites are being diagnosed in areas where the spiders sparsely occur, or do not occur at all. I personally am aware of two correctional facilities in Mississippi where dozens of medically documented brown recluse spider bites have occurred, yet pest control professionals and Cooperative Extension Service entomologists have failed to find even one brown recluse. From the entomological perspective, it borders on the ridiculous to claim that dozens of people have been bitten by BR spiders when none can be found. They are not *that* reclusive.

Most spider experts agree that BR spiders are not aggressive. Bites typically occur defensively, only when the spider is accidentally trapped against human skin while a person is dressing or sleeping.[2] Often there are no biting incidents even when hundreds of the spiders are present in a dwelling. Vetter[3] reported finding over 2000 BR spiders in a home in Kansas occupied by a family of four with no bites recalled or reported by the family. Bite reactions from BR spiders are also probably exaggerated. We must keep things in perspective. Certainly there is evidence that the venom may cause unsightly spots of necrosis on human skin. However, what is often not reported is that BR bite reactions can vary from no reaction, to a mild red wound, to a terrifying rotting flesh wound. In fact, Masters[4] says that cutaneous necrosis usually does not develop after untreated BR bites. Even deaths reported from BR bites (from hemolytic anemia) may not be strongly supported by evidence.

Diagnosis of BR bites is based on clinical presentation and history (unless of course, the patient brings in the offending specimen). Even though at least one researcher has developed a diagnostic immunoassay,[5] there is currently no widely available laboratory test to confirm that a patient has been bitten by a BR. The BR spider bite lesion is variable, but generally it is a dry blue-gray or blue-white irregular sinking patch with ragged edges surrounded by erythema — the *red, white, and blue sign*."[1,4] The lesion often is asymmetric and gravity-dependent owing to downward flow of venom through tissues. A perfectly symmetrical necrotic lesion is often not due to BR spider. Necrosis and sloughing of tissues may follow over a period of days or weeks, leading to an unsightly sunken scar. Many conditions may lead to spots of necrosis on human skin resembling BR spider bites and should be considered by physicians before assigning the diagnosis of BR spider bite. The most common cause of necrotic wounds misdiagnosed as BR spider bites are infections with *Staphylococcus* and *Streptococcus* species.

REFERENCES

1. Sandlin, N., Convenient culprit: myths surround the brown recluse spider, Amednews.com (E-news for America's physicians), American Medical Association, Chicago, IL, August 5, 2002.
2. Vetter, R.S. and Bush, S.P., The diagnosis of brown recluse spider bite is overused for dermonecrotic wounds of uncertain etiology, *Ann. Emergency Med.*, 39, 544, 2002.
3. Vetter, R.S. and Barger, D.K., An infestation of 2,055 brown recluse spiders and no envenomations in a Kansas home: implications for bite diagnoses in nonendemic areas, *J. Med. Entomol.*, 39, 948, 2002.
4. Masters, E.J. and King, L.E., Jr., Differentiating loxoscelism from Lyme disease, *Emergency Med.*, 36, 47, 1994.
5. Gomez, H.F., Krywko, D.M., and Stoecker, W.V., A new assay for the detection of *Loxosceles* species spider venom, *Ann. Emergency Med.*, 39, 469, 2002.

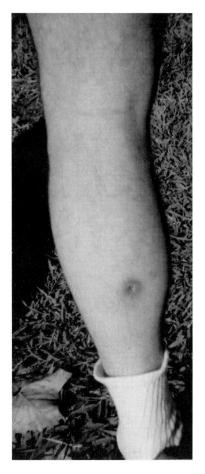

Figure 2
Sunken scar resulting from brown recluse bite — 4 months postbite. (From Mississippi Cooperative Extension Service Publication No. 2154.)

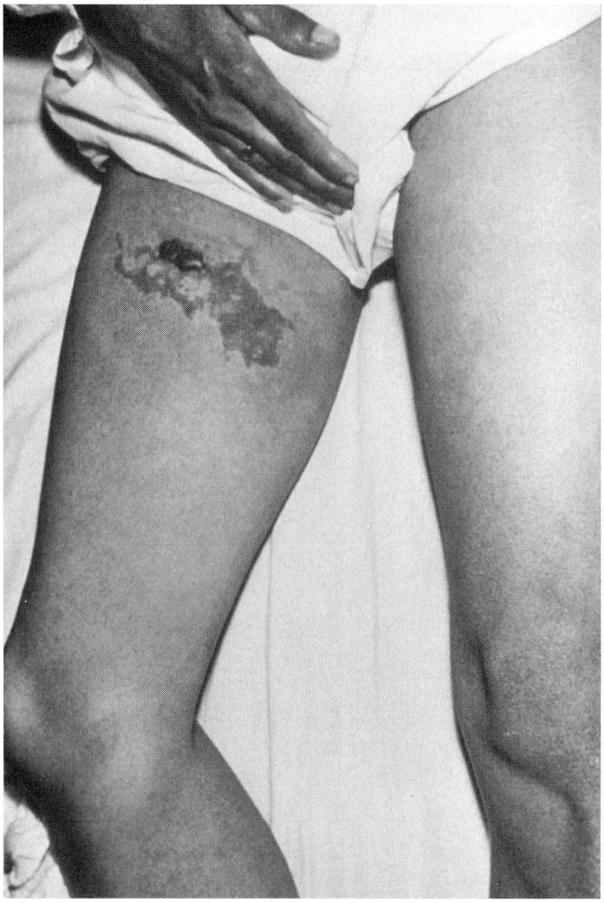

Figure 2a
Brown recluse spider bite showing asymmetrical lesion (CDC photo).

B.   General Description

The adult brown recluse is a medium-sized spider with a 20- to 40-mm leg span and a color range from tan to dark brown (Figure 3 and Color Figure 29.25). The most distinguishing characteristics are six eyes arranged in a semicircle of three pairs on top of the head and a violin-shaped marking (with base forward) extending from the area of the eyes to the beginning of the abdomen.

C.   Geographic Distribution

Although its distribution may be more widespread, the brown recluse spider is most commonly found in the southcentral U.S. (Figure 4). *L. deserta*, *L. rufescens*, *L. arizonica*, and *L. devia* are closely related species, occurring mostly in the southwestern U.S., that

VIOLIN SPIDERS____

Figure 3
Brown recluse spider (USDA photo).

Importance
Several closely related species cause necrotic lesions; possibly hemolytic anemia

Distribution
U.S., South America, parts of Africa and Australia

Lesion
Variable through time — papule with erythema, large central zone of pallor, mottling, then blackening of central zone, ulceration

Disease Transmission
None

Key References
Swanson, D. and Vetter, R., Bites of brown recluse spiders and suspected necrotic arachnidism, *New Engl. J. Med.*, 352, 700, 2005

Diaz, J.H., The global epidemiology, syndromic classification, management, and prevention of spider bites, *Am. J. Trop. Med. Hyg.*, 71, 239, 2004

Treatment
Controversial — may include ice packs, antibiotics, dapsone, or antivenin

may also cause necrotic arachnidism (Figure 4). In South America, there is a close relative of the brown recluse, *L. laeta*, that causes a severe necrotic lesion sometimes referred to as *gangrenous spot. L. laeta* has been introduced into the U.S. and is established in Los Angeles County, CA, and possibly at Harvard University, Cambridge, MA.

D. Biology and Behavior

The brown recluse is nocturnal in its feeding habits and most frequently found inside houses in bathrooms, bedrooms, cellars, attics, the folds of clothing, cardboard boxes, and storage areas; and also outdoors under rocks and rubble. Homes that are infested may easily contain hundreds of specimens. The spiders are not aggressive but will bite if pressure is placed on them. At night, they may crawl into garments and bite people attempting to dress the next morning. Common sites are under the arms or on the lower parts of the abdomen. Brown recluse spiders are very resilient, being active in temperatures ranging from 40 to 110°F, and they may live as long as 2 years.

E. Treatment of Bites

Persons bitten by a spider should make an effort to collect the spider and bring it in to a clinic for identification. Positive identification of the spider by an expert can be helpful, because bites of most spiders are temporarily painful but not dangerous. In the case of the brown recluse, prompt medical treatment may be needed to prevent severe reactions.

A specific antidote (antivenin) has shown success in patients prior to development of the necrotic lesion,[9] but currently it is not widely available. Also, some brown recluse bites are unremarkable, not leading to necrosis. Therefore, treatment is quite controversial. Some studies have indicated that application of ice is effective (the necrotic enzyme, sphingomyelinase D, increases in activity as temperature increases; so keep the temperature low.) There have been reports of successful treatment by

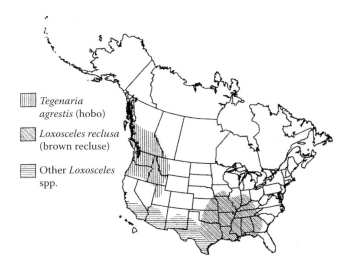

*Tegenaria
agrestis* (hobo)

Loxosceles reclusa
(brown recluse)

Other *Loxosceles*
spp.

Figure 4
*Approximate geographic distributions of spiders causing necrotic arachnidism in the U.S. and
Canada. (From CDC, MMWR, 45, 1, 1996.)*

early, total excision of the bite site followed by split thickness skin grafting; however,
more recent evidence argues against wound excision.[6] Systemic corticosteroids have
been associated with slower healing of brown recluse bites.[10] Nitroglycerin apparently
does not help.[11] Rees et al.[9] reported good success with the leukocyte inhibitor dapsone,
which eliminated the need for surgery in many brown recluse bites (Figure 5). Others
have also advocated dapsone.[12–14] Theoretical evidence supports its use — histological
examination of bites in animals shows evidence of extensive neutrophil infiltration at
the site of ultimate necrosis. Dapsone is contraindicated in persons with glucose-6-
phosphate dehydrogenase (G6PD) deficiency owing to potential massive hemolysis.
It may produce side effects (even in non-G6PD patients) that could be confused with
the systemic effects of brown recluse spider bite: malaise, nausea, and hemolysis.
Therefore, before dapsone is administered in patients with spider bites, a baseline
assessment of G6PD, a complete blood count, and a test of liver enzymes should be
performed and be repeated weekly while the patient is receiving the drug.[15] There is
some evidence that dapsone is ineffective. Randomized, blinded, controlled studies of
venom effects in rabbits failed to show any benefit from the use of the drug.[15a,16] However, at least one letter published in the *New England Journal of Medicine* (NEJM) cites
clinical experience indicating that dapsone is the most suitable treatment and leads to
a satisfactory resolution.[17] For systemic reactions to brown recluse spider bites, steroids
may minimize hemolysis (especially in children) and may protect renal function.[18]

III.   BLACK WIDOW AND OTHER WIDOW SPIDERS

A.   General and Medical Importance

Several species of widow spiders occur worldwide, but *Latrodectus mactans* is the
one most generally associated with the name *black widow spider*. It is also sometimes
referred to as the hourglass, shoe button, or po-komoo spider. The black widow spider

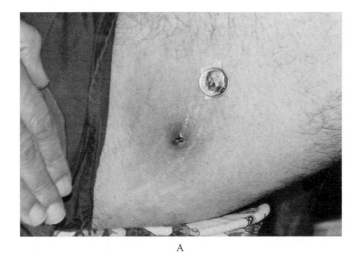

A

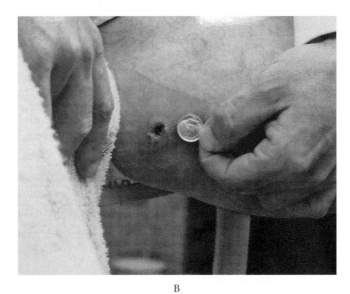

B

Figure 5
Brown recluse spider bite lesion development — treated with dapsone. (A) Lesion at 3 d postbite, (B) 8 d, (C) 12 d, (D), 15 d, (E) and 27 d.

injects a potent neurotoxin upon biting. Initially, the bite itself produces a mild burning or stinging pain; but more than half of the patients in one study did not know they had been bitten.[19] The venom exerts a specific effect on the central nervous system, causing depletion of acetylcholine at motor nerve endings and provoking the release of catecholamines at adrenergic nerve endings. Unlike the skin necrosis or systemic hemolysis associated with bites from the brown recluse spider, black widows produce pain in the regional lymph nodes, usually in the axilla or inguinal area, piloerection, increased blood pressure and white blood cell count, and profuse sweating and nausea. The bite site may appear as a bluish red spot with a white areola and sometimes

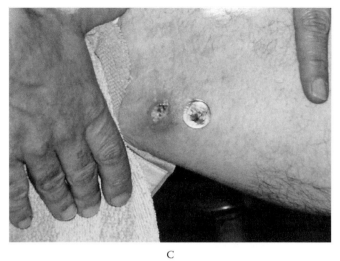

C

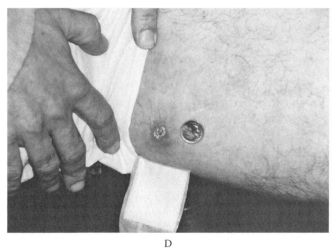

D

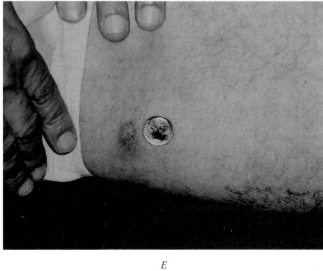

E

Figure 5
Continued.

WIDOW SPIDERS

Female black widow spider,
Latrodectus mactans

Importance
Neurotoxic venom

Distribution
Several species almost worldwide

Lesion
Minimal — two puncta

Disease Transmission
None

Key Reference
Diaz, J.H., The global epidemiology,
  syndromic classification, management,
  and prevention of spider bites, *Am. J.
  Trop. Med. Hyg.*, 71, 239, 2004

Treatment
Ice packs; analgesics; muscle relaxants;
sometimes antivenin; calcium gluconate

a urticarial rash. Significant envenomization may be accompanied by systemic symptoms of weakness, tremor, severe myalgia, muscular spasm, a rigid boardlike abdomen, and tightness in the chest. Bites on the torso are more likely to cause muscle cramps in the abdomen. Paralysis, stupor, and convulsions may occur in severe cases, and rarely death. The painful, rigid abdomen may be mistaken for appendicitis. Latrodectism can also be misdiagnosed as an alimentary toxic infection, acute psychosis, tabic crisis, pneumonia, tetanus, meningitis, acute renal failure, and various exanthematic diseases.[19] Wingo[20] reported that the mortality rate from black widow bites is less than 1%; Alexander[21] said it is 4 to 6%. Other, related widow spiders are perhaps more dangerous. The redback spider, *Latrodectus mactans hasselti*, ubiquitous in Australia, can occasionally cause death in humans, and its bite is the commonest envenomization requiring antivenom in Australia, with at least 250 cases per year receiving antivenom.[22] At least 17 deaths had been reported from bites by this species prior to development of antivenom.[23]

Males and immature spiders also bite but usually produce milder symptoms. Some references indicate that the males are harmless owing to the lesser amount of venom injected because of their smaller size and proportionately shorter chelicerae.

B.   General Description

Mature female black widow spiders are shiny black with a leg span of 30 to 40 mm (Figure 6 and Color Figure 29.26). On the underside of the abdomen is a characteristic red or orange hourglass-shaped marking. There is considerable variation among species and even subspecies (see Figure 6). Males are considerably smaller than females, averaging 16 to 20 mm in leg span, and have red and white marks on the dorsal side of the abdomen. Immature widow spiders (spiderlings) are tan to grayish in color with very little or no black. They have orange and white markings on their abdomens that resemble "racing stripes."

CASE HISTORY

UNKNOWN SPIDER BITE

An out-of-town emergency room (ER) physician called one night about a possible black widow (BW) spider bite in a young boy. The family had killed the spider and brought it in with the boy. By report, it somewhat resembled a black widow. The site of the bite was unremarkable, except for two small puncta. At the time of the call, the boy was not exhibiting any significant pain, nausea, sweating, or other symptoms usually associated with BW bites. The physician asked for identification advice and possible suggestions on what course of action to take.

Based on the description given over the telephone, the spider seemed to be an aberrant BW, a male BW, or possibly a closely related species, *Latrodectus variolus*. What seemed to be confusing the physician was the presence of several bright red hash marks on the dorsal side of the spider's abdomen (not present on a "normal" BW) in addition to the red hour-glass-shaped marking on its underside. The physician was informed that in all likelihood the specimen was a BW and the case should be treated as such (based on development of symptoms in that particular patient).

Comment: There are at least five species of widow spider in the U.S. Physicians should be aware that not all widow spiders fit the typical black widow appearance (Figure 1). Some specimens have red markings on the dorsal side of the body (in addition to the red hourglass-shaped mark on the underside of the abdomen). In addition, in some species, the hourglass marking is "broken" and not complete.

Source: Adapted in part from *J. Agromed.*, 2, 53, 1995. Copyright 1995, The Hayworth Press, Binghamton, New York. With permission.

Figure 1
Typical female black widow underside showing red hourglass marking.
(Photo courtesy of Dr. James Jarratt, Mississippi Cooperative Extension Service.)

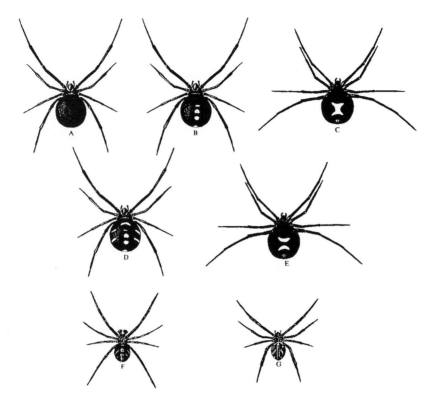

Figure 6
Widow spiders. Dorsal view of Latrodectus mactans *adult female (A) normal and (B) variant pattern; (C) ventral view.* L. variolus *(D) dorsal view and (E) ventral view. (F) Dorsal view of* L. mactans *male and (G)* L. mactans *immature female.*

C. Geographic Distribution

Five species of widow spiders occur in America, north of Mexico. Three (*L. mactans, L. variolus,* and *L. hesperus*) are similar in appearance. The black widow spider, *L. mactans,* is widely distributed in the U.S., occurring from southern New England to Florida, west to California and Oregon; it is more common in the southern part of the range. The northern black widow spider, *L. variolus,* occurs in the New England area and adjacent Canada, south to Florida and west to eastern Texas, Oklahoma, and Kansas; it is more common in the northern part of this range. The western black widow, *L. hesperus,* is found in western Texas, Oklahoma, and Kansas, north to the adjacent Canadian provinces and west to the Pacific Coast states. The brown widow spider, *L. geometricus,* has been reported from various parts of the southern U.S., and the coast of southern California; the red widow spider, *L. bishopi,* is limited to southern Florida.

Dangerous widow spiders in other parts of the world include *L. geometricus* (already mentioned occurring in the southern U.S.), occurring worldwide in the tropics and subtropics; the black widow subspecies, *L. mactans tredecimquttatus,* in southern Europe; *L. mactans hasselti,* a problem in Australia and India; *L. mactans cinctus,* a dangerous widow spider in South Africa; and *L. mactans menavodi,* one of medical significance on the island of Madagascar. The redback spider, *L. mactans hasselti,* is now established in Japan (probably imported from Australia).[24]

D. Biology and Behavior

Widow spiders are found in various habitats in the wild: in protected places such as crawl spaces under buildings, in water meter housings, holes in dirt embankments, piles of rocks, boards, bricks, or fire wood, and dense plant growth such as grain and cotton fields.

Inside buildings, they avoid strong light and favor dark corners behind or underneath appliances or furniture that is seldom moved, garage corners, deep closets, damp cabinets, etc. With the advent of indoor plumbing, the most frequent site of encounter in previous decades, the outdoor privy, has become less common. Widow spiders will establish a web wherever conditions seem favorable, and if there is a suitable food source, they may establish a long-term infestation.

Widow spiders are active primarily during the warm months of the year. Few adults survive cold winter weather, except indoors in heated places, but immature spiders may overwinter in large numbers. Mating takes place in the spring. Contrary to popular belief, which gave rise to the name black widow spider, the male is seldom killed and eaten by the female. If the female is well fed, the male usually leaves after mating and may mate again with other females. If the female is not well fed, she may consume the male for nutrients to produce the fertilized eggs, thus contributing to the success of the species. The female lays 250 to 750 eggs at a time and spins a spherical sac of strong silk around them.

The young hatch in 2 to 4 weeks. After emerging from the egg sac, they spin a strand of silk that enables them to be carried by the wind, dispersing them into the surrounding area. Male black widow spiders molt 4 to 7 times and females molt 7 to 9 times during a period of several weeks to several months, depending on the temperature, food availability, and other factors. More than one brood can be produced in a summer, and as many as 2000 offspring may be produced during a year. If they survive the winter, females may produce egg sacs the following year. Thus, they may live up to 3 years.

Widow spiders typically will be seen hanging upside down in their web, waiting for prey that includes many types of arthropods including insects and other spiders. The prey's contact with the web and subsequent struggle to free itself produces vibrations that the spider follows to locate the prey. The spider then seizes and bites the prey, injecting venom to paralyze and kill it so that it can be consumed. Venom is produced in two glands located in the basal segment of the chelicerae and is injected through a tiny hole in each of the two fangs.

Widow spiders are reclusive; they usually will not bite humans unless provoked. When their web is disturbed, they will attempt to escape, but if teased or pursued, they will bite in self-defense.

E. Treatment of Bites

First aid for *Latrodectus* envenomization consists of cleansing the wound thoroughly and applying ice packs to slow absorption of the venom. Incision of the wound and suction to remove venom should not be attempted.

Cramps may be helped with muscle relaxants. If pain is severe, there is often quick response to sustained i.v. calcium gluconate; i.m. calcium gluconate is not advisable, especially in infants and children. Calcium gluconate is used in preference to calcium chloride, because the gluconate lessens the danger of tissue necrosis from accidental extravasation. Pain may also be relieved by administration of oral, or occasionally i.v., analgesics such as morphine or fentanyl.[18] Tetanus vaccine status should be assessed. Physicians should monitor for shock and treat as necessary. Antivenin is available but rarely indicated.[25] It is sometimes used for severe cases (especially Australian cases involving the redback spider), but may be of little help once illness has progressed and unnecessarily risky if given routinely for mild bites. Russell et al.[26] recommend the muscle relaxant methocarbamol for uncomplicated adult cases. Children, the very elderly, and persons with hypertensive disease or other medical conditions of risk need special attention, including hospitalization and possible antivenin. If antivenin is given, the content of one vial (be sure to check package insert for instructions) is usually administered i.m. *after* tests for serum sensitivity are made. Symptoms should start to subside in 0.5 to 3 h after administration of the antivenin.

IV.   HOBO SPIDERS

A.   General and Medical Importance

Hobo spider bites have only recently been suspected of causing necrotic lesions. Necrotic spider bites in the states of Washington, Oregon, and Idaho have steadily increased in frequency during the last 30 years. At first, these were attributed to the brown recluse spider, but intensive efforts to find brown recluses in/near patients' homes were fruitless. Some scientists now believe that local effects of hobo spider (formerly called *the aggressive house spider*) envenomization may be similar to those of brown recluse bites, and that the hobo spider may be the cause of necrotic arachnidism in the northwestern U.S.[27,28] Akre and Myhre[29] recorded 52 serious bites from this species, *Tegenaria agrestis*, between 1989 and 1994. However, epidemiological and laboratory toxicity studies do not fully support common claims that hobo spiders cause necrotic lesions.[30,31]

Hobo spider bites are usually painless at first. Induration may appear within 30 min, surrounded by an area of expanding erythema that may reach a diameter of 5 to 15 cm. Blisters develop within 15 to 35 h, subsequently rupturing with a serous exudate encrusting the cratered wound. An eschar may develop with underlying necrosis and eventual sloughing of affected tissues. Lesions may require up to 3 years to heal if the bite occurred in fatty tissue.[27] Systemic symptoms include headache (often prolonged), nausea, weakness, fatigue, memory loss, and vision impairment. Protracted and severe systemic effects may rarely lead to death.[27]

B.   General Description

Hobo spiders are large, very fast-running specimens, about 45 mm in length (including legs), with distinct chevron stripes on the abdomen. They have eight eyes with both the anterior and posterior eye rows in straight lines and have solid, light brown legs (not banded).

C.   Geographic Distribution

Hobo spiders are native to Europe and were probably introduced into the Seattle area in the 1920s. They now occur from central Utah north to the Alaska Panhandle (Figure 4). The hobo spider will probably continue to spread east in Montana and Wyoming, and may or may not reach California.[32]

D. Biology and Behavior

Hobo spiders build funnel-shaped webs in or near houses. Common sites include rock walls, along house foundations, in garages, piles of debris, and stacks of firewood. Both males and females build webs, and both may be in a single web in the late summer and fall. Mature males leave their webs in the fall to search for females. Male hobo spiders enter houses in large numbers in the fall but are usually restricted to basements or ground floor rooms because they are poor climbers. Males are more venomous than females and are responsible for most bites.[29]

E. Treatment of Bites

Optimal treatment for necrotic spider bites is not well defined[27] (see previous discussion under violin spiders for current ideas).

V. FUNNEL WEB SPIDERS

A. General and Medical Importance

The Sydney funnel web spider, *Atrax robustus*, is Australia's most dangerous spider, capable of causing death in as little as 15 min.[33] There are several relatives of the Sydney funnel web that are also dangerous to humans such as *Hadronyche cerberea* and *H. formidabilis*.[34] In many cases, little or no venom is injected during the biting event and no symptoms develop. If envenomization does occur, the bite site becomes extremely painful. Systemic symptoms may develop within minutes owing to the toxin's direct effect on somatic and autonomic nerves, leading to widespread release of neurotransmitter. Progressive hypotension and apnea may ensue.

B. General Description

The Sydney funnel web is a large, dark-colored aggressive spider with prominent fangs (Figure 7). The carapace is glossy dark brown to black, whereas the abdomen is

Figure 7
Sydney funnel web spider.

usually dark plum to black. The spiders often range in size from 45 to 60 mm (including legs); body length alone ranges from 15 to 45 mm. Males are smaller than females. Spinnerets are obvious, fingerlike, and at the end of the abdomen.

C. Geographic Distribution

Sydney funnel web spiders are only found within a 160-km radius of Sydney, Australia.[33] Specifically, they occur from Newcastle to Nowra and west to Lithgow. Other related funnel webs occur all along the east coast of Australia.

D. Biology and Behavior

Sydney funnel webs live in burrows, rotting logs, tree holes, or crevices in rocks, where they build (as their name implies) a funnel-shaped web. *H. formidabilis* builds the funnel-shaped web in trees. The web characteristically contains irregular silken trap lines radiating out from the entrance. Colonies may consist of more than 100 spiders. Males may wander in search of females into houses during summer, especially during rainy weather.

E. Treatment of Bites

Immediate first aid should be administered for bites by any large black spider along the east coast of Australia, but especially in the Sydney area. A pressure bandage can be applied and the bitten limb immobilized using a splint. The spider should be captured, if possible, for identification.

At the hospital, patients are carefully monitored for signs and symptoms for 4 h[33]; little or no venom may have been injected during the bite. If signs and symptoms occur, such as mouth numbness, tongue spasms, nausea, vomiting, profuse sweating, salivation, and other muscle spasms, administration of antivenom is indicated (per package insert instructions). Patients should be monitored closely for development of allergic reactions; however, severe allergic reactions to funnel web spider antivenom are uncommon.[34]

VI. TARANTULA SPIDERS

A. General and Medical Importance

Spiders in the family Theraphosidae are commonly called *tarantulas* in the U.S.; however, in other parts of the world this name is shared with other spiders, leading to confusion. Because of their great size and reputation, tarantulas are sometimes feared. This fear is mostly unfounded. Tarantulas attack only when roughly handled or deliberately provoked, and their bite is relatively minor (except in the case of a few tropical species that have more toxic venom). Bites of the North American species vary from being almost painless to a deep, throbbing pain that may last for hours. Baerg[35] allowed himself to be bitten twice and reported mild pain (like that of pinpricks) lasting 15 to 30 min, not accompanied by inflammation or swelling. Hypersensitive individuals could have more severe reactions. Tarantula venom consists mainly of hyaluronidase and a protein that is toxic to cockroaches and other arthropods for which it is intended. Interestingly, tarantula bites are often fatal to domestic dogs.[36]

Many tarantulas occurring in the Western Hemisphere have urticarial hairs on the dorsal surface of the abdomen. These hairs are "flicked off" by the spiders as a defensive mechanism.[21] Urticarial hairs do not occur on African or Asian species. Contact with these urticarial hairs on the skin or mucous membranes may result in pruritic papular lesions that persist for weeks. If the hairs get into the eyes, complications can arise, with symptoms similar to ophthalmia nodosa.

B. General Description

Tarantulas are large hairy spiders. Their jaws (chelicerae) are attached in front of their head and can be moved up and down, opening parallel to the long axis of the body (Figure 8). Species native to the U.S. have a leg span of about 18 cm, but some South American jungle species have a span of 24 cm.

C. Geographic Distribution

About 30 species of tarantulas occur in the U.S., mostly in the southwestern states. None occur east of the Mississippi River. Owing to their popularity as pets, they can be found in captivity all across the U.S. The majority of tarantulas sold in pet stores in the U.S. are imported species of *Aphonopelma*, which are especially attractive as pets because of their bright colors.

D. Biology and Behavior

Tarantulas hatch from eggs into spiderlings that look like a miniature version of the adult. They mature in 10 to 12 years. A male tarantula usually does not live longer than a year after becoming sexually mature, but the female life span may exceed 15 to 20 years. In their native habitats, tarantulas dig burrows to rest in during the day. These burrows are often dug under large stones found in open hillside areas among mixed desert flora. Tarantulas are sluggish during daytime and may hibernate through the winter in colder areas. They emerge at night to hunt, usually ranging only

TARANTULA SPIDERS

Figure 8
Adult female tarantula.
(Photo courtesy of the Ross E. Hutchins Photograph Collection, Mississippi Entomological Museum, Mississippi State University.)

Importance
Painful bites; urticating hairs

Distribution
U.S., Central and South America, Africa

Lesion
Bites — generally just pinprick type lesions; abdominal hairs — may cause itchy pruritic lesions

Disease Transmission
None

Key Reference
Diaz, J.H., The global epidemiology, syndromic classification, management, and prevention of spider bites, *Am. J. Trop. Med. Hyg.*, 71, 239, 2004

Treatment
Bites — analgesics, tetanus prophylaxis; urticarial lesions — corticosteroids and antihistamines

a few yards from the burrow. During the mating season, males travel long distances from their burrows in search of females, but females and immatures remain near their burrow throughout their life. Tarantulas have extremely poor eyesight and detect their prey by vibrations. Their diet consists of arthropods and small vertebrates.

E. Treatment of Bites

Treatment of tarantula bites consists of washing the bite site thoroughly, keeping the affected area elevated, and administering systemic analgesics if the wound is painful. Because tarantula mouthparts are dirty, tetanus vaccine status should be assessed. Treatment for urticaria produced by the abdominal hairs includes topical corticosteroids and oral antihistamines. If severe and widespread lesions occur, a short course of 1 or 2 weeks of oral corticosteroids may be considered.

REFERENCES

1. Carpenter, T.L., Bernacky, B.J., and Stabell, E.E., Human envenomation by *Plectreurys tristis*, *J. Med. Entomol.*, 28, 477, 1991.

2. King, L.E., Jr., Spider bites, *Arch. Dermatol.*, 123, 41, 1987.

3. Atkins, J.A., Wingo, C.W., and Sodeman, W.A., Probable cause of necrotic spider bite in the Midwest, *Science*, 126, 1957, p. 73.

4. Atkins, J.A., Wingo, C.W., and Sodeman, W.A., Necrotic arachnidism, *Am. J. Trop. Med. Hyg.*, 7, 165, 1958.

5. Masters, E.J. and King, L.E., Jr., Differentiating loxoscelism from Lyme disease, *Emergency Med.*, August Issue, 47, 1994.

6. DeLozier, J.B., Reaves, L., King, L.E., Jr., and Rees, R.S., Brown recluse spider bites of the upper extremity, *South. Med. J.*, 81, 181, 1988.

7. Murray, L.M. and Seger, D.L., Hemolytic anemia following a presumptive brown recluse spider bite, *Clin. Toxicol.*, 32, 451, 1994.

8. Bey, T.A., Walter, F.G., Lober, W., Schmidt, J., Spark, R., and Schlievert, P.M., *Loxosceles arizonica* bite associated with shock, *Ann. Emergency Med.*, 30, 701, 1997.

9. Rees, R., Campbell, D., Rieger, E., and King, L.E., The diagnosis and treatment of brown recluse spider bites, *Ann. Emergency Med.*, 16, 945, 1987.

10. Mold, J.W. and Thompson, D.M., Management of brown recluse spider bites in primary care, *J. Am. Board Fam. Pract.*, 17, 347, 2004.

11. Lowry, B.P., Bradfiled, J.F., and Carroll, R.G., A controlled trial of topical nitroglycerin in a New Zealand white rabbit model of brown recluse spider envenomation, *Ann. Emergency Med.*, 37, 161, 2001.

12. King, L.E., Jr. and Rees, R.S., Dapsone treatment of a brown recluse bite, *JAMA*, 250, 648, 1983.

13. Nonavinakere, V.K., Stamm, P.L., and Early, J., A case study of brown recluse spider bite: role of the community pharmacist in achieving a successful outcome, *J. Agromed.*, 3, 37, 1996.

14. Rees, R.S., Altenbern, D.P., Lynch, J.B., and King, L.E., Jr., Brown recluse spider bites: a comparison of early surgical excision versus dapsone and delayed surgical excision, *Ann. Surg.*, 202, 659, 1985.

15. Swanson, D. and Vetter, R., Bites of brown recluse spiders and suspected necrotic arachnidism, *New Engl. J. Med.*, 352, 700, 2005.

15a. Elston, D.M., Miller, S.D., Young, R.J., Eggers, J., McGlasson, D., Schmidt, W.H., and Bush, A., Comparison of colchicine, dapsone, triamcinolone, and diphenhydramine therapy for the treatment of brown recluse spider envenomation: a double-blind, controlled study in a rabbit model, *Arch. Dermatol.*, 141, 595, 2005.

16. Phillips, S., Kohn, M., Baker, D., Vander Leest, R., Gomez, H., McKinney, P., McGoldrick, J., and Brent, J., Therapy of brown spider envenomation: a controlled trial of hyperbaric oxygen, dapsone, and cyproheptadine, *Ann. Emergency Med.*, 25, 363, 1995.

17. Masters, E.J., Loxoscelism, *New Engl. J. Med.*, 339, 379, 1998.

18. Buescher, L.S., Spider bites and scorpion stings, in *Conn's Current Therapy*, Rakel, R.E. and Bope, E.T., Eds., Elsevier, Philadelphia, PA, 2005, p. 1302.

19. Maretic, Z., Latrodectism: variations in clinical manifestations provoked by *Latrodectus* species of spiders, *Toxicon*, 21, 457, 1983.

20. Wingo, C.W., Poisonous Spiders, Missouri Agric. Exp. Stn. Bull. No. 738, 1960.

21. Alexander, J.O., *Arthropods and Human Skin*, Springer-Verlag, Berlin, 1984, chap. 14.

22. Sutherland, S.K., *Australian Animal Toxins: The Creatures, Their Toxins, and Care of the Poisoned Patient*, Oxford University Press, Melbourne, 1983, 275.

23. Wiener, S., Spider bite in Australia: an analysis of 167 cases, *Med. J. Aust.*, 48, 44, 1961.

24. Horton, P., Redback spider is now established in Japan, *Br. Med. J.*, 314, 1484, 1997.

25. Olson, K.R., Poisoning, in *Current Medical Diagnosis and Treatment*, 41st ed., Tierney, L.M., McPhee, S.J., Papadakis, M.A., Eds., Lange Medical Books/McGraw Hill, New York, 2002, pp. 1609–1643.

26. Russell, F.E., Wainsschel, J., and Gertsch, W.J., Bites of spiders and other arthropods, in *Current Therapy*, Conn, H.F., Ed., W.B. Saunders, Philadelphia, PA, 1973, p. 868.

27. CDC, Necrotic arachnidism — Pacific Northwest, 1988–1996, *MMWR*, 45, 433–436, 1996.

28. Vest, D.K., Necrotic arachnidism in the northwest United States and its probable relationship to *Tegenaria agrestis* spiders, *Toxicon*, 25, 175, 1987.

29. Akre, R.D. and Myhre, E.A., The Great Spider Whodunit, *PCT Magazine*, April issue, 1994 p. 44.

30. Vetter, R.S. and Isbister, G.K., Do hobo spider bites cause dermonecrotic injuries?, *Ann. Emergency Med.*, 44, 605, 2004.

31. Binford, G.J., An analysis of geographic and intersexual chemical variation in the venoms of the spider *Tegenaria agrestis*, *Toxicon*, 39, 955, 2001.

32. Vetter, R.S., Hobo spiders: are they everywhere?, *PCT Magazine*, December issue, 2003, pp. 46–48.

33. Hawdon, G.M. and Winkel, K.D., Spider bite: a rational approach, *Aust. Fam. Phys.*, 26, 1380, 1997.

34. Isbister, G.K., Gray, M.R., Balit, C.R., Raven, R.J., Stokes, B.J., Porges, K., Tankel, A.S., Turner, E., White, J., and Fisher, M.M., Funnel-web spider bite: a systematic review of recorded clinical cases, *Med. J. Aust.*, 182, 407, 2005.

35. Baerg, W.J., Regarding the habits of tarantulas and the effects of their poison, *Sci. Monthly*, 14, 482, 1922.

36. Diaz, J.H., The global epidemiology, syndromic classification, management, and prevention of spider bites, *Am. J. Trop. Med. Hyg.*, 71, 239, 2004.

CHAPTER 30

TICKS

TABLE OF CONTENTS

I. GENERAL AND MEDICAL IMPORTANCE

Ticks are blood-sucking ectoparasites that are efficient vectors of several different types of disease agents such as bacteria, viruses, rickettsiae, and protozoans (see Color Figure 30.27). In fact, they are second only to mosquitoes as arthropod vectors of human disease. In addition, tick bites cause a variety of acute and chronic skin lesions.[1]

Table 1
Major Tick-Borne Diseases Worldwide

Disease	Causative Agent	Where Occurs	Tick Vectors
Lyme disease	Spirochete	U.S., Europe, Japan, China, Australia	*Ixodes ricinus* complex
Rocky Mountain spotted fever	Rickettsia	U.S., Canada, Mexico, Central and South America	*Dermacentor variabilis, D. andersoni*, others
Siberian tick typhus	Rickettsia	Asiatic Russia, some islands in Sea of Japan	Primarily *D. marginatus, D. silvarum, D. nuttalli*
Boutonneuse fever	Rickettsia	Southern Europe, India, Mediterranean area, Africa	*Rhipicephalus sanguineus, R. appendiculatus, Haemphysalis leachi*, others
American boutonneuse fever[a]	Rickettsia	Southern and south central U.S., parts of South America	*Amblyomma maculatum* in U.S., *A. triste* in South America
Tularemia	Bacterium	U.S., Europe, Russia, Japan, Canada, Mexico	*Amblyomma americanum, D. variabilis, D. nuttalli, I. ricinus*
Colorado tick fever	Virus	Mountain states of western U.S., British Columbia, Canada	*D. andersoni*
Ehrlichiosis and Anaplasmosis	Rickettsiae (*Ehrlichia* and *Anaplasma*)	Mainly eastern and central U.S.	*A. americanum, I. scapularis*
American babesiosis	Protozoan	U.S., mostly in Massachusetts and New York areas	*I. scapularis*
Relapsing fever	Spirochete	Almost worldwide, in U.S., primarily in Washington, Oregon, California	*Ornithodoros* spp., U.S. cases mostly from *O. hermsi*
Tick-borne encephalitis	Virus	Eastern and western Europe, Powassan virus occurs in U.S.	*I. ricinus, I. persulcatus*, other *Ixodes* spp., *D. marginatus*
Crimean-Congo hemorrhagic fever	Virus	Europe, Asia, Africa	*Hyalomma marginatum, H. anatolicum*, others
Kyasanur forest disease	Virus	India, especially Karnataka State	*Haemaphysalis spinigera*
Tick paralysis	Salivary toxin	U.S., particularly Montana, British Columbia border, Australia	*D. andersoni, D. variabilis, I. holocyclus*

[a] *Newly described human infections caused by* Rickettsia parkeri *(see text).*

OFTEN-ASKED QUESTION

HOW DO TICKS GET ON PEOPLE?

There is considerable confusion among the public as to how ticks get on people. Folklore has it that they live in pine thickets, hide under bark, and/or fall out of trees on unsuspecting passersby. Frequently, park rangers or other outdoor workers erroneously recommend a hat for protection against ticks (presumably to guard against falling ticks). But ticks cannot fly, jump, or swim. They quest for host animals by climbing vegetation and passively waiting. The height of questing varies by species and life stage, but usually is 3 ft or less (often just a few inches). If environmental conditions are favorable, they may stay atop the vegetation for hours or even days. When a potential host approaches, signaled by vibrations or carbon dioxide, questing ticks may wave their legs (Figure 1) or move around on the plant trying to get on the animal. Once on the host, they crawl around for some time looking for a suitable place to feed. As ticks often attach to the scalp of humans, it is falsely assumed that they fall out of trees.

Figure 1
Ticks crawl up blades of grass to "quest" for passing hosts.

Listed in Table 1 are 14 of the common diseases produced by tick-transmitted agents, and a discussion of the major ones is provided in the following text. The reader is referred to other references[2-6] for more complete information on the natural history, epidemiology, and treatment of these diseases.

A.   Lyme Borreliosis

Lyme borreliosis (LB), caused by the spirochete, *Borrelia burgdorferi* (there are at least three "genospecies" of this organism), is a systemic tick-borne illness with many clinical manifestations that occurs over much of the world in temperate zones. Although rarely fatal, the disease may be long and debilitating with cardiac, neurologic, and joint involvement. Initial symptoms include a flulike syndrome with headache, stiff neck, myalgias, arthralgias, malaise, and low-grade fever. Often, a more or less circular,

CASE HISTORY

TICK ON MAN'S EARDRUM

One week after sleeping on the ground near Mexican Hat, Utah, a man began to have severe pain in his right ear. After a few days, lymphadenitis developed in the neck region surrounding the ear. He was seen by a family physician approximately 7 d later, who referred him to an ENT. After 4 days (and 18 days after camping) the ENT examined his eardrum, found a small, "crablike" creature attached, and removed it (Figure 1). As none of the physicians in the medical group knew exactly what the creature was, the patient — with specimen in hand — was referred to a parasitologist at the University of Mississippi Medical Center. The parasitologist identified the specimen as a tick, and referred the patient to the Mississippi Department of Health for more specific identification. The tick was identified as a nymphal spinose ear tick, *Otobius megnini*. The patient recovered uneventfully.

Comment: This tick belongs to the soft tick family. Most ticks seen by people are hard ticks: the ones that attach to a vertebrate host such as a dog or deer and remain attached for a week or so. Soft ticks are classified in another entire tick family as they differ from hard ticks in appearance and behavior. Soft ticks have little sexual dimorphism, have a wrinkled, granulated integument, and the mouthparts are generally not visible when the specimen is viewed from above. Soft ticks are especially adapted to dry climates or dry conditions within wet climates, and generally only feed on their vertebrate hosts for a short time (1 h or less), not remaining attached for days. Most soft tick species in the U.S. occur in the West. The tick involved in this particular case differs from other soft ticks in that the larval and nymphal stages invade ears of cattle, horses, sheep, deer, and other wild animals and remain attached for long periods of time. There have been records of this species remaining in the ears of animals for as long as 121 d. The scientific literature contains several records of this species being found in the ears of humans.

This case is interesting and illuminates several issues. For one thing, people travel and may bring back with them all sorts of parasites or microbes. Species that normally do not occur in an area may be seen from time to time by practicing physicians. Second, none of the physicians knew that this creature was a tick. To their credit, the physicians knew that the specimen was some sort of parasite, and referred the man to a parasitologist. The parasitologist — a broadly trained, organismal-level scientist knowing a wide range of internal and external parasites on sight — immediately knew it was a tick. The medical community needs such scientists, especially in light of rapid, modern travel methods and immigration

painless, macular dermatitis is present at the bite site, called *erythema migrans* (EM). The EM lesion is often said to be pathognomonic for LB, although not all patients develop it. EM lesions may steadily increase in size with subsequent central clearing. Untreated EM and associated symptoms usually resolve in 3 to 4 weeks. However, the disease often disseminates within weeks or months, resulting in cardiac, neurologic, and joint manifestations. Symptoms may include Lyme carditis, cranial neuropathy, radiculopathy, diffuse peripheral neuropathy, meningitis, and asymmetric oligoarticular arthritis. The number of reported LB cases in the U.S. continues to increase. There were 23,763 cases reported to the CDC in 2002, more than any previous year.[7] In the U.S., the vast majority of cases are from the northeastern and north central states (Figure 1).

Lyme borreliosis is solely tick-borne. In the U.S., *Ixodes scapularis* is the primary vector in the East, and *Ixodes pacificus* in the West. In Europe, *I. ricinus* and *I. persulcatus*

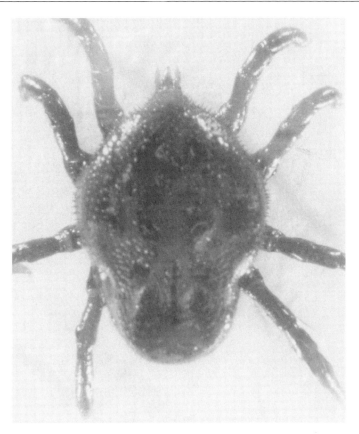

Figure 1
Tick removed from patient's eardrum.

(both legal and illegal). In the last 20 years or so, there has been a drift away from organismal-level training, with increasing emphasis on molecular biology. But there will always be a need for scientists who can "eyeball" a specimen and place it in its appropriate taxonomic group. A key to treatment strategies in such cases is knowledge of the biology and behavior of the parasite and what diseases, if any, it transmits.

serve as principal vectors of the *Borrelia*. In Asia, vectors include *I. persulcatus, I. ovatus, I. granulatus, I. moschiferi,* and a few *Haemaphysalis* species.

Two important aspects of vector-borne disease ecology are host availability and diversity. If immature ticks feed on hosts that are refractory to infection with the LB spirochete, then overall prevalence of the disease agent in an area will decline. On the other hand, if an abundant host is available that also is able to be infected with *B. burgdorferi* producing long and persistent spirochetemias, then prevalence of tick infection increases. This is precisely the case in the northeastern and upper midwestern U.S. In those areas the primary host for immature *I. scapularis* is the white-footed mouse, which is capable of infecting nearly 100% of larval ticks during feeding. As infection can be transferred from tick stage to tick stage, this obviously leads to high numbers of infected nymphs and adults. In the West and South, tick infection rates are much lower

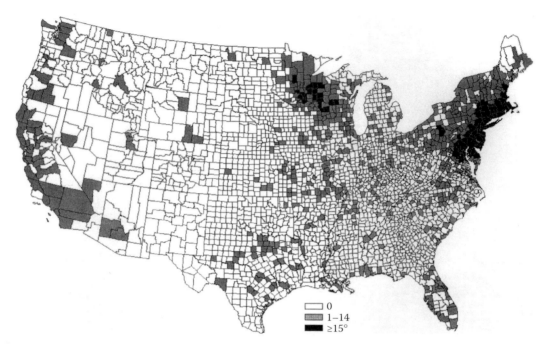

Figure 1
Reported number of Lyme disease cases by county — United States, 2000 (CDC figure).

(and hence, lower numbers of LB cases). This is attributed to the fact that immature stages of *I. scapularis* and *I. pacificus* feed primarily on lizards, which are incompetent as reservoirs and incapable of infecting ticks. Another factor affecting the dynamics of LB is the fact that nymphal *I. scapularis* are the stage primarily biting people and transmitting the disease agent in the Northeast, whereas in the South, nymphal *I. scapularis* rarely, if ever, bite humans.[8] Adult ticks are certainly capable of transmitting the LB agent in all areas — North, South, or West — but adult ticks are large enough to be easily seen and removed by people.[9] Nymphs, on the other hand, are about the size of a pinhead and may be easily overlooked or confused with a freckle.

Other tick species may be involved in the ecology of Lyme borreliosis in the U.S. Alternatively, there may be several, as yet undescribed, *Borrelia* species that cause Lymelike illness. In the southern U.S. there have been reports for years about an LB-like illness that researchers have voiced doubts about: doubts as to whether or not it is true LB. Also, it has been widely known for some time that 1 to 5% of lone star ticks, *Amblyomma americanum*, harbor spirochetes that react with reagents prepared against *B. burgdorferi*. These lone star tick spirochetes are a true *Borrelia* species (called *lonestari*), but distinct from the LB agent.[10] At least one case of erythema migrans has been attributed to *B. lonestari*;[11] however, recent data suggest that this agent is not the primary etiologic agent of Lymelike illnesses in the South.[12]

B. Rocky Mountain Spotted Fever

Ticks may transmit a wide variety of rickettsial organisms, classified by scientists into several distinct groups. The spotted fever group (SFG) contains rickettsial species related to the agent of Rocky Mountain spotted fever (RMSF), *Rickettsia rickettsii*. But there are

Table 2

Epidemiologic Information on Seven Spotted Fever Group Rickettsiae

Rickettsia	Disease	Tick Vectors	Distribution
R. rickettsii	RMSF	Primarily *Dermacentor variabilis* and *D. andersoni*	Western Hemisphere
R. conorii	Boutonneuse Fever	Primarily genera *Rhipicephalus, Hyalomma,* and *Haemaphysalis*	Africa, Mediterranean Area, Middle East
R. parkeri	American Boutonneuse Fever	*Amblyomma maculatum* (U.S.), *A. triste* (South America)	Southern and Central U.S., parts of South America
R. siberica	North Asian Tick Typhus (Siberian Tick Typhus)	Primarily genera *Dermacentor* and *Hyalomma*	Siberia, Central Asia, Mongolia
R. australis	Queensland Tick Typhus	*Ixodes holocyclus*	Australia
R. africae	African Tick Bite Fever	*Amblyomma hebraeum*	Sub-Saharan Africa
R. japonica	Japanese Spotted Fever	*Haemaphysalis flava, H. longicornis, Ixodes ovatus*	Japan

many other rickettsial species in the spotted fever group; it contains at least eight disease agents and 15 others with low or no pathogenicity to humans. Table 2 presents some distributional and epidemiologic information on seven human disease-causing SFG rickettsiae. RMSF is the most frequently reported rickettsial disease in the U.S. with about 600 cases reported each year.[13] Probably many more cases occur but go unreported. If an unusual febrile illness is treated successfully with one of the tetracyclines, there may be little interest in follow-up and reporting. At the time of initial presentation, there is often the classic triad of RMSF: fever, rash, and history of tick bite. Other characteristics are malaise, severe headache, chills, and myalgias. Sometimes, gastrointestinal symptoms such as abdominal pain and diarrhea are reported. More than one member of the family may be infected.[14] The rash, appearing on the third day or after, usually begins on the extremities and then spreads to the rest of the body. However, there have been confirmed cases without rash. Mental confusion, coma, and death may occur in severe cases. Untreated, the mortality rate is about 20%; even with treatment, the rate is 4%.

Laboratory findings may include hyponatremia (20 to 50% of cases), thrombocytopenia (30 to 50% of cases), anemia (5 to 25% of cases), and mildly elevated aminotransferase levels (40 to 60% of cases). There are no widely available, specific laboratory tests for the diagnosis of RMSF, although serological tests such as IFA may help confirm the disease later. It is vital to understand that diagnosis and treatment decisions for RMSF are based on clinical and epidemiologic clues and should *never* be delayed pending laboratory confirmation.

RMSF is usually transmitted by the bite of an infected tick (Figure 2). Not all tick species are effective vectors of the rickettsia and, even in the vector species, not all ticks are infected. Therefore, the presence of an infected tick in an area is like a needle in a haystack. Generally, only 1 to 5% of vector ticks in an area are infected. Several tick vectors may transmit RMSF organisms, but the primary ones are the American dog tick, *Dermacentor variabilis* (Color Figure 30.28) in the eastern U.S., and *D. andersoni* in the West. Adults of both species feed on a variety of medium-to-large mammals and

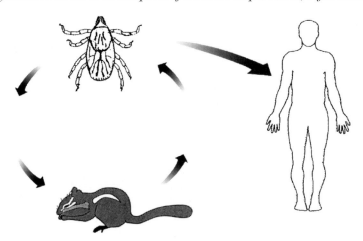

Figure 2
Life cycle of Rocky Mountain spotted fever agent.

humans. Recently, the brown dog tick, *Rhipicephalus sanguineus*, was found to be the vector in an outbreak of RMSF in Arizona.[15] Ticks are often brought into close contact with people via pet dogs or cats (dog ticks may also feed on cats).

C. Other Spotted Fever Group Rickettsioses

Boutonneuse fever.   Boutonneuse fever (BF), or Mediterranean spotted fever, caused by *Rickettsia conorii*, is widely distributed in Africa, areas surrounding the Mediterranean, southern Europe, and India. The name is derived from the black, buttonlike lesion (eschar) at the site of tick bite. BF resembles a mild form of RMSF, characterized by mild to moderately severe fever and a rash usually involving the palms and soles. Several tick species serve as vectors of the agent to humans, but especially *Rhipicephalus sanguineus, R. appendiculatus,* and *Amblyomma hebraeum.*

   American boutonneuse fever.   For nearly a century following the discovery of the agent of RMSF, *R. rickettsii* was considered the sole tick-borne rickettsia unequivocally associated with human disease in the U.S. Soon after the initial isolation of *R. parkeri* and during subsequent decades, investigators speculated about a possible role for this bacterium as an agent of disease in humans.[16,17] Parker and others identified similarity among "the maculatum agent" and the agents of other SFG rickettsioses, including boutonneuse fever, South African tick bite fever, and Siberian tick typhus by comparing clinical and serological characteristics of the infections in guinea pigs.[18–21] These observations have been validated by contemporary phylogenetic analyses showing relationships among SFG rickettsiae that position *R. parkeri* most closely with Old World, eschar-producing, SFG pathogens such as *R. africae, R. conorii, R. mongolotimonae,* and *R. sibirica*.[17,22] It follows that *R. parkeri* should also produce eschars. Goddard[23] reported eschar formation in guinea pigs infected with *R. parkeri,* and a human case of this eschar-characterized disease was recently described by Paddock et al.[24] Four other cases were subsequently identified serologically.[25] The tick vector of human *R. parkeri* infection is *Amblyomma maculatum* (see Table 2). Because of clinical similarities between *R. parkeri* infection and boutonneuse fever, the descriptive moniker for this newly recognized rickettsiosis is American Boutonneuse fever (ABF).[26] Only time will tell how prevalent ABF is in the U.S.

African tick-bite fever.   African tick-bite fever (ATBF), caused by a newly described spotted fever group rickettsia, *Rickettsia africae*, is clinically similar to BF with the exception that there is usually an absence of rash in ATBF patients.[27,28] ATBF also produces an eschar. The disease has been recognized in sub-Saharan Africa and the French West Indies, and is believed to be transmitted by *Amblyomma hebraeum* and *A. variegatum* ticks.[29,30]

Siberian tick typhus.   Siberian tick typhus (STT), or North Asian tick typhus, caused by *Rickettsia siberica*, is very similar clinically to RMSF with fever, headache, and rash. The disease can be mild to severe, but is seldom fatal. STT was first recognized in the Siberian forests and steppes in the 1930s, but now is known to occur in many areas of Asiatic Russia and on islands in the Sea of Japan. Various hard ticks are vectors of the agent, but especially *Dermacentor marginatus, D. silvarum, D. nuttalli,* and *Haemaphysalis concinna.*

Queensland tick typhus.   Queensland tick typhus (QTT), caused by *Rickettsia australis*, occurs along the east coast of Australia and is named after the northeastern Australian state of Queensland. It is primarily restricted to dense forests interspersed with grassy savanna or secondary scrub. Most patients have fever, headache, and rash that may be vesicular and petechial — even pustular. Commonly, there is an eschar at the site of the tick bite. The agent of QTT is transmitted to humans by the bite of an infected *Ixodes holocyclus* tick.

D.   Ehrlichiosis and Anaplasmosis

Ehrlichia organisms are rickettsiae that primarily infect circulating leukocytes. Much of the knowledge gained concerning ehrlichiae has come from the veterinary sciences, with intensive studies on *Anaplasma marginale* (cattle disease agent), *Ehrlichia (Cowdria) ruminantium* (cattle, sheep, goats), *Ehrlichia equi* (horses), and *Anaplasma (Ehrlichia) phagocytophilum* (sheep, cattle, deer). Canine ehrlichiosis, caused by *Ehrlichia canis*, wiped out 200 to 300 military working dogs during the Vietnam War.[31] In the U.S., human cases of ehrlichiosis were unknown until a report in March 1986 of a 51-year-old man who had been bitten by a tick in Arkansas and was sick for 5 d before being admitted to a hospital in Detroit.[32] He had malaise, fever, headache, myalgia, pancytopenia, abnormal liver function, renal failure, and high titers of *E. canis* antibodies that fell sharply during convalescence. The patient was thought to have the dog disease. It turned out not to be the case. For this reason, in the literature there are several reports of human infection with *E. canis*, when, in fact, human ehrlichiosis is caused by another closely related *Ehrlichia* organism. There are at least three ehrlichial disease agents infecting humans in the U.S.[33] One, *Ehrlichia chaffeensis*, the causative agent of human monocytic ehrlichiosis (HME), occurs mostly in the southern and southcentral U.S., and infects mononuclear phagocytes in blood and tissues.[34] Another, *Anaplasma* (formerly *Ehrlichia) phagocytophilum*, infects granulocytes and causes human granulocytic anaplasmosis (HGA); it is mostly reported from the upper midwest and northeastern U.S. The third, *E. ewingii*, causes a clinical illness similar to the other two, but thus far, has only been identified in a few patients, most of whom were immune-compromised.[35] Therefore, much is yet unknown about human infection with this agent.

Clinical and laboratory manifestations of infection with HME or HGA are similar.[33,36] The patient usually presents with fever, headache, myalgia, progressive leukopenia (often with a left shift), thrombocytopenia, and anemia. In addition, there may be moderate elevations in levels of hepatic transaminases. Sometimes there is a cough, gastroenteritis, or meningitis. Only about 2–40% of the time is there a rash, so ehrlichiosis and anaplasmosis are sometimes called *spotless* RMSF. Illness due to HGA is somewhat milder than with HME; reported fatality rates are about 1 and 2% for HGA and HME, respectively. Some research has indicated that both agents alter the patient's immune system, allowing opportunistic infections such as fungal pneumonia to occur. Diagnosis depends mainly on clinical findings, although IFA tests may be used to detect antibodies against the respective ehrlichial agent.

Ehrlichiosis and anaplasmosis are transmitted to humans via the bite of an infected tick. HME, primarily occurring within the geographic distribution of the lone star tick, *Amblyomma americanum,* seems to have a close association with that tick and the white-tailed (WT) deer. Lone star ticks (LSTs) generally occur from central Texas east to the Atlantic Coast and north to approximately Iowa and New York. WT deer, possibly along with dogs or small rodents, serve as reservoir hosts for the agent, and LSTs are the likely vectors. However, detection of the HME agent in other tick vectors and a few cases outside the distribution of LST may indicate that additional vectors occur.

The ecology of HGA is not well known at this time. It has been diagnosed mostly in patients from the upper Midwest and northeastern U.S., although cases have also been reported along the Pacific Coast. The tick vector is *Ixodes scapularis*, the same species that transmits the agent of Lyme borreliosis; thus, there is the possibility of co-infection with Lyme borreliosis and HGA (and even babesiosis).[37] Possible animal reservoirs of the HGA agent include deer and small rodents.

E. Babesiosis

Human babesiosis is a tick-borne disease primarily associated with two protozoa of the family Piroplasmordia: *Babesia microti* and *Babesia divergens,* although other newly recognized species may also cause human infection. The disease is a malaria-like syndrome characterized by fever, fatigue, and hemolytic anemia lasting from several days to a few months. In terms of clinical manifestations, babesiosis may vary widely, from asymptomatic infection to a severe, rapidly fatal disease. The first demonstrated case of human babesiosis in the world was reported in Europe in 1957.[38] Since then, there have been at least 28 additional cases in Europe.[38] Most European cases occurred in asplenic individuals and were with *Babesia divergens*, a cattle parasite. In the U.S. there have been hundreds of cases of babesiosis (most with intact spleens), mainly caused by *Babesia microti*, mostly from southern New England, and specifically Nantucket, Martha's Vineyard, Shelter Island, Long Island, and Connecticut.[39,40] The tick vector in Europe is believed to be the European castor bean tick, *Ixodes ricinus*, one of the most commonly encountered ticks in central and western Europe. In the U.S., cases of human infection by *Babesia microti* are caused by bites from the same tick that transmits the agent of Lyme borreliosis, *Ixodes scapularis.*

Babesiosis is very similar clinically to malaria; in fact, confusion between the two diseases is often reported in the scientific literature. Headache, fever, chills, nausea, vomiting, myalgia, altered mental status, disseminated intravascular coagulation, anemia with dyserythropoiesis, hypotension, respiratory distress, and renal insufficiency are common to both diseases. However, the symptoms of babesiosis do not show periodicity. The incubation period varies from 1 to 4 weeks. Physical examination of patients is generally unremarkable, although the spleen and liver may be palpable. Diagnosis of babesiosis is based on recognition of the organism within erythrocytes in Giemsa-stained blood smears.

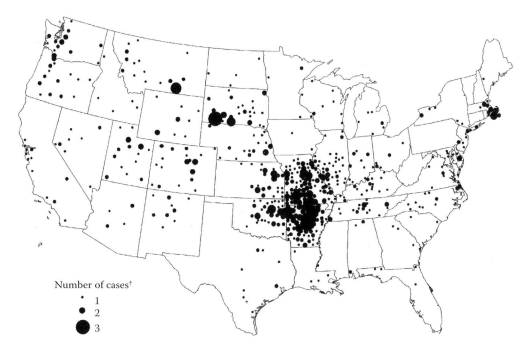

Number of cases†
- 1
- 2
- 3

Figure 3
Reported number of tularemia cases — United States, 1990–2000 (CDC figure).

Babesial parasites, along with members of the genus *Theileria*, are called *piroplasms* because of their pear-shaped intraerythrocytic stages. There are at least 100 species of tick-transmitted *Babesia*, parasitizing a wide variety of vertebrate animals. Some notorious ones are as follows: *Babesia bigemina*, the causative agent of Texas cattle fever; *B. canis* and *B. gibsoni*, canine pathogens; *B. equi*, a horse pathogen that occasionally infects humans; *B. divergens*, a cattle parasite that infects humans; and *B. microti*, a rodent parasite that infects humans. Recently, new *Babesia* species have been recovered from sick humans and have tentatively been variously designated as the WA1 agent, the CA1 agent, or the MO1 agent.[38,41] The WA1 agent, isolated from a patient in Washington state, was particularly interesting because the man was only 41 years old, had an intact spleen, and was immunocompetent.[38,41] Although the parasites were morphologically identical to *Babesia microti*, the patient did not develop a substantial antibody to *Babesia microti* antigens. Subsequent DNA sequencing of the organism indicated that it was most closely related to the canine pathogen *Babesia gibsoni*. There have been three reported cases of babesiosis in Washington state due to WA1, and a fourth from another, entirely different *Babesia*.[42] Obviously, there is much to be learned about the many and varied *Babesia* species, and their complex interactions in nature.

F.   Tularemia

Tularemia, sometimes called *rabbit fever* or *deer fly fever*, is a bacterial zoonosis that occurs throughout temperate climates of the Northern Hemisphere. Approximately 150 to 300 cases occur in the U.S. each year, but most cases occur in Arkansas, Missouri, and Oklahoma (Figure 3).[13] The causative organism, *Francisella tularensis*, is a small, Gram-negative, nonmotile coccobacillus named after Sir Edward Francis (who did the classical early studies on the organism) and Tulare, CA (where it was first isolated). The disease may be contracted in a variety of ways: food, water, mud, articles of clothing, and (particularly) arthropod bites. Arthropods involved in transmission of tularemia include ticks, biting flies, and possibly even mosquitoes. Ticks account for more than 50% of all cases, especially west of the Mississippi River. Tularemia may present as several different clinical syndromes, including glandular, ulceroglandular, oculoglandular, oropharangeal, pneumonic, and typhoidal. In general, the clinical course is characterized by an influenza-like attack with severe initial fever, temporary remission, and a subsequent febrile period of at least 2 weeks. Later, a local lesion with or without glandular involvement may occur. Additional symptoms vary depending on the method of transmission and form of the disease (see the following discussion). Untreated, the mortality rate for tularemia is about 8%; early diagnosis and treatment can reduce that to 1 to 2%.

Depending on the route of entry of the causative organism, tularemia may be classified in several ways. The most common is ulceroglandular — resulting from cutaneous inoculation — characterized by an ulcer with sharp undetermined borders and a flat base. Location of the ulcers may help identify the mode of transmission. Ulcers on the upper extremities are often a result of exposure to infected animals, whereas ulcers on the lower extremities, back, or abdomen most often reflect arthropod transmission. When there is lymphadenopathy without an ulcerative lesion, the classification glandular tularemia is used. If the tularemia bacterium enters via the conjunctivae, oculoglandular tularemia may result. Oropharyngeal tularemia results from ingestion of contaminated food or water. If airborne transmission of the agent is involved, the pneumonic form occurs. These patients often present with fever, a nonproductive cough, dyspnea, and chest pain. Finally, tularemia may be classified as typhoidal, characterized by disseminated infection mimicking typhoid fever, brucellosis, tuberculosis, or some of the RMSF-type infections.

In ticks, tularemia infection occurs in both the gut and body tissues and hemolymph fluid (tick blood). Infection is known to persist for many months and even years in some species. Tularemia organisms may be passed from tick stage to tick stage, and to the offspring of infected female ticks. The three major North American ticks involved in transmission of tularemia organisms are the lone star tick, *Amblyomma americanum*, the Rocky Mountain wood tick, *Dermacentor andersoni*, and the American dog tick, *Dermacentor variabilis*.[43] Both the lone star tick and the American dog tick occur over much of the eastern U.S.; the Rocky Mountain wood tick occurs in the West. All three tick species are avid human biters. In Central and Western Europe, *Ixodes ricinus* is probably a vector. *Dermacentor nuttalli* may be a vector in Russia. *Dermacentor dagestanicus (= niveus)* and *Rhipicephalus pumilio* may be vectors in Kazakhstan.

G. Colorado Tick Fever

Colorado tick fever (CTF) is a generally moderate, acute, self-limited, febrile illness caused by an *Orbivirus* in the Reoviridae. Typically, onset of CTF is sudden, with chilly sensations, high fever, headache, photophobia, mild conjunctivitis, lethargy, myalgias, and arthralgias. The temperature pattern may be biphasic, with a 2- to 3-d febrile period, a remission lasting 1- to 2-d, then another 2- to 3-d of fever, sometimes with more severe symptoms.[44] Rarely, the disease may become severe in children with encephalitis, myocarditis, or tendency to bleed. Infrequently, a transient rash may accompany infection. Recovery is usually prompt, but a few fatal cases have been reported. CTF occurs in areas above 4000 ft in at least 11 western states (South Dakota, Montana, Wyoming, Colorado, New Mexico, Utah, Idaho, Nevada, Washington, Oregon, and California) and in British Columbia and Alberta, Canada. Exact case numbers are hard to ascertain as many cases may be so mild that sick persons fail to seek medical care, but approximately 200 to 400 cases are reported in the U.S. annually. Peak incidence is during April and May at lower elevations and during June and July at higher elevations. The virus is maintained in nature by cycles of infection among various small mammals and the ticks that parasitize them. Infection in humans is by the bite of an infected tick. Several tick species have been found infected with the virus, but *Dermacentor andersoni* is by far the most common. This tick is especially prevalent where there is brushy vegetation to provide good protection for small mammalian hosts of immature ticks and yet with sufficient forage to attract large hosts required for the adults.

H. Relapsing Fever

Tick-borne (endemic) relapsing fever (TBRF) is a systemic spirochetal disease characterized by periods of fever lasting 2 to 9 d alternating with afebrile periods of 2 to 4 d. The total number of relapses can vary from 1 to 10 or more, lasting 2 or 3 weeks. Transitory petechial rashes are common during the initial febrile period. Untreated, the mortality rate is between 2 and 10%. Several hundred cases are reported worldwide each year, with approximately 30 to 50 of those being diagnosed in the U.S. (primarily in Washington, Oregon, and northern California).

TBRF is caused by *Borrelia recurrentis,* or various tick-adapted strains of this organism (many authors maintain that each tick-adapted strain is a distinct species; see Color Figure 30.29). The spirochetes are transmitted to humans by several species of soft ticks in the genus *Ornithodoros.* As soft ticks generally feed for only a short period of time (30 min or so), the victim may be unaware of any recent tick bites. Rodents serve as a natural source of infection for ticks, and transmission is by tick bite (saliva) and also sometimes through contamination of the bite wound with infective coxal fluid produced by feeding ticks just before they detach. Transstadial and transovarial transmission of the agent occurs readily; thus, the ticks are reservoirs of infection. The disease is endemic across central Asia, northern Africa, tropical Africa, parts of the Middle East, and North and South America.[45] Foci of infection are restricted to *Ornithodoros*-infested areas such as huts, caves, log cabins, cattle barns, and uninhabited houses. Outbreaks in the western U.S. have most often been associated with mountain cabins or rented state or federal park cabins.[46,47] A recent outbreak of 11 cases resulted from a 1-d family gathering in a remote New Mexico cabin.[48]

I. Tick-borne Encephalitis

Tick-borne encephalitis (TBE) should be considered a general term encompassing at least three diseases caused by similar flaviviruses spanning the British Isles (Louping ill), across Europe (Central European tick-borne encephalitis), to far-eastern Russia (Russian spring-summer encephalitis, RSSE). The three diseases also differ in severity, Louping ill being the mildest and RSSE the worst. In Central Europe the typical case has a biphasic course with an early, viremic, flulike stage, followed about a week later by the appearance of signs of meningoencephalitis.[49] CNS disease is relatively mild, but occasional severe motor dysfunction and permanent disability occur. The case fatality rate is 1 to 5%.[50] RSSE (sometimes referred to as the "far-eastern form") is characterized by violent headache, high fever, nausea, and vomiting. Delirium, coma, paralysis, and death may follow; the mortality rate is about 25 to 30%. A recent report showed that new variants of TBE virus in Russia may produce a hemorrhagic syndrome.[51,52] Louping ill — named after a Scottish sheep disease — in humans also displays a biphasic pattern and is generally mild.[2] As mentioned, the virus infects sheep; few cases are actually ever reported in humans. Reported case numbers for TBE (excluding the few Louping ill cases) is between 500 and 1000. Transmission to humans is mostly by the bite of an infected tick; however, infection may also be acquired by consuming infected milk and uncooked milk products. The distribution and seasonal incidence of TBE is closely related to the activity of the tick vectors: *Ixodes ricinus* in western and central Europe, and *I. persulcatus* in central and eastern Europe (there is overlap of the two species). *Ixodes ricinus* is most active in spring and autumn. Two peaks of activity may be observed: one in late March to early June, and one from August to October. *Ixodes persulcatus* is usually active in spring and early summer. Apparently, *I. persulcatus* is more cold-hardy than *I. ricinus,* thus inhabiting harsher, more northern areas.

Powassan encephalitis (POW) — also in the TBE subgroup — is a rare infection of humans that mostly occurs in the northeastern U.S. and adjacent regions of Canada. Characteristically, there is sudden onset of fever with temperature up to 40°C along with convulsions. Also, accompanying encephalitis is usually severe, characterized by vomiting, respiratory distress, and prolonged, sustained fever. Only about 20 cases of POW have been reported in North America.[53] Recognized cases have occurred in children and adults, with a case fatality rate of approximately 50%. POW is transmitted in an enzootic cycle among ticks (primarily *Ixodes cookei*) and rodents and carnivores. *Ixodes cookei* only occasionally bites people; this may explain the low case numbers. Antibody prevalence to POW in residents of affected areas is less than 1%, indicating that human exposure to the virus life cycle is a rare event.

J. Tick Paralysis

Tick paralysis is characterized by an acute, ascending, flaccid motor paralysis that may terminate fatally if the tick is not located and removed. The causative agent is believed to be a salivary toxin produced by ticks when they feed. In the strictest sense, tick paralysis is not a zoonosis; however, many contend that zoonoses should include not only infections that humans acquire from animals, but also diseases induced by non-infective agents such as toxins and poisons.[54] The disease is more common than one might think. In North America, hundreds of cases have been documented from the Montana–British Columbia region.[55,56] It occurs in the southeastern U.S. as well. Tick paralysis is also especially common in Australia. Sporadic cases may occur in Europe, Africa, and South America.

The site of tick bite in a case of tick paralysis looks no different from that in cases without paralysis. There is a latent period of 4 to 6 d before the patient becomes restless and irritable. Within 24 h there is an acute ascending lower motor neuron paralysis of the Landry type. It usually begins with weakness of the lower limbs, progressing in a matter of hours to falling down and obvious incoordination, which is principally due to muscle weakness, although rarely there may also be true ataxia. Finally, cranial nerve weakness with dysarthria and dysphagia leads to bulbar paralysis, respiratory failure, and death. In children, presenting features may include restlessness, irritability, malaise, and sometimes anorexia and vomiting. A tick may usually be found attached to the patient, usually on the head or neck. Some controversy exists over whether or

CASE HISTORY

TICK PARALYSIS

A 4-year-old female presented to a Mississippi hospital emergency department on July 3 with a 24-h history of worsening weakness that began as difficulty in walking. Upon admission, the child was unable to stand, sit up, or even raise herself from bed. The physical exam revealed normal vital signs, mild lethargy, total body weakness, and some drooping of the eyelids. Tick paralysis was suspected, and upon closer examination, a large tick was found attached on her scalp. The tick was removed and identified as a partially engorged female American Dog tick (Figure 1). This tick is known to cause paralysis in humans and animals in the eastern U.S. The patient remained stable, but continued to exhibit total body motor weakness and difficulty swallowing resulting in drooling when she attempted to drink liquids. When the child failed to improve after 4 h, she was transferred to a regional pediatric intensive care unit. After arrival in the ICU, she quickly began regaining motor strength over the next 2 to 3 h. Residual motor weakness lasted for a total of about 7 to 8 h after tick removal. One interesting aspect of this case is how long it took the patient to recover. In most cases of tick paralysis in North America (this is not the case in Australia), the patient quickly recovers after tick removal.

Figure 1
Female American dog tick, Dermacentor variabilis, *involved in a case of tick paralysis.*

not severity of symptoms is related to the proximity of the attached tick to the patient's brain. In one study, the case fatality rate in patients with ticks attached to the head or neck was higher than that in patients with ticks attached elsewhere; however, the difference was not statistically significant. Although ticks causing paralysis are often attached to the head or neck, it must be noted that cases of paralysis may occur from tick bites anywhere on the body (published examples: external ear, breast, groin, and back[57]). Once the tick is found and removed, all symptoms usually disappear rapidly

TICKS

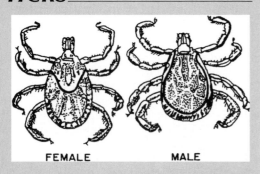

FEMALE   **MALE**

Figure 4
Female and male hard ticks.
(From USAF Publication USAFSAM-SR-89-2.)

Importance

Annoyance; disease transmission; tick paralysis

Distribution

Numerous species worldwide

Lesion

Red papule with erythema; sometimes nodular

Disease Transmission

Lyme disease, Rocky Mountain spotted fever, tularemia, relapsing fever, tick-borne encephalitis, ehrlichiosis, anaplasmosis, and others

Key Reference

Arthur, D. R., Pergamon Press, New York, 445 pp., 1962

Treatment

Generally none needed for bites other than palliatives after tick removal; watch for development of tick-borne disease

(there are exceptions to this, especially in Australia with *Ixodes holocyclus*).

As many as 43 tick species in 10 genera have been incriminated in tick paralysis in humans, other mammals, and birds.[58] However, human cases of the malady mostly occur in only a few geographic regions, caused by three main tick species. In the Northwestern U.S. and British Columbia region of North America, the Rocky Mountain wood tick, *Dermacentor andersoni*, is the principal tick involved. In the southeastern U.S. a related species, *Dermacentor variabilis,* known as the American dog tick, is the main cause of tick paralysis. Human cases in Australia are due to the Australian paralysis tick, *Ixodes holocyclus.*

Interestingly, not all feeding female ticks — even of the species known to cause paralysis — produce paralysis. Why, out of hundreds of tick bites, does one result in paralysis? There is some evidence that in cattle, sheep, and dogs, numerous ticks feeding simultaneously (to reach a minimum dose) is necessary to elicit paralysis. In humans, however, one tick is usually involved. Most researchers believe that tick paralysis is caused by a toxin, but its nature is not well characterized. Generally, it is thought that the toxin is produced in the salivary glands of the female tick as she feeds. One alternative view would be that the toxin is produced in tick ovaries and subsequently passes to the salivary glands during later stages of tick engorgement. Although the vast majority of cases are due to female ticks, there are reports of male ticks causing limited paralysis. This fact seems to argue against the ovary toxin theory. There are other theories for the cause of the paralysis such as host reactions to components of the tick saliva or possibly symbiotic rickettsial organisms commonly found in tick salivary glands.

II. GENERAL BIOLOGY AND ECOLOGY

There are three families of ticks recognized today: (1) Ixodidae (hard ticks), (2) Argasidae (soft ticks), and (3) Nuttalliellidae (a small,

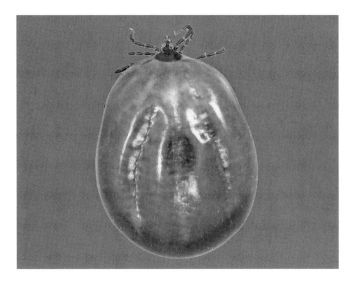

Figure 4a
Engorged female tick.
(From Dr. Blake Layton, Mississippi Cooperative Extension Service, Mississippi State University.)

curious, little-known group with some characteristics of both hard and soft ticks). The terms *hard* and *soft* refer to the presence of a dorsal scutum or "plate" in the Ixodidae, which is absent in the Argasidae.

Hard ticks display sexual dimorphism; males and females look obviously different (Figure 4), and the blood-fed females are capable of enormous expansion (Figure 4a). Their mouthparts are anteriorly attached and visible from dorsal view (Figures 5 and 6B). When eyes are present, they are located dorsally on the sides of the scutum.

Soft ticks are leathery and nonscutate, without obvious sexual dimorphism. Their mouthparts are subterminally attached (in adult and nymphal stages) and not visible from dorsal view (Figure 6A). Eyes, when present, are located laterally in folds above the legs.

There are major differences in the biology of hard and soft ticks. Some hard tick species have a one-host life cycle, wherein engorged larvae and nymphs remain on the host after feeding; they then molt, and the subsequent stages reattach and feed. The adults mate on the host, and only engorged females drop off to lay eggs on the ground. Although some hard ticks complete their development on only one or two hosts, most commonly encountered ixodids have a three-host life cycle. In this case, adults mate on a host (except for some *Ixodes* spp.), and the fully fed female drops from a host animal to the ground and lays from 2,000 to 18,000 eggs, after which she dies. The eggs hatch in about 30 d into a 6-legged seed tick (larval) stage, which feeds predominantly on small animals. The fully fed seed ticks drop to the ground and transform into 8-legged nymphs. These nymphs seek an animal host, feed, and drop to the ground. They then molt into adult ticks, thus completing the life cycle. Figure 7A shows all the motile life stages of hard ticks.

Many hard tick species *quest* for hosts, whereby they climb blades of grass or weeds and remain attached, forelegs outstretched, awaiting a passing host. For this reason, dragging a white flannel cloth through brushy areas works well for collecting ticks (Figure 8). They may travel up a blade of grass (to quest) and back down to the leaf

Ticks

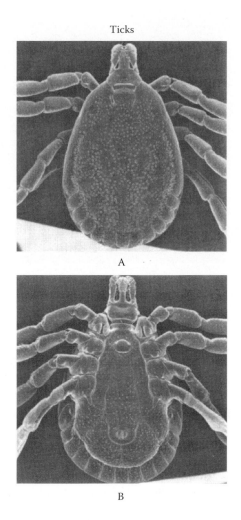

A

B

Figure 5
Hard tick showing (A) dorsal and (B) ventral aspects.
(National Institutes of Health. Photo courtesy of Dr. Jim Keirans.)

litter where humidity is high (to rehydrate) several times a day. Also, hard ticks will travel toward a carbon dioxide source. Adult ticks are more adept at traveling through vegetation than the minute larvae. Studies have shown that adult lone star ticks may travel up to 10 m (33 ft) to a carbondioxide source, but other species such as *I. scapularis* will only travel distances of 1 to 2 m (3.3 to 6.6 ft) toward a carbon dioxide source. Goddard[59] demonstrated minimal (less than 2 ft) lateral movement by questing adult *I. scapularis* in a mark–release–recapture study.

Ticks feed by cutting a small hole into the host epidermis with their chelicerae and inserting the hypostome into the cut, thereby attaching to the host. Blood flow is presumably maintained with the aid of an anticoagulant from the salivary glands. Some hard ticks secure their attachment to the host by forming a cement cone around the mouthparts and surrounding skin. Two phases are recognized in the feeding of nymphal and female hard ticks: (1) a growth feeding stage characterized by slow

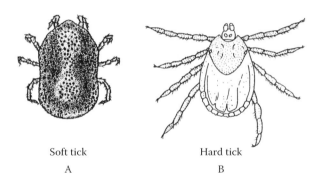

Figure 6
Comparison of (A) soft and (B) hard ticks. (From USAF Publication USAFSAM-SR-89-2.)

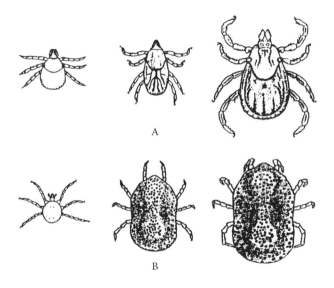

Figure 7
Motile life stages of (A) hard ticks and (B) soft ticks. (From USAF Publication USAFSAM-SR-89-2.)

continuous blood uptake and (2) a rapid engorgement phase occurring during the last 24 h or so of attachment.

The biology of soft ticks differs from hard ticks in several ways (see Figure 7B for life stages). Adult female soft ticks feed and lay eggs several times during their lifetime. Soft tick species may also undergo more than one nymphal molt before reaching the adult stage. With the exception of larval stages of some species, soft ticks do not firmly attach to their hosts for several days like the Ixodidae. They are adapted to feeding rapidly and leaving the host promptly.

The expansion capability of hard ticks sometimes causes confusion among non-specialists. I have often been sent fully engorged hard ticks removed from dogs with instructions to "identify enclosed soft tick;" this misconception is because engorged ixodids do sometimes appear "soft." Another common misconception is that flat, unengorged hard ticks and engorged hard ticks represent different species. Ranchers often

Figure 8
Tick collection by dragging or flagging with a white flannel cloth.

speak of two "species" on their cattle; "the large swollen species" and the "small flat, brown species."

Hard ticks and soft ticks occur in different habitats. In general, hard ticks occur in brushy, wooded, or weedy areas containing numerous deer, cattle, dogs, small mammals, or other hosts. Soft ticks are generally found in animal burrows or dens, bat caves, dilapidated or poor-quality human dwellings (huts, cabins, etc.), or animal-rearing shelters. Many soft tick species thrive in hot and dry conditions, whereas ixodids are more sensitive to desiccation (the genus *Hyalomma* may be an exception) and, therefore, usually found in areas protected from high temperatures, low humidities, and constant breezes.

Most hard ticks, being sensitive to desiccation, must practice water conservation and uptake. Their epicuticle contains a wax layer that prevents water movement through the cuticle. Water can be lost through the spiracles; therefore, resting ticks keep their spiracles closed most of the time (opening them only once or twice an hour). Tick movement and the resultant rise in carbon dioxide production cause the spiracles to open about 15 times an hour with a corresponding water loss.

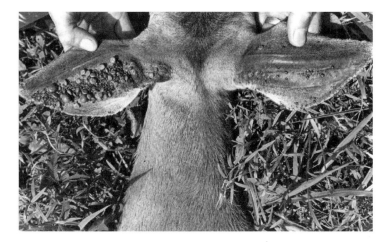

Figure 9
Deer heavily infested with lone star ticks. (Tennessee Valley Authority. Photo by Denise Schmittou.)

Development, activity, and survival of hard ticks is influenced greatly by temperature and humidity within the tick microhabitat. Lancaster[60] found that lone star tick eggs reared in an environment of less than 75% humidity would not hatch. Lees[61] demonstrated that *I. ricinus* died within 24 h if kept in a container of 0% RH, but survived 2 to 3 months at 90% RH. Because of their temperature and humidity requirements, as well as host availability, hard ticks tend to congregate in areas providing those factors. Ecotonal areas (ecological interface areas such as between forests and fields) are excellent habitats for hard ticks. Open meadows and prairies, along with climax forest areas, support the fewest lone star ticks. Ecotone areas and small openings in the woods are usually heavily infested. In a study by Semtner et al.,[62] lone star tick populations decreased with an increase in distance from the ecotone. Studies in Virginia demonstrated that American dog ticks tend to be especially abundant along trails, roadsides, and forest boundaries surrounding old fields or other clearings.[63–65]

Deer and small mammals thrive in ecotonal areas, thus providing blood meals for ticks. In fact, deer are often heavily infested with hard ticks in the spring and summer months (Figure 9). The optimal habitat of white tail deer has been reported to be the forest ecotone, as the area supplies a wide variety of browse and frequently offers the greatest protection from their natural enemies. Many favorite deer foods are also found in the low trees of an ecotone, including greenbrier, sassafras, grape, oaks, and winged sumac.

Ticks are not evenly distributed in the wild; instead, they are localized in areas providing their necessary temperature, humidity, and host requirements. These biologic characteristics of ticks, when known, may enable us to avoid the parasites.

III. TICK IDENTIFICATION

Ticks look like large mites, with both having eight legs and one (apparent) disk-shaped body region. However, ticks differ from mites in having a toothed hypostome and no claws on their palps. Like those of many other arthropod groups, tick morphological structures have been assigned rather long and somewhat confusing names. Figure 10

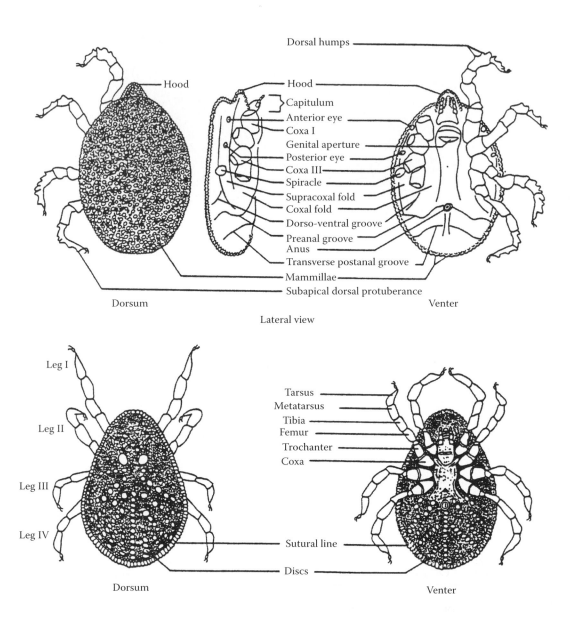

Figure 10
Diagnostic characters of soft ticks. (From Strickland et al. Ticks of Veterinary Importance, *USDA, APHIS, Agri. Hnbk. No. 485, Washington, DC, 1976.)*

and Figure 11 are provided to acquaint the reader with those terms. Some identification keys are well written and relatively easy to use (a good example is the pictorial key to adult hard ticks east of the Mississippi River[66]). However, tick identification to the species level is probably best accomplished by an expert to avoid misinformation. In a "study" that the author was asked to review, more than 30% of the ticks collected and identified from Mississippi did not even occur in this country.

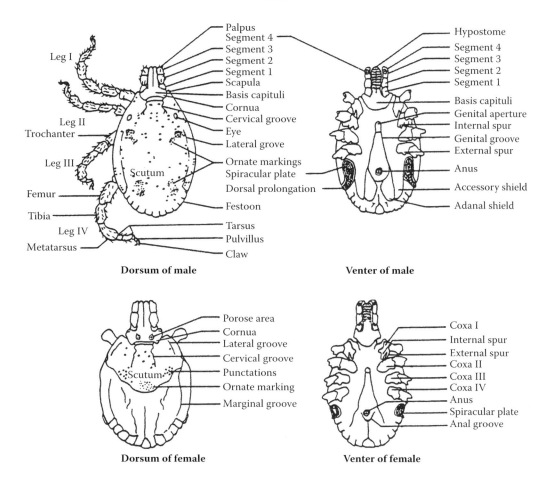

Dorsum of male

Leg I
Leg II
Trochanter
Leg III
Femur
Tibia
Leg IV
Metatarsus

Palpus
Segment 4
Segment 3
Segment 2
Segment 1
Scapula
Basis capituli
Cornua
Cervical groove
Eye
Lateral grove
Ornate markings
Spiracular plate
Dorsal prolongation
Festoon
Tarsus
Pulvillus
Claw
Scutum

Venter of male

Hypostome
Segment 4
Segment 3
Segment 2
Segment 1
Basis capituli
Genital aperture
Internal spur
Genital groove
External spur
Anus
Accessory shield
Adanal shield

Dorsum of female

Porose area
Cornua
Lateral groove
Cervical groove
Punctations
Ornate marking
Marginal groove
Scutum

Venter of female

Coxa I
Internal spur
External spur
Coxa II
Coxa III
Coxa IV
Anus
Spiracular plate
Anal groove

Figure 11
Diagnostic characters of hard ticks. (From Strickland et al. Ticks of Veterinary Importance, *USDA, APHIS, Agri. Hnbk. No. 485, Washington, DC, 1976.)*

IV.   DISCUSSION OF SOME OF THE COMMON U.S. SPECIES*

Ornithodoros hermsi Wheeler, Herms, and Meyer

Medical Importance: Primary vector of tick-borne relapsing fever (TBRF) spirochetes in the Rocky Mountain and Pacific Coast states, U.S; implicated in several TBRF outbreaks[46,47]

Description: Typical-looking soft tick about 10 mm long, gray in color, and covered with numerous bumplike projections (mammillae); as with all soft ticks, head (mouthparts in this case) not visible from dorsal view; foreleg depicted in Figure 12

---

* *The descriptions for each tick species are only general comments about their macroscopic appearance. Tick identification to the species level is quite difficult, utilizing a number of microscopic characteristics. Specific identifications should be performed by specialists at institutions that routinely handle such requests (universities, extension services, state health departments, the military, etc.).*

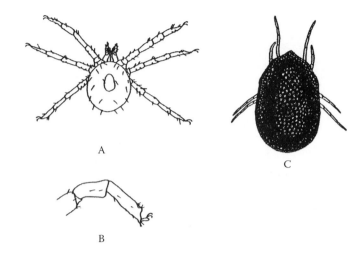

Figure 12
(A) Larva, (B) foreleg, and (C) adult of Ornithodoros hermsi.
(From USAF Publication USAFSAM-SR-89-2.)

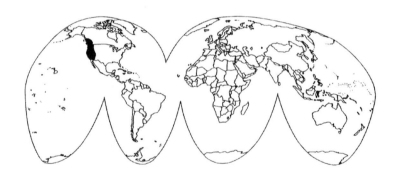

Figure 13
Approximate geographic distribution of Ornithodoros hermsi.
(From USAF Publication USAFSAM-89-2.)

Distribution: California, Nevada, Idaho, Oregon, Utah, Arizona, Washington, and Colorado, as well as in British Columbia, Canada (Figure 13)

Host: Rodents and humans

Seasonality: Varies with geographic location, hosts, and habitat

Remarks: Often found infesting corners and crevices of vacation or summer cabins; larvae only remain attached to a host for about 15 to 20 min; usually 4 nymphal molts; cycle from egg to egg takes about 4.5 months; often found in coniferous forests at elevations above 1000 m

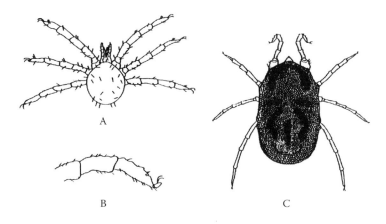

Figure 14
(A) Larva, (B) foreleg, and (C) adult, of Ornithodoros turicata.
(From USAF Publication USAFSAM-89-2.)

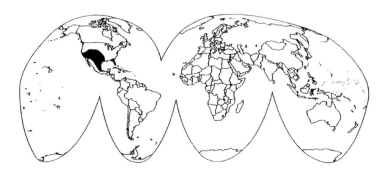

Figure 15
Approximate geographic distribution of Ornithodoros turicata.
(From USAF Publication USAFSAM-89-2.)

Relapsing Fever Tick
Ornithodoros turicata (Duges)

Medical Importance: May produce an intense irritation and edema at the bite site in humans; serves as a vector of relapsing fever spirochetes in portions of Kansas, Oklahoma, Texas, and other southwestern states

Description: About 10 mm long, gray in color, and covered with bumplike projections (mammillae); foreleg depicted in Figure 14

Distribution: Texas, New Mexico, Oklahoma, Kansas, California, Colorado, Arizona, Florida, and Utah; reported from Mexico in the states of Aguascalientes, Coahuila, Gunanjuato, Morelos, Queretaro, San Luis Potosi, and Sinaloa (Figure 15); also reportedly found in Venezuela, Honduras, Bolivia, Chile, and Argentina; records from Central and South America probably incorrect

Hosts: Collected from rattlesnakes, turtles, birds, rodents, rabbits, sheep, cattle, horses, pigs, and humans

Seasonality: Varies with geographic location, hosts, and habitat; may be active in warmer geographic areas throughout year

Remarks: Often found in burrows used by rodents or burrowing owls; bite is painless but may be followed in a few hours by intense local irritation and swelling; subsequently, subcutaneous nodules may form and persist for months; 3 to 5 nymphal stages; time required for development from larva to adult is approximately 6 months

Lone Star Tick
Amblyomma americanum (Linnaeus)

Medical Importance: Transmits the pathogen of tularemia to humans; a known vector of agent of human ehrlichiosis — *Ehrlichia chaffensis*;[31] reported to transmit the agent of Rocky Mountain spotted fever (RMSF) (Goddard[67] provided a detailed review of the disease potential of this species; however, studies indicate that *A. americanum* may not be an important vector of RMSF[68,69]); Lyme-borreliosis-like spirochetes have been recovered from this species;[10] found naturally infected with *Rickettsia parkeri*[69] and *Ehrlichia ewingii*[70,71]

Description: Palps long, with segment 2 at least twice as long as segment 3; mouthparts visible from above (in contrast to the soft ticks); eyes present, but not in sockets; reddish-brown tick species; adult females have distinct single white spot on their back (scutum); males have no single spot, instead have inverted horseshoe-shaped markings at the posterior edge of their dorsal side (Figure 16)

Distribution: Central Texas east to the Atlantic Coast and north to approximately Iowa and New York; reported from Mexico in the northern states of Coahuila, Nuevo Leon, and Tamaulipas (Figure 17); also occasionally reported from Panama, Venezuela, Argentina, Guatemala, Guayana, and Brazil (however, Central and South American records of this species may not be valid)

Hosts: Extremely aggressive and nonspecific in its feeding habits; all three motile life stages will feed on a wide variety of mammals (including humans) and ground-feeding birds

Seasonality: Adults and nymphs generally active from early spring through midsummer, with larvae being active from late summer into early fall

Remarks: Probably the most annoying and commonly encountered tick occurring in the southern U.S.; in some rural areas almost every person has been bitten by these ticks at one time or another; the "seed ticks" occurring in late summer in the southern U.S. are most often this species (Figure 18); lone star ticks especially found in interfacing zones between forested and open (meadow) areas, especially where there is an abundance of deer or other hosts; they seldom occur in high numbers in the middle of pastures or meadows because of low humidities and high daytime temperatures

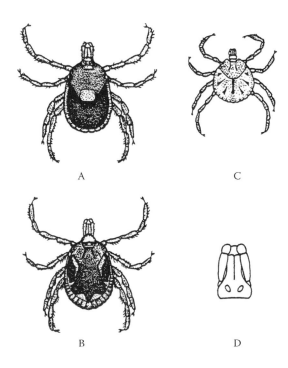

Figure 16
(A) Adult female, (B) adult male, (C) nymph, and (D) dorsal view of capitulum of Amblyomma americanum. *(From USAF Publication USAFSAM-89-2.)*

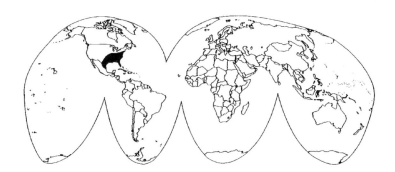

Figure 17
Approximate geographic distribution of Amblyomma americanum.
(From USAF Publication USAFSAM-89-2.)

present in those areas; larvae may survive from 2 to 9 months, nymphs and adults 4 to 15 months each; females usually deposit 3000 to 8000 eggs

Female often falsely referred to as the "spotted fever tick" because of the single white spot visible on its back; however, this spot has nothing to do with the presence or absence of RMSF organisms; adults have very long mouthparts and can produce painful bites

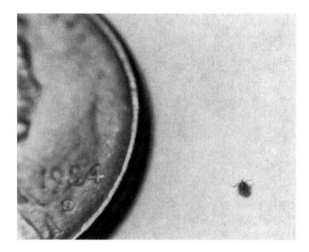

Figure 18
Larval lone star tick showing extremely small size.

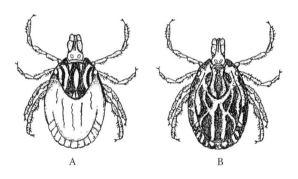

A   B

Figure 19
(A) Adult female and (B) male Amblyomma maculatum.

Gulf Coast Tick
Amblyomma maculatum (Koch)

Medical Importance: Known vector of American boutonneuse fever (infection with *R. parkeri*);[24,26] nuisance effects due to painful bites

Description: Macroscopically, somewhat similar to the American dog tick, except that it has metallic markings (instead of white) and long mouthparts typical of all *Amblyomma*; large tick species with long mouthparts visible from above; adult females with metallic white or gold markings on scutum (Figure 19); males with numerous, mostly connected, linear spots of golden white

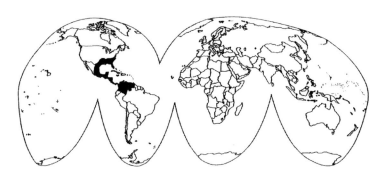

Figure 20
Approximate geographic distribution of Amblyomma maculatum.

Distribution: Portions of Atlantic and Gulf coast areas (generally 100 to 200 mi inland) and south into Mexico and portions of Central and South America (Figure 20)

Hosts: Adults on large animals including deer, cattle, sheep, and humans; larvae and nymphs on small mammals and ground-feeding birds such as rabbits, fox, meadow larks, and bobwhite quail

Seasonality: Variable depending on geographic location; larvae generally out from June through October; nymphs active from July through October; adults can be found from March through September, usually peaking in activity during August

Remarks: Increasingly a pest in the southern U.S.; large ticks often found in the ears of cattle, producing great irritation, destruction of cartilage, and drooping, called *gotched* ears

Rocky Mountain Wood Tick
Dermacentor andersoni Stiles

Medical Importance: Primary vector of RMSF in the Rocky Mountain states and also known to transmit the causative agents of Colorado tick fever and tularemia; produces cases of tick paralysis in the U.S. and Canada each year[72]

Description: Adults have shorter mouthparts than the *Amblyomma* species and are usually dark brown or black with bright white markings on the scutum (see Figure 21 for pattern); basis capituli rectangular when viewed from above; a pair of medially directed spurs on the first pair of coxae.

Distribution: Found from the western counties of Nebraska and the Black Hills of South Dakota to the Cascade and Sierra Nevada Mountains; also reported from northern Arizona and northern New Mexico to British Columbia, Alberta, and Saskatchewan, Canada (Figure 22)

Hosts: Immatures prefer many species of small mammals such as chipmunks and ground squirrels, whereas adults feed mostly on cattle, sheep, deer, humans, and other large mammals

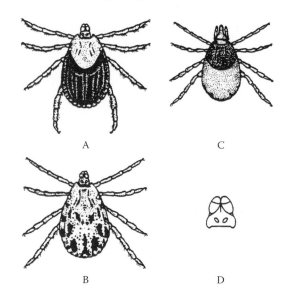

Figure 21
(A) Adult female, (B) adult male, (C) nymph, and (D) dorsal view of capitulum of Dermacentor andersoni. *(From USAF Publication USAFSAM-89-2.)*

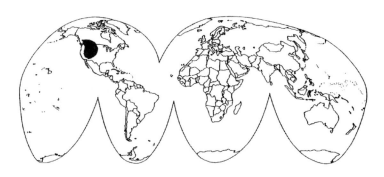

Figure 22
Approximate geographic distribution of Dermacentor andersoni.
(From USAF Publication USAFSAM-89-2.)

Seasonality: Larvae feed throughout the summer and adults usually appear in March, disappearing by July; nymphs may continue to be present, although in diminishing numbers, until late summer

Remarks: Especially prevalent where there is brushy vegetation to provide good protection for small mammalian hosts of immatures and with sufficient forage to attract large hosts required by adults; unfed larvae may live for 1 to 4 months, nymphs for 10 months or more, and adults 14 months or longer; females deposit about 4000 eggs

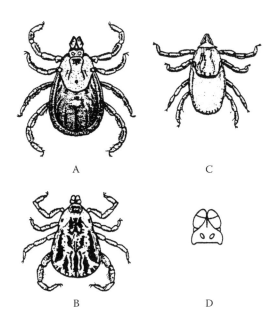

Figure 23
(A) Adult female, (B) adult male, (C) nymph, and (D) dorsal view of capitulum of Dermacentor variabilis. *(From USAF Publication USAFSAM-89-2.)*

American Dog Tick
Dermacentor variabilis (Say)

Medical Importance: One of the most medically important ticks in the U.S.; primary vector of RMSF in the East; also transmits tularemia and causes tick paralysis[72,73]

Description: Adults are dark brown or black with short, thick mouthparts; dull or bright white markings on the scutum (see Figure 23 for pattern); basis capituli rectangular when viewed from above; a pair of medially directed spurs on the first pair of coxae.

Distribution: Throughout the U.S. except in parts of the Rocky Mountain region; also established in Nova Scotia, Manitoba, and Saskatchewan, Canada; reported in Mexico from Chiapas, Gunajuato, Hidalgo, Oaxaca, Puebla, San Luis Potosi, Sonora, Tamaulipas, and Yucatan (Figure 24)

Hosts: Immatures feed primarily on small mammals (particularly rodents); adults prefer the domestic dog, but will readily bite humans

Seasonality: Adults active from about mid-April to early September; nymphs predominate from June to early September; larvae active from about late March through July

Remarks: Principal vector of RMSF in the central and eastern U.S. (should be avoided whenever possible); deticking dogs important mode of RMSF transmission that may be

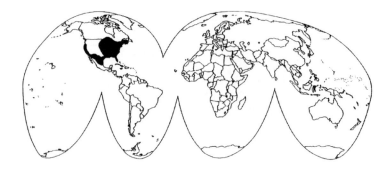

Figure 24
Approximate geographic distribution of Dermacentor variabilis.
(From USAF Publication USAFSAM-89-2.)

overlooked (handpicking *D. variabilis* from dogs is dangerous because infected tick secretions on the hands may be transmitted through contact with the eyes, mucous membranes, etc.); unfed larvae may live up to 15 months, nymphs 20 months, and adults up to 30 months or longer; females deposit 4000 to 6500 eggs

Western Black-Legged Tick
Ixodes pacificus Cooley and Kohls

Medical Importance: Known to be a vector of Lyme borreliosis spirochetes; most, if not all, cases of Lyme borreliosis occurring in California transmitted by this tick; transmits agent of human granulocytic anaplasmosis; there are reports of Type I (IgE-mediated) hypersensitivity reactions in humans as a result of bites by this species

Description: No white markings on their dorsal side, no eyes or festoons (Figure 25) (like other members of genus); anal groove that encircles the anus anteriorly; males with sclerotized ventral plates; adults generally dark brown in color with moderately long mouthparts; looks almost identical to *I. scapularis*

Distribution: Along the Pacific coastal margins of British Columbia, Canada, and U.S., possibly extending into Baja, California and other parts of Mexico (Figure 26); also reported from at least one area in Arizona

Hosts: Immatures feed on numerous species of small mammals, birds, and lizards; in certain areas of California, predominance of feeding on lizards; adults feed primarily on Columbian black-tailed deer

Seasonality: Adults primarily active from fall to late spring with immatures active in the spring and summer

Remarks: Adults, like *I. scapularis*, have long mouthparts, enabling them to be especially painful parasites of humans; adults most abundant in the early spring; infection rates with Lyme borreliosis spirochetes usually in the range of 1 to 5% compared to rates of 25 to 75% in the northern form of *I. scapularis* (effect may be related to vector competence or host preferences of the immatures); immatures will bite people

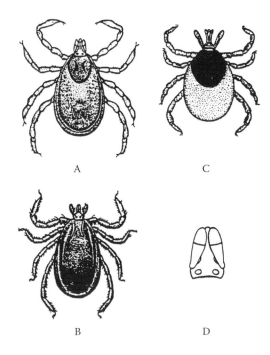

Figure 25
(A) Adult female, (B) adult male, (C) nymph, and (D) dorsal view of capitulum of Ixodes *pacificus.*
(From USAF Publication USAFSAM-89-2.)

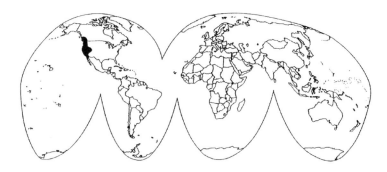

Figure 26
Approximate geographic distribution of Ixodes *pacificus. (From USAF Publ. USAFSAM-89-2.)*

Black-Legged Tick
Ixodes scapularis Say

Taxonomic Note: For several years, the northern form of *Ixodes scapularis* was thought to be a distinct species named *Ixodes dammini* (see Goddard[74] for a discussion of this issue). Subsequently, evidence was produced indicating that the two are one species.[75] Morphologically, they are almost identical. However, there are important behavioral differences, especially in the immature stages. In this book, *Ixodes dammini* is considered a synonym and is included under *I. scapularis*

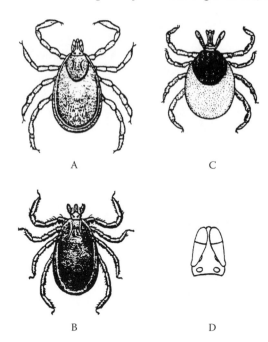

Figure 27
(A) Adult female, (B) adult male, (C) nymph, and (D) dorsal view of capitulum, of Ixodes scapularis.
(From USAF Publ. USAFSAM-89-2.)

Medical Importance: Northern-form primary vector of the causative agent of Lyme borreliosis, especially in the northeastern and upper midwestern areas of the U.S.; vector of the protozoan, *Babesia microti*, also in the Northeast and Upper Midwest; also a vector of agent of human granulocytic anaplasmosis (HGA); southern form of *Ixodes scapularis* a vector of Lyme borreliosis spirochetes,[76] however, infection rates extremely low

Description: Adults have no eyes, festoons, or white markings on their dorsal side (Figure 27), anal groove that encircles the anus anteriorly; males with sclerotized ventral plates; dark brown in color (occasionally the abdomen from dorsal view is light brown or orangish)

Distribution: Northern form in the New England states and New York, south into New Jersey, Virginia, and Maryland; also found in Upper Midwest and Ontario, Canada; southern form found in the southern Atlantic Coast states and throughout the South including Texas and Oklahoma; also reported from the Mexican states of Jalisco and Tamaulipas (Figure 28)

Hosts: Immatures feed on lizards, small mammals, and birds; adults prefer deer but will bite people; in Mexico, additional host records from dogs, cattle, and jaguar

Seasonality: In U.S., adults active in fall, winter, and spring (Figure 29) (seasonal activity pattern interesting, as most people do not think of ticks being active in the dead of winter); immatures active in spring and summer

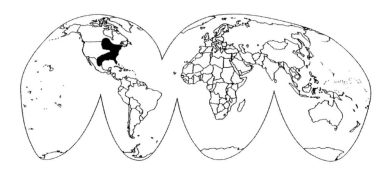

Figure 28
Approximate geographic distribution of Ixodes scapularis. *(From USAF Publication USAFSAM-89-2.)*

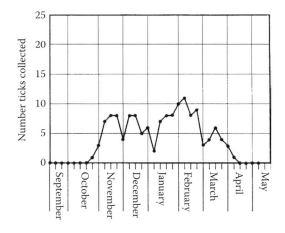

Figure 29
Seasonal activity of adult I. scapularis *in Mississippi, as determined by dragging vegetation with a white flannel cloth.*

Remarks: Congregates along paths, trails, and roadways in various types of forested areas such as those with mature pine hardwoods with dogwood, wild blueberry, privet, blackberry, huckleberry, and sweetgum; inflicts a painful bite; adult males rarely bite; nymphs of northern form bite people aggressively during summer months and can be collected with a drag cloth; nymphs of southern form rarely, if ever, bite people and can rarely be collected with drag cloths; most hard ticks acquired by persons in the southcentral and southeastern states in the winter months are of this species; in one study,[77] they were most often collected questing at around 20°C, but were collected on days as cold as 6.9°C

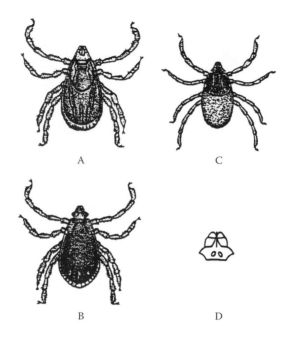

Figure 30
(A) Adult female, (B) adult male, (C) nymph, and (D) dorsal view of capitulum of Rhipicephalus
sanguineus. *(From USAF Publication USAFSAM-89-2.)*

Brown Dog Tick
Rhipicephalus sanguineus (Latreille)

Medical Importance: Recently reported to transmit the agent of RMSF;[15] in southern
Europe and Africa, known vector of *Rickettsia conorii*, the causative agent of boutonneuse fever

Description: Light to dark brown in color, with no white markings on the dorsum
(Figure 30); hexagonal-shaped basis capituli; festoons and eyes both present

Distribution: Probably the most widely distributed of all ticks, being found almost
worldwide (Figure 31); in the Western Hemisphere, records from most of U.S. and
southeastern and southwestern parts of Canada; reported from most of Mexico, Argentina, Venezuela, Colombia, Brazil, Nicaragua, Panama, Uruguay, Paraguay, Galapagos
Islands, Surinam, British Guiana, French Guiana, Peru, Costa Rica, Caribbean islands
of Cuba, Jamaica, and Bahamas; also widely distributed throughout Eurasia, Africa, and
Australian region

Hosts: The dog is the principal host, although in immature stages sometimes attacks
numerous other animals; humans historically only occasionally bitten in U.S.; could be
becoming more anthropophilic[78,79]

Seasonality: May be active in the warmer parts of its range year-round; however, in
temperate zones adults and immatures primarily active from late spring to early fall

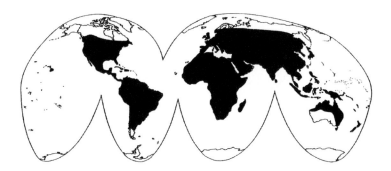

Figure 31
Approximate geographic distribution of Rhipicephalus sanguineus.
(From USAF Publication USAFSAM-89-2.)

Remarks: Most often found indoors in and around pet bedding areas; strong tendency to crawl upward and often seen climbing the walls of infested houses; associated with homes and yards of pet owners and seldom found out in the middle of a forest or uninhabited area; unfed larvae may survive as long as 8.5 months, nymphs 6 months, and adults 19 months; females usually lay 2000 to 4000 eggs

V. DISCUSSION OF SOME MAJOR PEST SPECIES IN OTHER AREAS OF THE WORLD

Eyeless Tampan
Ornithodoros moubata (Murray)

Medical Importance: Known vector of African tick-borne relapsing fever spirochetes in eastern, central, and southern Africa

Description: Like many other soft ticks, about 9 to 12 mm long in adult stage; bumpy integument (mammillated) and protuberances on the tarsi (Figure 32)

Distribution: Throughout eastern Africa and the northern portions of southern Africa, extending into the drier parts of central Africa (Figure 33)

Hosts: Humans, warthogs, domestic pigs, antbears, and porcupines

Seasonality: Varies with geographic location, hosts, and habitat

Remarks: Often found in cracks in walls and in earthen floors of huts; female usually lays 6 to 7 batches of eggs (several hundred per batch) during her lifetime; larvae do not feed; nymphs engorge in about 20 to 25 min; usually 4 nymphal molts for males and 5 for the females; able to live up to 5 years without feeding

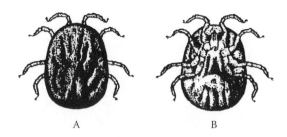

Figure 32
(A) Dorsal and (B) ventral view of Ornithodoros moubata.
(From USAF Publication USAFSAM-89-2.)

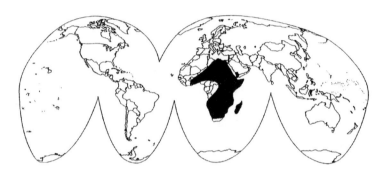

Figure 33
Approximate geographic distribution of Ornithodoros moubata.
(From USAF Publication USAFSAM-89-2.)

Carios rudis (includes *venezuelensis*) Karsch
Klompen[80] has determined that several
Ornithodoros species should be placed in the genus *Carios*

Medical Importance: Most important vector of relapsing fever spirochetes in Panama, Colombia, Venezuela, and Ecuador

Description: Unlike some of the other soft ticks, has no dorsal humps on legs (see Figure 34)

Distribution: Panama, Paraguay, Colombia, Venezuela, Peru, and Ecuador (Figure 35)

Hosts: Domestic birds and humans

Seasonality: Varies with geographic location, hosts, and habitat; may be active in warmer areas throughout year

Remarks: Appears especially adapted as parasite of humans but feeds on other animals; night feeder with the larval stages engorging rapidly; 3 to 4 nymphal stages; developmental time from larvae to adult is about 3 months

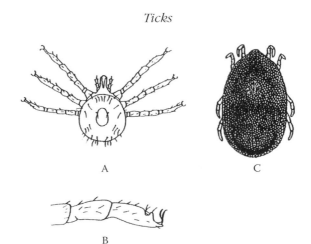

Figure 34
(A) Larva, (B) foreleg, and (C) adult of Carios (= Ornithodoros) rudis.
(From USAF Publication USAFSAM-89-2.)

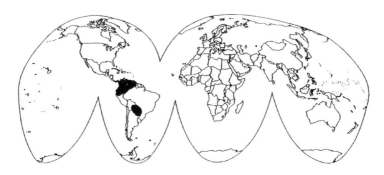

Figure 35
Approximate geographic distribution of Carios (= Ornithodoros) rudis.
(From USAF Publication USAFSAM-89-2.)

Carios talaje (Guerin-Meneville)[80]

Medical Importance: Transmits the agent of relapsing fever to humans in Guatemala, Panama, and Colombia

Description: Typical-looking soft tick with large disks (round spots mostly on the dorsal side of the integument) and no humps on tarsi (Figure 36)

Distribution: Reported in Florida, Texas, Arizona, Nevada, Kansas, New Mexico, and California (however, Hoogstraal[81] maintains that in the U.S. it has only been reported from Kansas and California); occurs in Mexico in the states of Baja California, Chiapas, Guerrero, Morelos, Oaxaca, Puebla, Sinaloa, Sonora, Veracruz, and Yucatan; also reported from Venezuela, Uruguay, Brazil, French Guiana, Panama, Ecuador, and Chile. Hoffman[82] notes this species also reported from Guatemala, Colombia, Argentina, and Galapagos Islands (although according to Keirans et al.,[83] the *O. talaje* reported from Galapagos Islands actually is *O. galapagensis*; Figure 37)

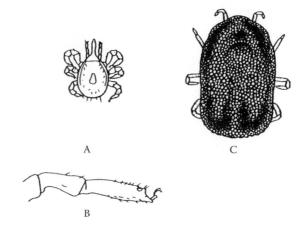

Figure 36
(A) Larva, (B) foreleg, and (C) adult of Carios talaje. *(From USAF Publication USAFSAM-89-2.)*

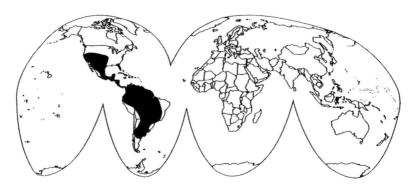

Figure 37
Approximate geographic distribution of Carios talaje. *(From USAF Publication USAFSAM-89-2.)*

Hosts: Rodents (principally) and humans, as well as birds, bats, pigs, cattle, horses, opossums, and snakes

Seasonality: Varies with geographic location, hosts, and habitat; may be active in warmer geographic areas throughout year

Remarks: Adults seldom observed in dwellings and not avid parasites of humans; larvae remain attached to a host for several days; 3 to 4 nymphal stages; developmental time from larva to adult is about 8 months

Cayenne Tick
Amblyomma cajennense (Fabricius)

Medical Importance: Probably most commonly encountered and aggressive of all Central and South American ticks; considered vector of RMSF rickettsiae in Mexico, Panama, Colombia, and Brazil

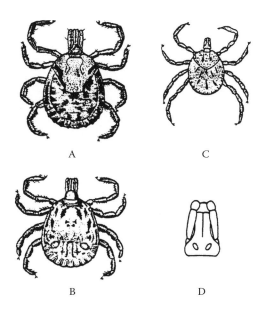

Figure 38
(A) Adult female, (B) adult male, (C) nymph, and (D) dorsal view of capitulum of Amblyomma cajennense. *(From USAF Publication USAFSAM-89-2.)*

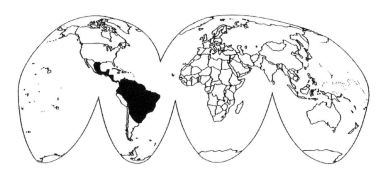

Figure 39
Approximate geographic distribution of Amblyomma cajennense.
(From USAF Publication USAFSAM-89-2.)

Description: Long mouthparts, eyes, and festoons similar to that of *A. americanum*; males have weblike ornamentation radiating from the center of the scutum; females also have extensive ornamentation and festoons with tubercules at the posterior edge (Figure 38)

Distribution: Extreme southern Texas, south throughout Mexico and Central America into parts of South America; most of Mexico, Panama, several Caribbean islands including Cuba and Jamaica, Brazil, Honduras, Venezuela, Costa Rica, Uruguay, Ecuador, Nicaragua, and Bolivia (Figure 39); Hoffman[82] states it also occurs in Guatemala, Colombia, Guayana, Paraguay, and Argentina

Hosts: All active stages commonly attack people, domestic and wild animals, and ground-frequenting birds

Figure 40
(A) Adult female and (B) male Amblyomma hebraeum. *(From USAF Publication USAFSAM-89-2.)*

Seasonality: May be active in tropical areas year-round; may be reduced activity in midwinter in the cooler areas at the northernmost and southernmost extent of distribution

Remarks: Very similar to *A. americanum* in aggressiveness and nonspecific feeding habits; basically, where southernmost distribution of *A. americanum* stops, *A. cajennense* picks up and continues southward throughout Central and South America; longevity of larvae, nymphs, and adults, as well as numbers of eggs laid by engorged females, similar to that of *A. americanum*; as with *A. americanum, A. cajennense* have long mouthparts and produce painful bites

Bont Tick
Amblyomma hebraeum Koch

Medical Importance: One of several ixodid vectors of *Rickettsia conorii*, the agent of boutonneuse fever; vector of African tick-bite fever organisms

Description: Long mouthparts, eyes, and festoons typical of other *Amblyomma* spp. (Figure 40); males have black or brown stripes and spots on a pale greenish white background; females have dark markings confined to the scutum

Distribution: Distributed throughout southern Africa, specifically Angola, Botswana, Cameroon, Mozambique, Madagascar, Kenya, Nigeria, Somalia, South Africa, Swaziland, Tanzania, Zaire, Zambia, and Zimbabwe; accidentally introduced into the U.S. on several occasions (primarily on rhinoceroses), but each time has been successfully eliminated (Figure 41)

Hosts: Immatures feed on many medium- and large-sized mammals, particularly wild hares; adults parasitize a variety of domestic and wild mammals but seem to prefer cattle and antelopes; all life stages will bite people

Seasonality: Active in spring, summer, and fall months; in South Africa, adults most abundant on hosts during the late summer and autumn[84]

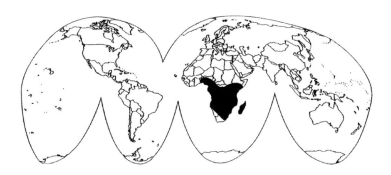

Figure 41
Approximate geographic distribution of Amblyomma hebraeum.
(From USAF Publication USAFSAM-89-2.)

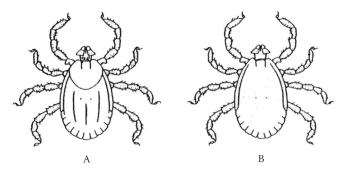

A   B

Figure 42
(A) Female and (B) male Haemaphysalis concinna. *(From USAF Publication USAFSAM-89-2.)*

Remarks: Larvae, like their U.S. cousins, the lone star ticks, are troublesome pests of people; attach themselves in large numbers on legs and about waist, causing intense irritation, rashlike lesions, and occasional pustules; unfed larvae may live for up to 11 months, nymphs for 8 months or more, and adults 22 months or longer; females deposit about 15,000 eggs

Haemaphysalis concinna Koch

Medical Importance: Vector of Siberian tick typhus rickettsia and viruses in the tick-borne encephalitis complex; may be a vector of Lyme borreliosis spirochetes in Asia.

Description: As a group, small inornate ticks with festoons but without eyes (Figure 42); second palpal segment projecting beyond lateral margin of basis capituli (gives palpi appearance of being triangular — should not be confused with basis capituli being angular, as in the *Rhipicephalus* ticks); males have pincerlike palps; females have a hypostome dentition of 6/6 and scutum is broadest in middle

Distribution: Widely distributed in forests of temperate Eurasia, including most of Central Europe, Estonia, Latvia, Belarus, Ukraine, Russia, Kazakhstan, Uzbekistan, Turkmenistan, China, Japan, Korea, and Vietnam (Figure 43)

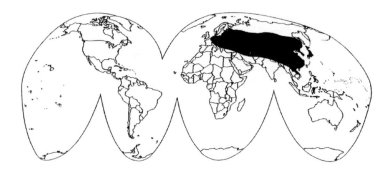

Figure 43
Approximate geographic distribution of Haemaphysalis concinna.
(From USAF Publication USAFSAM-89-2.)

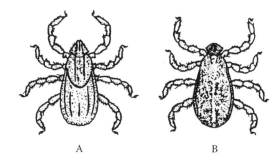

A   B

Figure 44
(A) Female and (B) male Haemaphysalis leachi. *(From USAF Publication USAFSAM-89-2.)*

Hosts: Adults feed on large wild and domestic mammals; immatures infest smaller mammals and birds, sometimes even reptiles; both adults and nymphs bite people

Seasonality: All stages active from spring to autumn; peak adult activity in June

Remarks: Found chiefly in deciduous and mixed forests, grass tussock swamps, birch–aspen groves, and alpine taiga forests; reported abundant in low-lying areas with high humidities

Yellow Dog Tick
Haemaphysalis leachi (Audouin) (including *H. l. muhsami*)

Medical Importance: Vector of boutonneuse fever rickettsia (human infection with rickettsia may also be acquired by contamination of skin and eyes with infectious tick fluids from crushing while deticking dogs)

Description: Typical-looking *Haemaphysalis* sp. (Figure 44) characteristically appears to have large, wedge-shaped mouthparts as viewed from above (persons identifying these specimens should be careful not to confuse them with brown dog tick, *R. sanguineus*)

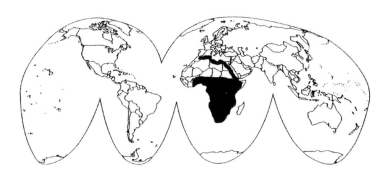

Figure 45
Approximate geographic distribution of Haemaphysalis leachi.
(From USAF Publication USAFSAM-89-2.)

Distribution: Primarily in tropical and southern Africa (although there are records from Algeria, Libya, and Egypt); records from India and Southeast Asia probably represent related but distinct species (Figure 45)

Hosts: Immatures usually parasitize field rodents; adults commonly bite domestic dogs but will also bite people readily; subspecies *H. leachi muhsami* prefers small carnivores (mongooses, wildcats, etc.) instead of canines

Seasonality: Most active from late spring to early fall

Remarks: Very common on dogs; in some areas more prevalent on dogs than *R. sanguineus*; usually 2 generations produced each year; unfed larvae may survive at least 169 d, nymphs 52 d, and adults 210 d; females lay up to 5000 eggs

Haemaphysalis spinigera Neumann

Medical Importance: Primary vector of the virus of Kyasanur Forest disease (KFD) in India

Description: Apart from usual generic characteristics of *Haemaphysalis* (inornate, no eyes, festoons present, triangular-shaped second palpal segment), males have long spurs on coxae I and IV and poorly developed spurs on coxae II and III (Figure 46); females have no spurs on the trochanters

Distribution: Widely distributed in central and southern India; also reported from southeast Asia and Indonesia (Figure 47)

Hosts: Immatures parasitize wide range of small mammals and birds; adults prefer large mammals such as cattle, monkeys, bears, and tigers; nymphs avidly bite humans

Seasonality: Generally active in the spring, summer, and fall; immatures peak in numbers from September to November

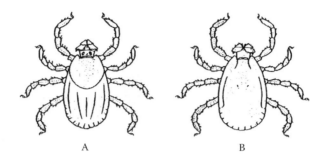

Figure 46
(A) Female and (B) male Haemaphysalis spinigera. *(From USAF Publication USAFSAM-89-2.)*

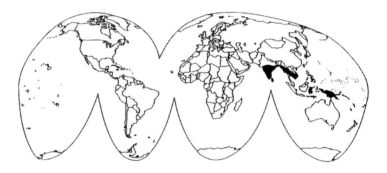

Figure 47
Approximate geographic distribution of Haemaphysalis spingera.
(From USAF Publication USAFSAM-89-2.)

Remarks: People in the KFD endemic areas (Mysore, Karnataka state, India) turn from agricultural pursuits to wood gathering in the forests during the season of peak immature activity; activity greatly increases human–tick contact; Hoogstraal[85] reported tick population increases in India because of recent increased cattle-grazing practices in and beside forests; immatures thrive on numerous small vertebrates hiding in dense lantana thickets where sections of forests were cleared

Asiatic Hyalomma
Hyalomma asiaticum Schulze and Schlottke

Medical Importance: Vector of agent of Siberian tick typhus

Description: In general, scant ornamentation, long mouthparts, eyes present in sockets, and festoons (although festoons not always clearly delineated); often legs appear banded (Figure 48); females have scutum longer than wide; both sexes have white or yellowish bands on legs

Distribution: Generally found in Russia, Georgia, Azerbaijan, Kazakhstan, Uzbekistan, Turkmenistan, China, Afghanistan, Pakistan, Iran, and Iraq (Figure 49)

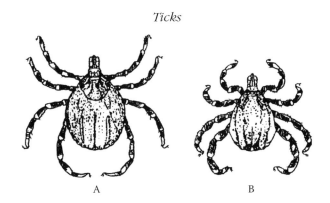

Figure 48
(A) Female and (B) male Hyalomma asiaticum. *(From USAF Publication USAFSAM-89-2.)*

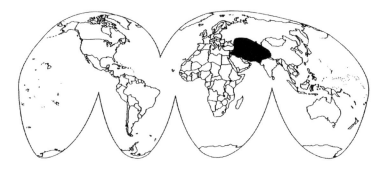

Figure 49
Approximate geographic distribution of Hyalomma asiaticum.
(From USAF Publication USAFSAM-89-2.)

Hosts: Adults parasitize all domestic animals, especially camels, cattle, horses, and sheep; people, hares, boars, and hedgehogs less frequently attacked; immatures feed on hedgehogs, rodents, hares, cats, and dogs

Seasonality: Most active in spring and summer throughout range

Remarks: In southwestern Kirghiz (former U.S.S.R.), foci of Siberian tick typhus occur where southern steppes give way to foothill semidesert zone; in these areas, associated with red-tailed jirds along dry waterways overgrown with shrubs and along irrigation canals[6]

Small Anatolian Hyalomma
Hyalomma anatolicum Koch (including *H. a. anatolicum* and *H. a. excavatum*)

Medical Importance: Vector of virus of Crimean–Congo hemorrhagic fever

Description: Usual characteristics of all *Hyalomma* ticks (scant ornamentation, eyes present in sockets, long mouthparts, banded legs); small and similar in appearance to *H. asiaticum* (Figure 50)

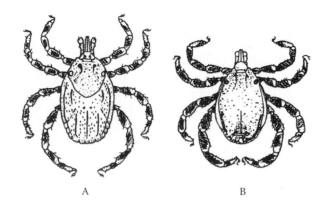

Figure 50
(A) Female and (B) male Hyalomma anatolicum. *(From USAF Publication USAFSAM-89-2.)*

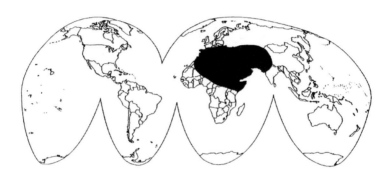

Figure 51
Approximate geographic distribution of Hyalomma anatolicum.
(From USAF Publication USAFSAM-89-2.)

Distribution: Throughout northern Africa, portions of the Near East, Asia Minor, southern Europe, Russia, Ukraine, Georgia, Azerbaijan, Armenia, Kazakhstan, Uzbekistan, Turkmenistan, and India (Figure 51)

Hosts: All stages observed feeding on hares in forest near Casablanca;[86] also avid parasite of humans and many domestic animals

Seasonality: Varies with latitude throughout range; in general, adults infest domestic animals from March to October, and larvae and nymphs from July to September; all stages most abundant in early August

Remarks: Engorged larvae and unfed adults usual overwintering stages; hibernate in cracks and crevices in wooden animal shelters in Russian climate and in rodent burrows in African desert conditions; larvae may survive up to 241 d, nymphs up to 246 d, and adults over 1 year

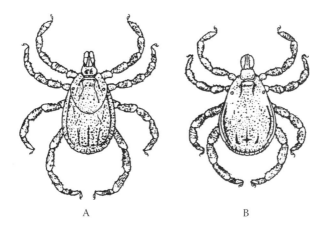

Figure 52
(A) Female and (B) male Hyalomma marginatum. *(From USAF Publication USAFSAM-89-2.)*

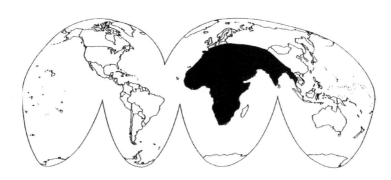

Figure 53
Approximate geographic distribution of Hyalomma marginatum.
(From USAF Publication USAFSAM-89-2.)

Hyalomma marginatum Koch
(including several subspecies)

Medical Importance: Ticks of this complex efficient vectors of the Crimean–Congo hemorrhagic fever virus

Description: Complex includes several subspecies; in general, large with banded legs (Figure 52)

Distribution: Most common in southeastern Europe, including Russia, Ukraine, Belarus, Georgia, Azerbaijan, Armenia, Kazakhstan, Uzbekistan, Turkmenistan; also occurs in India and Indochina, westward throughout southern Europe, into the Near East and Africa (Figure 53)

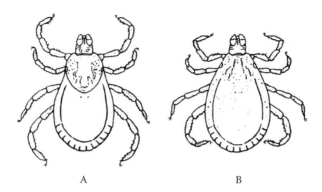

A   B

Figure 54
(A) Female and (B) male Dermacentor marginatus. *(From USAF Publication USAFSAM-89-2.)*

Hosts: Adults attack humans and most domestic animals, especially cattle and horses; immatures may also be found on domestic animals but prefer small wild mammals and birds

Seasonality: According to Hoogstraal,[86] rarely seen in winter throughout much of distribution, but begin to appear in March and continue until October; maximum densities reached in April, May, and June; nymphs active throughout summer

Remarks: Extremely hardy, often existing under varied conditions of cold, heat, and aridity; often occurs in high numbers; an aggressive human parasite; may act as either a 2-host or 3-host tick; unfed adults can survive over 2 years; females deposit between 4,000 and 15,000 eggs

Dermacentor marginatus Sulzer

Medical Importance: Primary vector of Siberian tick typhus rickettsia in Eurasia; vector of viruses in the tick-borne encephalitis complex; possible vector of Omsk hemorrhagic fever virus

Description: Usually ornate specimens with both eyes and festoons present (Figure 54); rectangular basis capituli dorsally; males have weakly defined spur on posterodorsal margin of second palpal segment and mixture of both large and small punctations over scutum; females have small ventral spurs on trochanters II and III

Distribution: Many areas of western and central Europe; specific countries include Afghanistan, Albania, Bulgaria, Czech Republic, Portugal, Slovakia, Spain, France, Germany, Greece, Hungary, Romania, Switzerland, Iran, Iraq, Poland, Italy, Turkey, Yugoslavia, Russia, Belarus, Ukraine, Kazakhstan, Uzbekistan, Kyrgyzstan, Georgia, Azerbaijan, and Armenia (Figure 55)

Hosts: Adults parasitize horses, cattle, sheep, people, dogs, buffalo, swine, camels, and hedgehogs; immatures feed most frequently on small mammals, especially rodents

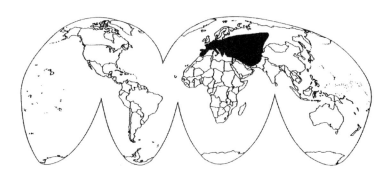

Figure 55
Approximate geographic distribution of Dermacentor marginatus.
(From USAF Publication USAFSAM-89-2.)

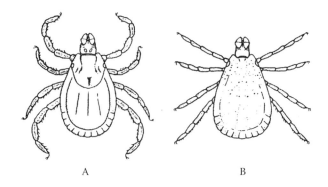

A   B

Figure 56
(A) Female and (B) male Dermacentor nuttalli. *(From USAF Publication USAFSAM-89-2.)*

Seasonality: Adults generally active in spring and again in autumn; larvae usually peak in activity in June and July, nymphs in July and August

Remarks: Inhabits shrubby areas, low forests, marshes, lowlands, alpine steppes, and semidesert areas; in southeastern France, found in close association with woods where oaks, *Quercus pubescens,* predominate[87]

Dermacentor nuttalli Olenev

Medical Importance: One of several known vectors of Siberian tick typhus; also vector of agent of tularemia in northern Eurasia

Description: Characteristics common to all members of genus *Dermacentor*; females have no internal spurs on coxa IV and no cornua (Figure 56)

Distribution: Through central and eastern Siberia, Asiatic Russia, northern Mongolia, and China (Figure 57); occasionally reported from Ukraine and Kazakhstan

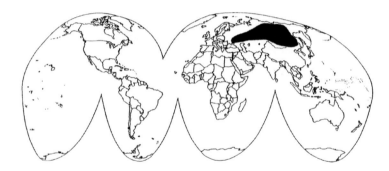

Figure 57
Approximate geographic distribution of Dermacentor nuttalli.
(From USAF Publication USAFSAM-89-2.)

Hosts: Immatures generally parasitize small mammals such as field mice, rats, marmots, hamsters, hares, cats, and dogs; adults feed predominantly on larger hosts such as horses, cattle, camels, sheep, dogs, and humans

Seasonality: Larvae and nymphs active from mid-June to mid-August with adults active primarily in the spring (peaking in mid-May); Splisteser and Tyron[88] reported high population numbers of adult *D. nuttalli* in steppe regions of Mongolia from mid-March to late May

Remarks: Seems especially associated with high grasslands; generally not found in dense forests, river lowlands, or hilly wooded country; unfed adults usually overwinter in cracks in soil and occasionally in burrows of rodents; cease questing and become inactive at temperatures below 10°C (50°F)[88]

Dermacentor silvarum Olenev

Medical Importance: Vector of Siberian tick typhus rickettsia in Eurasia and Asia, as well as viruses in the tick-borne encephalitis complex; may be a vector of Lyme borreliosis spirochetes in Asia

Description: Males have ornamentation similar to that in Figure 58B; prominent dorsal spur on trochanter I; females have coxa IV without internal spurs; cornua present and ventral spurs lacking on basis capituli

Distribution: Primarily in eastern and far eastern Russia and northern Mongolia; also reported from Belarus, Ukraine, Lithuania, Latvia, Estonia, Georgia, Kazakhstan, Uzbekistan, Turkmenistan, Kyrgyzstan, Romania, and Yugoslavia (Figure 59)

Hosts: Adults collected from people, horses, cattle, sheep, dogs, fox, and deer; larvae and nymphs feed on numerous species of small mammals

Seasonality: Bimodal pattern of seasonal activity of adults with one peak in early June and another in early September[89]; larvae most active in June and July and nymphs from June to mid-August

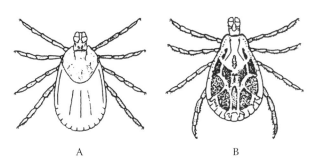

A   B

Figure 58
(A) Female and (B) male Dermacentor silvarum.
(From USAF Publication USAFSAM-89-2.)

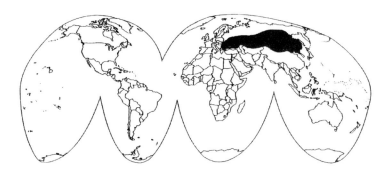

Figure 59
Approximate geographic distribution of Dermacentor silvarum.
(From USAF Publication USAFSAM-89-2.)

Remarks: Inhabitant of forest steppe zones; most numerous in birch–aspen marshes, glades in mixed forests, cultivated areas in taiga forests, and other localized dense shrub areas and secondary growth forest

Australian Paralysis Tick
Ixodes holocyclus Neumann

Medical Importance: Primary cause of tick paralysis cases in Australia;[72] bite also known to cause Type I hypersensitivity reactions in humans

Description: Inornate with no eyes or festoons (Figure 60); males have legs I and IV reddish, legs II and III yellowish; females have scutum broadest posterior to middle, with numerous punctations of varying sizes; coxae are large and trapezoid shaped

Distribution: Primarily in New Guinea and along eastern coastal areas of Australia (Figure 61)

Hosts: Parasitizes humans, other mammals, and birds; seems to especially prefer sheep, cattle, dogs, cats, and bandicoots

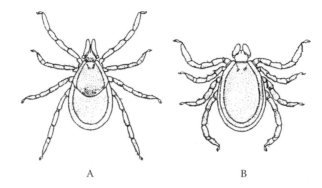

Figure 60
(A) Female and (B) male Ixodes holocyclus. *(From USAF Publication USAFSAM-89-2.)*

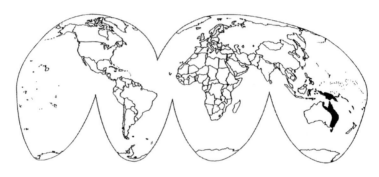

Figure 61
Approximate geographic distribution of Ixodes holocyclus.
(From USAF Publication USAFSAM-89-2.)

Seasonality: Active in warmer months of year

Remarks: Primarily in heavily vegetated rain forest areas of eastern coastal Australia; bandicoot is natural host (bandicoot populations are increasing near urban areas owing to control campaigns against dingoes and foxes[90])

Taiga Tick
Ixodes persulcatus Schulze

Medical Importance: Vector of virus of Russian spring–summer encephalitis and Lyme borreliosis spirochete in Europe and Asia

Description: Characteristics common to all species in genus *Ixodes*; very similar in appearance to commonly encountered European castor bean tick, *I. ricinus* (Figure 62)

Distribution: Central and eastern Europe, Russia, Ukraine, Belarus, Kazakhstan, Uzbekistan, Kyrgyzstan, China, and Japan (Figure 63)

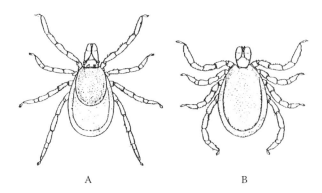

Figure 62
(A) Female and (B) male Ixodes persulcatus. *(From USAF Publication USAFSAM-89-2.)*

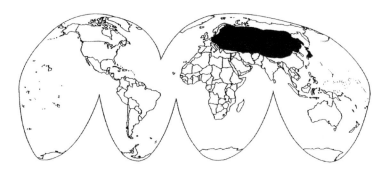

Figure 63
Approximate geographic distribution of Ixodes persulcatus.
(From USAF Publication USAFSAM-89-2.)

Hosts: Larvae and nymphs feed on wide variety of small forest mammals and birds; adults parasitize larger wild and domestic mammals; readily bites people

Seasonality: Found on hosts in late spring and summer months (Zemskaya[91] found that, after overwintering, adults usually resumed activity in eastern part of Russian plain during last 10 d of April, when upper soil layers warmed up to 5 to 10°C (41 to 50°F) and remained active for 65 to 95 d

Remarks: Apparently more cold-hardy than *I. ricinus*, thus inhabiting harsher, more northern areas; inhabits small-leaved forests near primary coniferous forests, such as spruce–basswood combinations (commonly referred to as *taiga*)

European Castor Bean Tick
Ixodes ricinus (Linnaeus)

Medical Importance: Primary vector of Lyme borreliosis spirochetes in Europe; also known to transmit viruses of tick-borne encephalitis complex

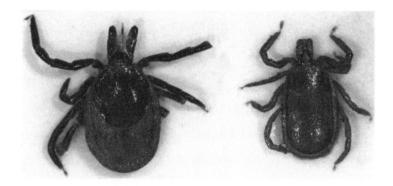

Figure 64
Female and male Ixodes ricinus. *(USAF photo).*

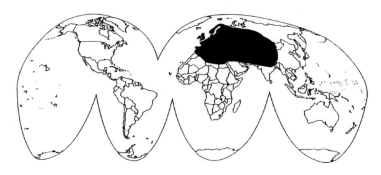

Figure 65
Approximate geographic distribution of Ixodes ricinus. *(From USAF Publication USAFSAM-89-2.)*

Description: Long mouthparts and dark brown to black in color (females may have portion posterior to scutum orangish) (Figure 64); resembles and closely related to American species *I. scapularis*

Distribution: Common throughout most of Europe, including the British Isles (Figure 65); also found in scattered locations in northern Africa and parts of Asia

Hosts: Immatures recorded from lizards, small mammals, and birds; adults feed mostly on sheep, cattle, dogs, horses, and deer; adults are avid parasites of humans

Seasonality: In temperate regions of range, most active in spring and autumn; two peaks of activity may be observed: late March to early June, August to October; in northern Africa most active in winter[84]

Remarks: One of the most commonly encountered ticks in central and western Europe; more than 90% of tick bites in England and Ireland are from *I. ricinus* nymphs; as long as 3 years usually required to complete the life cycle: larvae feed first year, nymphs second year, and adults third year; females deposit 2000 to 3000 eggs

Figure 66
Male Rhipicephalus appendiculatus.
(Specimen provided by Dr. K.Y. Mumcuoglu, Hebrew University, Jerusalem.)

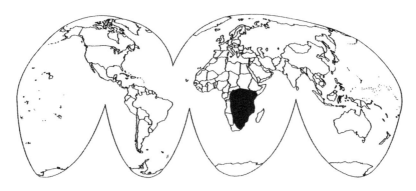

Figure 67
Approximate geographic distribution of Rhipicephalus appendiculatus.
(From USAF Publication USAFSAM-89-2.)

Brown Ear Tick
Rhipicephalus appendiculatus Neumann

Medical Importance: One of several vectors of boutonneuse fever rickettsia

Description: Usually inornate with both eyes and festoons present (Figure 66); look similar to cosmopolitan *R. sanguineus*; males have distinctly pointed dorsal projection on coxa I; females have flat eyes (not obviously convex)

Distribution: Throughout southern Africa up to about 10°N latitude (Figure 67)

Hosts: Cattle primarily (for adult ticks as well as immatures); also humans, domestic animals, and wild game such as antelope and buffalo

Seasonality: Larvae most abundant on hosts from May to July; nymphs occur on hosts from June to September; adults most active from November to March

Remarks: May be 1 to 3 generations occurring annually, depending on length and number of rainy seasons in range; seems especially sensitive to desiccation; during consecutive dry years tends to die out; unfed larvae may survive as long as 10 months, nymphs 15 months, and adults 24 months; females lay up to 6000 eggs

REFERENCES

1. McGinley-Smith, D.E. and Tsao, S.S., Dermatoses from ticks, *J. Am. Acad. Dermatol.*, 49, 363, 2003.

2. Benenson, A.S., Ed., *Control of Communicable Diseases Manual*, American Public Health Association, Washington, D.C., 1995.

3. Burgdorfer, W., A review of Rocky Mountain spotted fever: its agent, and its vectors in the U.S., *J. Med. Entomol.*, 12, 269, 1975.

4. Goddard, J., *Ticks and Tick-borne Diseases Affecting Military Personnel*, USAF, School of Aerospace Medicine, San Antonio, TX, 1989.

5. Hoogstraal, H., Ticks in relation to human diseases caused by viruses, *Annu. Rev. Entomol.*, 11, 261, 1966.

6. Hoogstraal, H., Ticks in relation to human diseases caused by *Rickettsia* species, *Annu. Rev. Entomol.*, 12, 377, 1967.

7. CDC, Lyme disease, United States, 2001–2002, *MMWR*, 53, 365–368, 2004.

8. Piesman, J.F., Ecology of *Borrelia burgdorferi* sensu lato in North America, in *Lyme Borreliosis — Biology, Epidemiology, and Control*, Gray, J.S., Kahl, O., Lane, R.S., and Stanek, G., Eds., CABI International, Trowbridge, England, 2002, p. 223.

9. Falco, R.C., Fish, D., and Piesman, J., Duration of tick bites in a Lyme disease-endemic area, *Am. J. Epidemiol.*, 143, 187, 1996.

10. Barbour, A.G., Maupin, G.O., Teltow, G.J., Carter, C.J., and Piesman, J., Identification of an uncultivable *Borrelia* species in the hard tick *Amblyomma americanum*: possible agent of a Lyme disease-like illness, *J. Infect. Dis.*, 173, 403, 1996.

11. James, A.M., Liveris, D., Wormser, G., Schwartz, I., Montecalvo, M.A., and Johnson, B., *Borrelia lonestari* infection after a bite by an *Amblyomma americanum* (L.), *J. Infect. Dis.*, 183, 1810, 2001.

12. Wormser, G., Masters, E.J., Liveris, D., Nowakowaski, J., Nadelman, R., Holmgren, D., Bittker, S., Cooper, D., Wang, G., and Schwartz, I., Microbiologic evaluation of patients from Missouri with erythema migrans, *Clin. Infect. Dis.*, 40, 423, 2005.

13. Spach, D.H., Liles, W.C., Campbell, G.L., Quick, R.E., Anderson, D.E.J., and Fritsche, T.R., Tick-borne diseases in the United States, *New Engl. J. Med.*, 329, 936, 1993.

14. CDC, Fatal cases of Rocky Mountain spotted fever in family clusters — three states, 2003, *MMWR*, 53, 407–410, 2004.

15. Demma, L.J., Traeger, M.S., Nicholson, W.L., Paddock, C.D., Blau, D.M., Eremeeva, M.E., Dasch, G.A., Levin, M.L., Singleton, J., Jr., Zaki, S.R., Cheek, J.E., Swerdlow, D.L., and McQuiston, J.H., Rocky Mountain spotted fever from an unexpected tick vector in Arizona, *New Engl. J. Med.*, 353, 587, 2005.

16. Anigstein, L. and Bader, M. N., Investigations on rickettsial diseases in Texas, *Tex. Rep. Biol. Med.*, 1, 105, 1943.

17. Stothard, D.R. and Fuerst, P.A., Evolutionary analysis of the spotted fever and typhus groups of *Rickettsia* using 16sRNA gene sequences, *Syst. Appl. Microbiol.*, 18, 52, 1995.

18. Bell, E.J. and Stoenner, H.C., Immunologic relationships among the spotted fever group of rickettsias determined by toxin neutralization tests in mice with convalescent animal serums, *J. Immunol.*, 84, 171, 1960.

19. Lackman, D.B., Bell, E.J., Stoemmer, H.G., and Pickens, E.G., The Rocky Mountain spotted fever group of rickettsias, *Health Lab. Sci.*, 2, 135, 1965.

20. Parker, R.R., A pathogenic rickettsia from the Gulf Coast tick, *Amblyomma maculatum*, in *Proc. 3rd Int. Congr. Microbiol.*, New York, 1940, pp. 390–391.

21. Parker, R.R., Kohls, G.M., Cox, G.W., and Davis, G.E., Observations on an infectious agent from *Amblyomma maculatum*, *Public Health Rep.*, 54, 1482, 1939.

22. Xu, W. and Raoult, D., Taxonomic relationships among spotted fever group rickettsias as revealed by antigenic analysis with monoclonal antibodies, *J. Clin. Microbiol.*, 36, 887, 1998.

23. Goddard, J., Experimental infection of lone star ticks, *Amblyomma americanum* (L.), with *Rickettsia parkeri* and exposure of guinea pigs to the agent, *J. Med. Entomol.*, 40, 686, 2003.

24. Paddock, C.D., Sumner, J.W., Comer, J.A., Zaki, S.R., Goldsmith, C.S., Goddard, J., McLellan, S.L.F., Tamminga, C.L., and Ohl, C.A., *Rickettsia parkeri* — a newly recognized cause of spotted fever rickettsiosis in the United States, *Clin. Infect. Dis.*, 38, 805, 2004.

25. Raoult, D. and Paddock, C.D., *Rickettsia parkeri* infection and other spotted fevers in the U.S., *New Engl. J. Med.*, 353, 626, 2005.

26. Goddard., J., American Boutonneuse Fever — a new spotted fever rickettsiosis, *Infect. Med.*, 21, 207, 2004.

27. Kelly, P.J., Beati, L., Mason, P.R., Matthewman, L.A., Roux, V., and Raoult, D., *Rickettsia africae* sp. nov.: the etiological agent of African tick bite fever, *Int. J. Syst. Bacteriol.*, 46, 611, 1996.

28. Kelly, P.J., Beati, L., Matthewman, L.A., Mason, P.R., Dasch, G.A., and Raoult, D., A new pathogenic spotted fever group rickettsia from Africa, *J. Trop. Med. Hyg.*, 97, 129, 1994.

29. Kelly, P.J., *Rickettsia africae* in the West Indies, *Emerging Infect. Dis.*, 12, 224, 2006.

30. Jensenius, M., Fournier, P.E., Kelly, P.J., Myrvang, B., and Raoult, D., African tick bite fever, *Lancet Infect. Dis.*, 3, 557, 2003.

31. Walker, D.H. and Dumler, J.S., Emergence of the ehrlichioses as human health problems, *Emerging Infect. Dis.*, 2, 18, 1996.

32. Maeda, K., Markowitz, N., Hawley, R.C., Ristic, M., Cox, D., and McDade, J.E., Human infection with *Ehrlichia canis* a leukocytic rickettsia, *New Engl. J. Med.*, 316, 853, 1987.

33. Bakken, J.S. and Dumler, J.S., Ehrlichiosis and anaplasmosis, *Infect. Med.*, 21, 433, 2004.

34. Dumler, J.S. and Bakken, J.S., Ehrlichial diseases of humans: emerging tick-borne infections, *Clin. Infect. Dis.*, 20, 1102, 1995.

35. Buller, R.S., Ariens, M., Hmiel, S.P., Paddock, C.D., Sumner, J.W., Rikihisa, Y., Unver, A., Gaudreault-Keener, M., Manian, F.A., Liddell, A.M., Schmulewitz, N., and Storch, G.A., *Ehrlichia ewingii,* a newly recognized agent of human ehrlichiosis, *New Engl. J. Med.*, 341, 148, 1999.

36. Dumler, J.S., Choi, K.-S., Garcia-Garcia, J.C., Barat, N.S., Scorpio, D.G., Garyu, J.W., Grab, D.J., and Bakken, J.S., Human granulocytic anaplasmosis and *Anaplasma phagocytophilum, Emerging Infect. Dis.*, 11, 1828, 2005.

37. Holman, M.S., Caporale, D.A., Goldberg, J., Lacombe, E., Lubelczyk, C., Rand, P.W., and Smith, R.P., *Anaplasma phagocytophilum, Babesia microti,* and *Borrelia burgdorferi* in *Ixodes scapularis* in southern coastal Maine, *Emerging Infect. Dis.*, 10, 744, 2004.

38. Gorenflot, A., Moubri, K., Precigout, E., Carcy, B., and Schetters, T.P., Human babesiosis, *Ann. Trop. Med. Parasitol.*, 92, 489, 1998.

39. CDC, Babesiosis — Connecticut, *MMWR*, 38, 649–650, 1989.

40. Markell, E., Voge, M., and John, D., *Medical Parasitology*, W.B. Saunders, Philadelphia, PA, 1992.

41. Thomford, J.W., Conrad, P.A., Telford, S.R., III, Mathiesen, D., Eberhard, M.L., Herwaldt, B.L., Quick, R.E., and Persing, D.H., Cultivation and phylogenetic characterization of a newly recognized human pathogenic protozoan, *J. Infect. Dis.*, 169, 1050, 1994.

42. Herwaldt, B.L., de Bruyn, G., Pieniazek, N.J., Homer, M., Lofy, K.H., Slemenda, S.B., Fritsche, T.R., Persing, D.H., and Limaye, A.P., Babesia divergens-like infection, Washington State, *Emerging Infect. Dis.*, 10, 622, 2004.

43. Farlow, J., Wagner, D.M., Dukerich, M., Stanley, M., Chu, M., Kubota, K., Petersen, J., and Keim, P., *Francisella tularensis* in the United States, *Emerging Infect. Dis.*, 11, 1835, 2005.

44. Emmons, R.W., Colorado tick fever, in *Viral Zoonoses*, Vol. 1, Steel, J.H., Ed., CRC Press, Boca Raton, FL, 1979, p. 113.

45. Varma, M.G.R., Ticks and mites, in *Medical Insects and Arachnids*, Lane, R.P. and Crosskey, R.W., Eds., Chapman and Hall, London, 1993, chap. 18.

46. CDC, Outbreak of relapsing fever — Grand Canyon National Park, Arizona, *MMWR*, 40, 296–297, 1991.

47. Thompson, R.S. and Russell, R., Outbreak of tick-borne relapsing fever in Spokane County, Washington, *JAMA*, 210, 1045, 1969.

48. CDC, Tick-borne relapsing fever outbreak after a family gathering — New Mexico, August 2002, *MMWR*, 52, 809–812, 2003.

49. Monath, T.P. and Johnson, K.M., Diseases transmitted primarily by arthropod vectors, in *Public Health and Preventive Medicine*, 13th ed., Last, J.M. and Wallace, R.B., Eds., Appleton and Lange, Norwalk, CT, 1992, p. 223.

50. Gresikova, M. and Calisher, C.H., Tick-borne encephalitis, in *The Arboviruses: Epidemiology and Ecology*, Vol. 4, Monath, T.P., Ed., CRC Press, Boca Raton, FL, 1989, p. 177.

51. Ternovoi, V.A., Kurzhukov, G.P., Sokolov, Y.V., Ivanov, G.Y., Ivanisenko, V.A., Loktev, A.V., Ryder, R.W., Netesov, S.V., and Loktev, V.B., Tick-borne encephalitis with hemorrhagic syndrome, Novosibirsk region, Russia, 1999, *Emerging Infect. Dis.*, 9, 743, 2003.

52. Loktev, V.B., Ternovoy, V.A., Kurgukov. G.P., Sokolov, Y.V., Ivanov, G.Y., Loktev, A.U., Ryder, R., and Netesov, S.V., New variants of tick-borne encephalitis discovered by retrospective investigation of fatal cases of tick-borne encephalitis with hemorrhagic syndrome occurring in Novosibirsk Region (Russia) during summer of 1999, in (Program containing abstracts), *Int. Conf. Emerging Infect. Dis.*, Atlanta, March 24–27, 2002, p. 11 (supplement).

53. Nuttall, P.A. and Labuda, M., Tick-borne encephalitis subgroup, in *Ecological Dynamics of Tick-borne Zoonoses*, Sonenshine, D.E. and Mather, T.N., Eds., Oxford University Press, New York, 1994, p. 351.

54. Kocan, A.A., Tick paralysis, *JAVMA*, 192, 1498, 1988.

55. Gregson, J.D., Tick paralysis: an appraisal of natural and experimental data, Canada Dep. Agric. Monograph No. 9, 48, 1973.

56. Schmitt, N., Bowmer, E.J., and Gregson, J.D., Tick paralysis in British Columbia, *Can. Med. Assoc. J.*, 100, 417, 1969.

57. Stanbury, J.B. and Huyck, J.H., Tick paralysis: a critical review, *Medicine*, 24, 219, 1945.

58. Gothe, R., Kunze, K., and Hoogstraal, H., The mechanisms of pathogenicity in the tick paralysis, *J. Med. Entomol.*, 16, 357, 1979.

59. Goddard, J., Ecological studies of *Ixodes scapularis* in Mississippi: lateral movement of adult ticks, *J. Med. Entomol.*, 30, 824, 1993.

60. Lancaster, J.L., Jr., Control of the Lone Star Tick, University of Arkansas Agric. Exp. Stn. Rep. Ser. No. 67, 1957, 39 pp.

61. Lees, A.D., The water balance in *Ixodes ricinus* and certain other species of ticks, *Parasitology*, 37, 1, 1946.

62. Semtner, P.J., Howell, D.E., and Hair, J.A., The ecology and behavior of the lone star tick. I. The relationship between vegetative habitat type and tick abundance and distribution in Cherokee Co., Oklahoma, *J. Med. Entomol.*, 8, 329, 1971.

63. Sonenshine, D.E., The Ticks of Virginia, Virginia Polytechnic Institute and State University Res. Div. Bull. No. 139, 1979, 42 pp.

64. Sonenshine, D.E., Atwood, E.L., and Lamb, J.T., The ecology of ticks transmitting Rocky Mountain spotted fever in a study area in Virginia, *Ann. Entomol. Soc. Am.*, 59, 1234, 1966.

65. Sonenshine, D.E. and Levy, G.F., Ecology of the American dog tick, *Dermacentor variabilis* in a study area in Virginia II. Distribution in relation to vegetative types, *Ann. Entomol. Soc. Am.*, 65, 1175, 1972.

66. Keirans, J.E. and Litwak, T.R., Pictorial key to the adults of hard ticks, family Ixodidae, *J. Med. Entomol.*, 26, 435, 1989.

67. Goddard, J., A review of the disease agents harbored and transmitted by the lone star tick, *Southwest. Entomol.*, 12, 158, 1987.

68. Burgdorfer, W., Tick-borne diseases in the United States: Rocky Mountain spotted fever and Colorado tick fever, *Acta Trop.*, 34, 103, 1977.

69. Goddard, J. and Norment, B.R., Spotted fever group rickettsiae in the lone star tick, *J. Med. Entomol.*, 23, 465, 1986.

70. Long, S.W., Pound, J.M., and Yu, X., *Ehrlichia* prevalence in *Amblyomma americanum* in central Texas, *Emerging Infect. Dis.*, 10, 1342, 2004.

71. Wolf, L., McPherson, T., Harrison, B.A., Engber, B.R., Anderson, A., and Whitt, P., Prevalence of *Ehrlichia ewingii* in *Amblyomma americanum* in North Carolina, *J. Clin. Microbiol.*, 38, 2795, 2000.

72. Goddard, J., Tick paralysis, *Infect. Med.*, 15, 28, 1998.

73. Goddard, J., Arthropod transmission of tularemia, *Infect. Med.*, 15, 306, 1998.

74. Goddard, J., Ticks and Lyme disease, *Infect. Med.*, 14, 698, 1997.

75. Oliver, J.H., Owsley, M.R., Hutcheson, H.J., James, A.M., Chunsheng, C., Irby, W.S., Dotson, E.M., and McLain, D.K., Conspecificity of the ticks *Ixodes scapularis* and *Ixodes dammini*, *J. Med. Entomol.*, 30, 54, 1993.

76. Piesman, J. and Sinksky, R.J., Ability of *Ixodes scapularis, Dermacentor variabilis,* and *Amblyomma americanum* to acquire, maintain, and transmit Lyme disease spirochetes, *J. Med. Entomol.*, 25, 336, 1988.

77. Goddard, J., Ecological studies of adult *Ixodes scapularis* in central Mississippi: questing activity in relation to time of year, vegetation type, and meteorologic conditions, *J. Med. Entomol.*, 29, 501, 1992.

78. Carpenter, T.L., McMeans, M.C., and McHugh, C.P., Additional instances of human parasitism by the brown dog tick, *J. Med. Entomol.*, 27, 1065, 1990.

79. Goddard, J., Focus of human parasitism by the brown dog tick, *Rhipicephalus sanguineus*, *J. Med. Entomol.*, 26, 628, 1989.

80. Klompen, J.S.H. and Oliver, J.H., Jr., Systematic relationships in the soft ticks, *Syst. Entomol.*, 18, 313, 1993.

81. Hoogstraal, H., Argasid and nuttalliellid ticks as parasites and vectors, *Adv. Parasitol.*, 24, 135, 1985.

82. Hoffman, A., Monografia de los Ixodoidea de Mexico, I parte, *Rev. Soc. Mex. Hist. Nat.*, 23, 191, 1962.

83. Keirans, J.E., Clifford, C.M., and Hoogstraal, H., *Ornithodoros yunkeri*, new species from seabirds and nesting sites in the Galapagos Islands, *J. Med. Entomol.*, 21, 344, 1984.

84. Strickland, R.K., Gerrish, R.R., Hourrigan, J.L., and Schubert, G.O., Ticks of veterinary importance, in USDA, APHIS, Agric. Handbook No. 485, U.S. Department of Agriculture, Washington, D.C., 1976, 122 pp.

85. Hoogstraal, H., Changing patterns of tick-borne diseases in modern society, *Annu. Rev. Entomol.*, 26, 75, 1981.

86. Hoogstraal, H., African Ixodoidea. I. Ticks of the Sudan, in U.S. Navy Bur. Med. Surg., Washington, D.C., Res. Rep. NM 005-050, 29.07, 1956, 89 pp.

87. Gilot, B. and Pautou, G., Distribution and ecology of *Dermacentor marginatus* in the French Alps and their piedmont, *Acarologia*, 24, 261, 1983.

88. Splisteser, H. and Tyron, U., Studies on the ecology and behavior of *Dermacentor nuttalli* in the Mongolian Republic, *Monatsschr. Vet.*, 41, 126, 1986.

89. Pomerantzev, B.I. and Serdyukova, G.V., Ecological observations of ticks of the family Ixodidae, vectors of spring-summer encephalitis in the Far East, *Parazitol. Sb. Zool. Inst. Akad. Nauk. USSR*, 9, 47, 1947.

90. Bagnall, B.G. and Doube, B.M., The Australian paralysis tick, *Ixodes holocyclus*, *Aust. Vet. J.*, 51, 151, 1975.

91. Zemskaya, A.A., Seasonal activity of adult ticks *Ixodes persulcatus* in the eastern part of the Russian plain, *Folia Parasitol.*, 31, 269, 1984.

CHAPTER 31

WASPS (YELLOWJACKETS, HORNETS, AND PAPER WASPS)

TABLE OF CONTENTS

_____ *YELLOWJACKETS*

Figure 1
Worker yellowjacket.
(From U.S. DHEW, PHS, CDC Pictorial Keys.)

Importance
Painful stings; allergic RXNs

Distribution
Numerous species almost worldwide

Lesion
Central white spot with erythematous halo; amount of local swelling is variable

Disease Transmission
None

Key Reference
Akre, R.D. et al., USDA, SEA, *Agriculture Handbook,* No. 552, 1981

Treatment
Pain relievers, antipruritic lotions for local RXNs; systemic RXNs may require antihistamines, epinephrine, and other supportive measures

I. YELLOWJACKETS

A. General and Medical Importance

Venomous wasps in the genera *Vespula, Paravespula,* and *Dolichovespula* are called *yellowjackets* and comprise about 26 species (some entomologists still do not recognize the generic status of *Paravespula* and would group them with *Vespula*). Some species nest near (in the ground) or in human dwellings and can be a nuisance and health threat to people (Figure 1). Yellowjackets produce painful stings and may cause death due to sting allergy (see Chapter 2). Cellulitis may occur following stings by some scavenger yellowjackets. Foraging yellowjackets are particularly numerous around recreation areas and refuse collection sites, where they are attracted to meats, sweet carbohydrates, soda pop, etc.

B. General Description

The name "yellowjacket" refers to the typical yellow and black bands on the abdomen (Figure 2), although some species are actually black and white, such as the bald-faced hornet. The pattern of markings on the yellowjacket gaster (most prominent portion of the abdomen) is often diagnostic as to species (Figure 3). Most species are smaller than paper wasps (1.5 to 2.0 cm) and more robust in appearance like honey bees. Wasps, yellowjackets, and hornets have inconspicuous hairs on their bodies that are not feathered (when observed under magnification).

C. Geographic Distribution

Paravespula maculifrons, the eastern yellowjacket, occurs from Minnesota to Texas and eastward. It is probably the most troublesome yellowjacket in the eastern and southeastern U.S. *P. vulgaris,* the common yellowjacket, is Holarctic, being widely distributed across Europe, Asia, and North America. *Vespula squamosa*, the southern yellowjacket, occurs from Wisconsin to Texas and southeastward to the Atlantic Coast. Yellowjackets that have

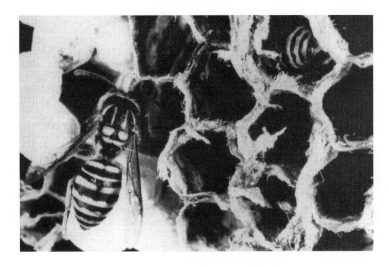

Figure 2
Yellowjacket workers tending nest.
(Photo courtesy of Dr. James Jarratt, Entomology Department, Mississippi State University.)

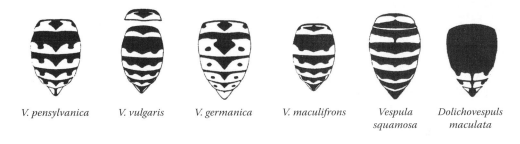

| *V. pensylvanica* | *V. vulgaris* | *V. germanica* | *V. maculifrons* | *Vespula squamosa* | *Dolichovespuls maculata* |

Figure 3
Gaster patterns of some yellowjackets; not drawn to scale.
(From USDA Agriculture Handbook No. 552.)

been introduced into nonnative habitats seem to be especially troublesome pests. The introduced German yellowjacket, *P. germanica*, is rapidly becoming a problem in the northeastern U.S. as well as the Midwest. Also, the western yellowjacket, *P. pennsylvanica*, is a severe pest in the western U.S. In certain years this species appears in such great numbers that it surpasses mosquitoes and flies as an annoyance to campers.[1] In addition, this species has been introduced into Hawaii, where it builds perennial colonies and is a pest in sugarcane fields.

D. Biology and Behavior

Yellowjackets of the genus *Dolichovespula* build aerial nests. This is the case for the large black and white species called the *bald-faced hornet* (*D. maculata*). Actually, this species is not a hornet and should be called an aerial yellowjacket (see Section II). The other genera of yellowjacket, *Vespula* and *Paravespula*, usually build nests in underground sites. Yellowjacket nests are multicombed with a surrounding paper envelope (Figure 4 and Color Figure 31.30). The mature size of yellowjacket colonies in

Figure 4
Yellowjacket nest showing multiple combs and paper envelope.
(From Mississippi Cooperative Extension Service.)

temperate regions ranges from 500 cells in 2 combs in some *Vespula* sp. to 15,000 cells in 8 to 10 combs in some *Paravespula* sp.

Mature colonies may have from as few as 75 to over 5000 worker yellowjackets. Most yellowjackets typically have annual nesting cycles, but in some of the southernmost areas of their distribution, perennial colonies of *V. squamosa* and *P. maculifrons* have been reported. This has led to discovery of 6 to 9 ft tall nests in Florida containing over 100,000 cells. (See Color Figure 31.31 for a picture of a 4-ft tall yellowjacket nest found in Mississippi.) A single overwintered queen yellowjacket initiates and builds a colony without aid from other queens. After the first brood of workers emerge, the queen ceases foraging and building activities and becomes the egg layer. At that point she rarely leaves the colony. The queen is much larger than her workers (1.5 to 3 times the workers' size). A yellowjacket colony usually survives until late fall (September to November), depending on the species and locality. There is rapid colony growth during late summer. The hundreds of new queens that emerge from the colony mate with males (which were also produced in the fall), and each inseminated queen hibernates in a protected place until the following spring when the cycle is repeated. Some species, typically in the genus *Paravespula*, scavenge for decaying protein and carbohydrates at carrion, garbage cans, rotting fruit, picnic areas, and meat-processing plants. This seems to be a problem especially in the fall, resulting in a serious stinging hazard to people.

E. Treatment of Stings

The treatment recommendations given for yellowjacket stings are generally those recommended for all stinging wasps.

Local treatment of yellowjacket stings involves using ice packs and pain relievers to minimize pain, washing the wound to lessen the chances of secondary infection, and administering oral antihistamines to counteract the direct release of histamine (not IgE-mediated). In the case of a large local reaction, rest and elevation of the affected arm or leg may also be needed.

For allergic reactions, administration of epinephrine, antihistamines, and other supportive treatment may be required (see Chapter 2).

II. HORNETS

A. General and Medical Importance

In the U.S., the term *hornet* is often misapplied to the yellowjacket, *D. maculata,* because it is large with white and black markings and it builds large aerial nests. True hornets are represented in North America by only one species, *Vespa crabro,* the brown hornet or European hornet (Figure 5). This large hornet produces a very painful sting and may produce allergic reactions in sensitive individuals (see also Chapter 2).

B. General Description

Again, it is important to mention that the commonly called *bald-faced hornet* that produces a large, egg-shaped aerial nest is not a true hornet — it is a yellowjacket. *V. crabro* (the only true U.S. hornet) is a distinctive wasp with a large, robust body (2.5 to 3.5 cm long) and characteristic brown, orange, and red coloration. They have the head swollen behind the eyes and ocelli (small, simple eyes) remote from the margin of the head.

C. Geographic Distribution

The true hornets belong to the genus *Vespa* and occur in Europe and Asia. The only species in the U.S. is *V. crabro,* which was accidentally introduced into the eastern U.S. around 1850. It now occurs sparsely in the Atlantic Seaboard states into the South.

D. Biology and Behavior

V. crabro is very similar in its biology to yellowjackets. They produce annual, single-female

HORNETS

Figure 5
Worker hornet, Vespa crabro.

Importance
Painful stings; allergic RXNs

Distribution
Several species, almost worldwide; only one U.S. species: *Vespa crabro*

Lesion
Central white spot with erythematous halo; amount of local swelling is variable

Disease Transmission
None

Key Reference
Akre, R. D. and Davis, H. G., *Annu. Rev. Entomol.,* 23, 215, 1978

Treatment
Pain relievers, antipruritic lotions for local RXNs; systemic RXNs may require antihistamines, epinephrine, and other supportive measures

OFTEN-ASKED QUESTION

HOW DO YELLOWJACKETS BUILD A NEST IN THE GROUND?

People often encounter yellowjackets when cutting grass, hiking, playing golf, or other outdoor activities during late summer. Unfortunately, they may fly up the pants legs when emerging from their holes in the ground. Stinging events can be quite severe with multiple stings possible. Why do these little pests come out of a hole in the ground? How do they construct a nest underground? First of all, the hole in the ground is just an entrance to a nest underground. Even though the nests are underground, they are still paper nests with the familiar combs and a paper envelope surrounding them. Raw materials gathered by yellowjackets to construct the paper include wood or vegetable fibers mixed with salivary secretions. They may even utilize human products such as blankets, newspapers, or cardboard. The queen yellowjacket selects a site (usually already a hole, crack, or cavern in the ground) and builds the first few nest cells. Almost simultaneously, she constructs a paper envelope around the developing nest (Figure 1 and Figure 2). When the first batch of workers emerge, the queen then switches from provisioning or building the nest to solely egg laying. Workers then take over the jobs of foraging for food and nest building. Nest building is continuous from the time the first workers emerge until the colony declines (usually fall). Accordingly, underground nests may get quite large in late summer, containing thousands of workers. As the nest grows, yellowjackets excavate dirt to enlarge the underground cavity. Most of the evacuated earth is carried outside the burrow and dropped. As summer ends and fall approaches, yellowjackets generally (there are exceptions) produce new queens and the colony dies. Newly formed queens spend the winter in hollow trees, stumps, wall voids of houses, etc., waiting to emerge in the spring and start the process all over again.

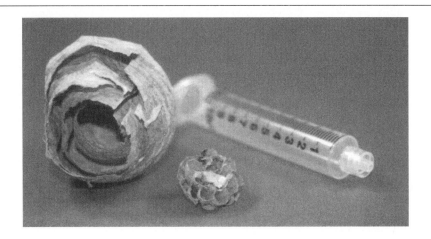

Figure 1
Yellowjacket nest in early stages of development. (With individual block of cells removed from inside.)

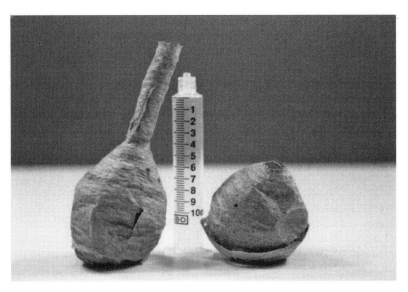

Figure 2
Early stages of yellowjacket nest (left) and bald-faced hornet nest (right).

colonies usually housed in a multiple-combed nest with a thick brown envelope in sheltered aerial locations such as hollow trees, attics, and wall voids of houses. This brown paper nest is easily distinguishable from the typical gray paper nests of other wasps and bald-faced hornets (Color Figure 31.32). The nests are generally large because of the large individual cells and contain about 1500 to 3000 cells in 6 to 9 combs with 200 to 400 workers. Hornets hunt insects, and several species attack honey bee colonies.[2] Although not usually scavengers, they do occasionally puncture the skin of ripe fruits and feed on them.

E.   Treatment of Stings

Hornet stings are treated the same as yellowjacket stings.

III.   PAPER WASPS

A.   General and Medical Importance

This large and diverse wasp group is usually divided into two major divisions depending on lifestyle: solitary or social. Of the 15,000 or so species of stinging wasps worldwide, 95% are solitary species that are not aggressive toward people. The social wasps, on the other hand, form large colonies and can pose serious threats to humans. The most common social groups in temperate regions are the yellowjackets and hornets (subfamily Vespinae) and the paper wasps (*Polistes*). Paper wasps have a great affinity for building their nests on or around buildings; therefore, human encounters and stinging incidents are common. Wasp stings may lead to large local and systemic reactions (see Chapter 2 for in-depth coverage), as well as unusual manifestations such as acute renal failure and polyradiculoneuropathy.[3,4]

B.   General Description

Paper wasps have elongate, slim bodies (2 to 2.5 cm long) and are variously colored yellow, black, brown, and red depending on the species. Some are striped, yellowish or white and brown, and are occasionally confused with yellowjackets. The introduced pest species, *Polistes dominulus,* especially looks like a yellowjacket.[5] Males of many paper wasps have more yellow on the face (frons and clypeus) than females of the same species.

C.   Geographic Distribution

Paper wasps in the genus *Polistes* include about 200 species distributed worldwide in temperate and tropical regions. *P. rubiginosa* (now divided into two species, *P. carolina* and *P. perplexus*) is a bright red-orange wasp that is common throughout the southern U.S. (Figure 6A). *P. exclamans* (Figure 6B), also commonly found in the southern U.S., is the striped species most often mislabeled as yellowjackets. *P. annularis* is a large, dark red paper wasp that nests near permanent bodies of water. It is common in the southern U.S. In the western U.S. and California, *P. aurifer* and *P. apachus* are significant pests. *P. fuscatus* is a commonly encountered wasp found throughout Canada and the U.S. (Figure 6D). *P. gallicus* is a species occurring widely throughout southern France and Italy. *Polistes* species common in Japan include *P. smelleni* and *P. jadwigae.*

D. Biology and Behavior

Paper wasps build a nest consisting of a single, open-faced comb of gray paper. The nest is attached to various substrates (buildings, trees, shrubs) by a single petiole. *Polistes* wasps feed on any extrafloral juices, plant sap, sweets, and various arthropods. They catch caterpillars to feed their brood. In temperate regions each inseminated queen hibernates in a protected site and emerges in the spring to initiate nest building. Usually a single female, called a *foundress*, initiates the nest, but in some species other individual females, called *cofoundresses*, may join the original foundress and help build and provision the nest. These multiple foundresses (usually two to six) compete for reproductive dominance in this small colony. Eventually a dominance hierarchy or pecking order develops with one foundress (i.e., queen), and it lays more of the eggs than the other foundresses. After emergence of the first brood of workers, the other foundresses lose their reproductive status and the colony usually develops into a single queen (monogynous) society. The workers perform most colony tasks, especially nest building and foraging; however, reproduction remains the primary function of the queen. Although the queen is distinct in behavior from the other colony members, she is not usually discernibly different in appearance from the workers.

The colony grows rapidly through the summer, and mature colony size is usually reached in August or September. Typical mature *Polistes* colonies probably contain 30 to 75 adults and 100 to 200 cells, although very large colonies of 150 to 200 adults and nests of 1000 to 1900 cells have been reported. In late summer and early fall, males and reproductive females (future queens) are produced. The colony declines as workers die and the reproductives leave the nest to mate. After mating, males soon die or do not usually survive the winter, whereas inseminated females seek out and survive in winter hibernation sites. Large aggregations of males and females are sometimes seen near hibernation sites or in tall trees, towers, and buildings during late fall,

PAPER WASPS

Paper wasp with nest.
(Photo courtesy of the Ross E. Hutchins photograph collection, Mississippi Entomological Museum, Mississippi State University.)

Importance
Painful stings; allergic RXNs

Distribution
Numerous species almost worldwide

Lesion
Central white spot with erythematous halo; amount of local swelling variable

Disease Transmission
None

Key Reference
Akre, R. D. and Davis, H. G., *Annu. Rev. Entomol.*, 23, 215, 1978

Treatment
Pain relievers, antipruritic lotions for local RXNs; systemic RXNs may require antihistamines, epinephrine, and other supportive measures

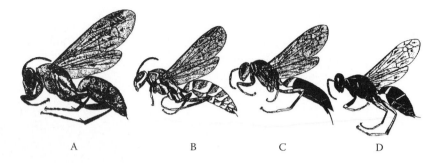

Figure 6

Some common paper wasps: (A) Polistes perplexus, *(B)* P. exclamans, *(C)* P. annularis, *(D)* P. fuscatus. *(From U.S. DHEW, PHS, CDC Pictorial Keys.)*

on warm days of winter, or in early spring. This "house-dwelling" hibernation behavior of *Polistes* may lead to human stings any time of the year.[6] The following spring, the annual cycle is repeated as females emerge from hibernation to initiate a new nest.

E.   Treatment of Stings

Paper wasp stings are treated in the same way as yellowjacket stings.

REFERENCES

1. Ebeling, W., *Urban Entomology*, University of California Press, Berkeley, CA, 1978.
2. Harwood, R.F. and James, M.T., *Entomology in Human and Animal Health*, Macmillan, New York, 1979.
3. Ridolo, E., Albertini, R., Borghi, L., Meschi, T., Montanari, E., and Dall'Aglio, P.P., Acute polyradiculoneuropathy occurring after hymenoptera stings: a clinical case study, *Int. J. Immunopathol. Pharmacol.*, 18, 385, 2005.
4. Vikrant, S., Pandey, D., Machhan, P., Gupta, D., Kaushal, S.S., and Grover, N., Wasp envenomation-induced acute renal failure: a report of three cases, *Nephrology (Carlton)*, 10, 548, 2005.
5. White, G., Hunting *Polistes dominulus*, *PCT Magazine*, September Issue, 2004, pp. 58–64.
6. Alexander, J.O., *Arthropods and Human Skin*, Springer-Verlag, Berlin, 1984.

PART IV

PERSONAL PROTECTION MEASURES AGAINST ARTHROPODS

PROS AND CONS OF INSECT REPELLENTS

TABLE OF CONTENTS

I.   INTRODUCTION

Insect repellents are chemicals that cause insects to make directed, oriented movements away from the source of repellent. In light of disease transmission by insects and other arthropods, chemical substances that have repellent effects or interfere with biting are wonderful because they enable us to go places and do things in insect- or disease-infested areas. Undoubtedly, repellents have prevented thousands of cases of malaria, dengue fever, encephalitis, and other mosquito-borne diseases. For example, a recent study demonstrated that persons practicing two or more personal protective behaviors against mosquito bites, including repellents, reduced the risk of West Nile Virus (WNV) infection by half.[1] However, in recent years, concerns about the potential adverse health effects of insect repellents have increased, especially for those containing the active ingredient DEET. *N,N*-diethyl-3-methylbenzamide (DEET) is the most effective and widely used insect repellent available.[2] It repels a variety of mosquitoes, chiggers, ticks, fleas, and biting flies, and an estimated 50 to 100 million people in the U.S. use it each year.[3,4] This chapter discusses various chemical repellents, their modes of action, possible side effects, and precautions necessary to prevent adverse reactions.

II.   MOSQUITO REPELLENTS

A.   DEET Products

Previously called *N,N*-diethyl-m-toluamide, DEET remains the gold standard of currently available insect repellents. The chemical was discovered by USDA scientists and patented by the U.S. Army in 1946. It was registered for use by the public in 1957. Twenty years of empirical testing of more than 20,000 other chemical compounds has not resulted in another marketed product with the duration of protection and broad-spectrum effectiveness of DEET.[2] DEET is sold under numerous brand names, and is formulated in various ways and concentrations — creams, lotions, sprays, extended-release formulations, etc. (Figure 1). Concentrations of DEET range from about 5 to 100%, and, generally, products with higher concentrations of DEET have longer repellence times.[5] However, at some point the direct correlation between concentration and repellency breaks down. For example, in one study, 50% DEET provided about 4 h of protection against *Aedes aegypti* mosquitoes, but increasing the concentration to 100% provided only 1 additional hour of protection.[6] When used, DEET is absorbed through the skin into the systemic circulation; one study showed that about 10 to 15% of each dose can be recovered from the urine.[3] Other studies have shown lower skin absorption values in the range of 5.6 to 8.4%.[2] Possibly, the solvent used in the product enhances absorption. One study showed that ethanol may increase permeation of DEET.[7] Regardless, the lowest concentration of DEET providing the longest repellency should be chosen for use. Products containing 10 to 35% DEET will provide adequate protection from biting insects under most circumstances (Figure 2). Small children should probably not be exposed to concentrations higher than 10%.[2] The U.S. military uses a polymer-based extended-release formulation containing 35% DEET, which is available to the general public through the 3M Corporation under the brand name Ultrathon®.

Figure 1
Several commercially available insect repellents; all but Natrapel® contain the active ingredient DEET.

B. Picaridin

Picaridin, also known as Bayrepel® or KBR 3023, is an effective alternative to DEET products[8] that provides long-lasting protection against mosquito bites. This relatively new repellent has been used worldwide since 1998. As opposed to DEET, Picaridin is nearly odorless, does not cause skin irritation, and has no adverse effect on plastics. However, even though the product is long lasting and effective against mosquitoes, in some cases it does not provide protection for as long as DEET.[9,10] One field study demonstrated 5-h protection against *Culex annulirostris* mosquitoes with Picaridin vs. 7-h protection with DEET.[9]

C. Plant-Derived Substances

Plant-derived substances that provide some repellency against mosquitoes include citronella, cedar, verbena, lemon eucalyptus, pennyroyal, geranium, lavender, pine, cajeput, cinnamon, rosemary, basil, thyme, allspice, garlic, and peppermint. However, some of these products only provide temporary protection, if any at all. One study testing DEET-based products against Buzz Away® (containing citronella, cedarwood, eucalyptus, lemongrass, alcohol, and water) and Green Ban® (containing citronella, cajuput, lavender, safrole-free sassafras, peppermint, bergaptene-free bergamot, calendula, soy, and tea tree oils) demonstrated essentially no repellency against *A. aegypti*.[5] However, other studies with Buzz Away® indicated that the product does have repellency for about 2 h.[2] One plant-based repellent that was released in the U.S. in 1997, Bite Blocker® (containing soybean oil, geranium oil, and coconut oil), has shown good repellency against *Aedes* mosquitoes for up to 3.5 h.[2] In addition, oil of lemon eucalyptus *p*-methane 3,8-diol (PMD) has performed well in a number of recent scientific

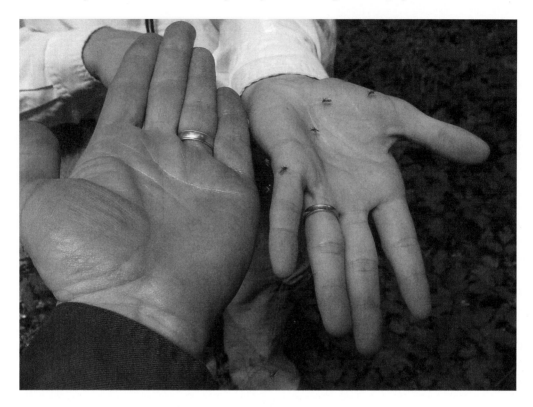

Figure 2
Repellency of DEET against mosquitoes. Hand on the right has no DEET on it.
(Photo copyright 2004 by Jerome Goddard.)

studies and is now listed on the CDC Web site as one possible alternative to DEET.[8] Citronella candles have been marketed as backyard mosquito repellents for years. One study compared the ability of commercially available 3% citronella candles, 5% citronella incense, and plain candles to prevent bites by *Aedes* mosquitoes under field conditions.[11] Persons near the citronella candles had 42% fewer bites than controls, but ordinary candles provided a 23% reduction. The efficacy of plain candles and citronella incense did not differ. The ability of plain candles to decrease biting may be because they serve as a decoy source of warmth, moisture, and carbon dioxide.

D. Permethrin

Permethrin, actually a pesticide rather than a repellent, is a synthetic pyrethroid available for use against mosquitoes, but can only be used on clothing. The product is sold in lawn, garden, or sporting goods stores as an aerosol under the name Permanone® Repel or something similar (Figure 3). It is nonstaining, nearly odorless, and resistant to degradation by light, heat, or immersion in water.[3] Interestingly, it can maintain its potency for at least 2 weeks, even through several launderings.[2,12] Permethrin can be applied to clothing, tent walls, and mosquito nets. In fact, sleeping under permethrin-

Figure 3
Permanone® tick repellent.

impregnated mosquito nets has been tried extensively in malaria prevention campaigns in Africa, New Guinea, Pakistan, and Malaysia.[13] For personal protection, the combination of permethrin-treated clothing and DEET-treated skin creates almost complete protection against mosquito bites. In field trials conducted in Alaska, persons wearing permethrin-treated uniforms and 35% DEET (on exposed skin) had more than 99.9% protection (1 bite/h) over 8 h, whereas unprotected persons received an average of 1188 bites/h.[14]

E. Skin-So-Soft®

The bath oil, Avon Skin-So-Soft®, is often used as a mosquito repellent and is discussed here because of its widespread use. Apparently, Skin-So-Soft does have some transient repellency for mosquitoes. Rutledge et al.[15] reported that the bath oil exhibits repellency for *A. aegypti*, but the effect is short-lived. They reported that Skin-So-Soft is not nearly as effective as DEET (gram for gram). In another study, the Avon product provided 0.64 h of protection from *A. albopictus* mosquito bites compared to greater than 10 h protection provided by 35% DEET.[12] Avon now markets products under the Skin-So-Soft label that contain an EPA-recognized repellent.

III. TICK REPELLENTS

A. DEET Products

One study[16] demonstrated that DEET on military uniforms provided between 10 and 87.5% protection against ticks, depending on species and life stage of the tick. There was an average of 59.8% protection against all species of ticks. Obviously, protection levels in the 50% range are less than desirable, considering the fact that trust one tick can transmit a tick-borne disease. In a U.S. Army repellent rating system, DEET is assigned a 2X value, whereas, permethrin is given a 3X rating.[17] DEET products are simply not as effective in protecting from tick infestation as permethrin products.[16,18] However, the advantage of DEET is that it can be applied to human skin in places likely to be encountered by ticks: ankles, legs, and arms.

B. Permethrin

By far, the most effective tick repellent is permethrin (sold under various brand names, but especially Permanone; Figure 2), a synthetic pyrethroid pesticide with very low mammalian toxicity. It is for use on clothing only, not to be applied directly to human skin. In one study, a pressurized spray of 0.5% permethrin was compared to 20 and 30% DEET products on military uniforms in a highly infested tick area. A 1-min application of permethrin provided 100% protection, compared to 86 and 92% protection with the two DEET products, respectively.[18] Additionally, permethrin has been shown to remain in clothing, providing 100% protection against ticks after several washings.[2,12,17] Generally, application of permethrin to clothing is done by a slow, sweeping application of the aerosol spray until clothing is slightly wet. Label instructions should be followed. Some people take the clothing to be worn (needing protection from ticks), hang it on a clothes line, spray it, and put it on after drying. When ticks subsequently crawl on the clothing treated with permethrin they are either killed or repelled. Permethrin is extremely effective against New World ticks, but there is some evidence that not all tick species are equally repelled by permethrin. One Old World species, the camel tick, actually showed a high tolerance to permethrin, and an increased biting response when exposed to the product.[19]

IV. HEALTH CONCERNS ASSOCIATED WITH DEET PRODUCTS

A. Background

DEET has been used for over 30 years by millions of people worldwide. Although it has an excellent safety record, there have been sporadic reports of adverse reactions associated with its use. Most of these have resulted from accidental exposure, such as swallowing, spraying into the eye, or repeated application, although at least one case occurred in an 18-month-old boy following brief exposure to low-strength (17.6%) DEET.[20] Although most complaints have involved transient minor skin or eye irritation, rare cases of toxic encephalopathy have been reported, especially in children. Adverse reactions have included headache, nausea, behavioral changes, disorientation, muscle incoordination, irritability, confusion, difficulty sleeping, respiratory distress, and even convulsions and death. In one report, 6 girls, ranging in age from 17 months to 8 years, developed behavioral changes, ataxia, encephalopathy, seizures, and coma after repeated cutaneous exposure to DEET; 3 later died.[4] However, if DEET products (preferably not the high concentration products such as 50 to 100%) are properly applied and used according to their label directions, they are generally considered safe. Use of DEET products according to EPA guidelines (next section) will greatly reduce the possibility of toxicity.

B. Safe Application Rates and Methods

Except under extraordinary conditions, high concentrations of DEET should not be used. Products with 10 to 35% DEET will provide adequate protection under most conditions. For children, even lower concentrations may be warranted. The American Academy of Pediatrics recommends that repellents used on children contain no more than 10% DEET.[2] The following guidelines will help ensure safe use of DEET-based repellents. Remember, repellents should only be applied to clothing and exposed skin according to the product label directions.

DO:
- Use aerosols or pump sprays for skin and for treating clothing. These products provide even application.
- Use liquids, creams, lotions, or sticks for more precise application on exposed skin.
- After outdoor activity, wash DEET-covered skin with soap and water.
- Always keep insect repellents out of the reach of small children.

DON'T:
- Apply to eyes, lips or mouth, or over cuts, wounds, or irritated skin.
- Overapply or saturate skin or clothing.
- Apply to skin under clothing.
- Apply more often than directed on the product label.

REFERENCES

1. Loeb, M., Protective behavior and west Nile virus risk, *Emerging Infect. Dis.*, 11, 1433, 2005.
2. Fradin, M.S., Mosquitoes and mosquito repellents: a clinician's guide, *Ann. Intern. Med.*, 128, 931, 1998.
3. Abramowicz, M., Insect Repellents, *Medical Letter*, 31, 45, 1989.
4. CDC, Seizures temporarily associated with use of DEET insect repellents — New York and Connecticut, *MMWR*, 38, 678–680, 1989.
5. Chou, J.T., Rossignol, P.A., and Ayres, J.W., Evaluation of commercial insect repellents on human skin against *Aedes aegypti*, *J. Med. Entomol.*, 34, 624, 1997.
6. Buescher, M.D., Rutledge, L.C., Wirtz, R.A., and Nelson, J.H., The dose-persistence relationship of DEET against *Aedes aegypti*, *Mosq. News*, 43, 364, 1983.
7. Stinecipher, J. and Shah, J., Percutaneous permeation of N,N-diethyl-m-toluamide from commercial mosquito repellents and the effect of solvent, *J. Toxicol. Environ. Health*, 52, 119, 1997.
8. CDC, Updated information regarding mosquito repellents, in CDC Information sheet, draft issued April 7, 2005, 2 pp.

9. Frances, S.P., Waterson, D.G., Beebe, N.W., and Cooper, R.D., Field evaluation of repellent formulations containing deet and picaridin against mosquitoes in Northern Territory, Australia, *J. Med. Entomol.*, 41, 414, 2004.

10. Klun, J.A., Khrimian, A., Margaryan, A., Kramer, M., and Debboun, M., Synthesis and repellent efficacy of a new chiral piperidine analog: comparison with Deet and Bayrepel activity in human-volunteer laboratory assays against *Aedes aegypti* and *Anopheles stephensi, J. Med. Entomol.*, 40, 293, 2003.

11. Lindsay, R.L., Surgeoner, G.A., Heal, J.D., and Gallivan, G.J., Evaluation of the efficacy of 3% citronella candles and 5% citronella incense for protection against field populations of *Aedes* mosquitoes, *J. Am. Mosq. Control Assoc.*, 12, 293, 1996.

12. Schreck, C.E. and McGovern, T.P., Repellents and other personal protection strategies against *Aedes albopictus, J. Am. Mosq. Control Assoc.*, 5, 247, 1989.

13. Service, M.W., Mosquitoes, in *Medical Insects and Arachnids*, Lane, R.P. and Crosskey, R.W., Eds., Chapman and Hall, London, 1993, p. 120.

14. Lillie, T.H., Schreck, C.E., and Rahe, A.J., Effectiveness of personal protection against mosquitoes in Alaska, *J. Med. Entomol.*, 25, 475, 1988.

15. Rutledge, L.C., Wirtz, R.A., and Buescher, M.D., Repellent activity of a proprietary bath oil, Skin-So-Soft, *Mosq. News*, 42, 557, 1982.

16. Evans, S.R., Korch, G.W., Jr., Lawson, M.A., Comparative field evaluation of permethrin and DEET-treated military uniforms for personal protection against ticks, *J. Med. Entomol.*, 27, 829, 1990.

17. Evans, S.R., Personal Protective Techniques Against Insects and Other Arthropods of Military Significance, U.S. Army Environmental Hygiene Agency, Aberdeen Proving Ground, MD, TG No. 174, 1991, 90 pp.

18. Schreck, C.E., Snoddy, E.L., and Spielman, A., Pressurized sprays of permethrin or deet on military clothing for personal protection against *Ixodes dammini, J. Med. Entomol.*, 23, 396, 1986.

19. Fryauff, D.J., Shoukry, M.A., and Schreck, C.E., Stimulation of attachment in a camel tick, *Hyalomma dromedarii:* the unintended result of sublethal exposure to permethrin-treated fabric, *J. Med. Entomol.*, 31, 23, 1994.

20. Briassoulis, G., Narlioglo, M., and Hatzis, T., Toxic encephalopathy associated with use of DEET insect repellents: a case analysis of its toxicity in children, *Hum. Exp. Toxicol.*, 20, 8, 2001.

CHAPTER 33

ARTHROPOD-SPECIFIC PERSONAL PROTECTION TECHNIQUES

TABLE OF CONTENTS

I. PROTECTION FROM MOSQUITOES

A. Avoidance

Practical nonchemical measures for mosquito avoidance include limiting outdoor activity after dark and avoiding known mosquito-infested areas (e.g., swamps, marshes, etc.) during the peak mosquito season. In addition, people who have to be outdoors after dark in the mosquito season should wear long sleeves and long pants.

B. Screening

Probably one of the most basic and effective sanitation measures to limit arthropod–human contact is that of screen wire windows and doors. Screens are constructed of various metals or plastic and are ordinarily 16 × 16 × 20 mesh. They should be tight fitting over window openings. Screen doors should be hung so that they open outward. Being such a basic protection measure, screens are sometimes overlooked. However, their importance cannot be overemphasized. I personally investigated a fatality due to eastern equine encephalitis in which the family had no screens on the house. By their own testimony, the family members stated, "Mosquitoes eat us up every night!"

C. Netting

During the mosquito season, people may choose to use protective jackets made of netting, when outdoors in heavily infested areas (Figure 1). Also, those camping may sleep under mosquito nets for protection from mosquitoes (Figure 2). This becomes mandatory on safaris or other trips to the tropics or subtropics for protection against disease-carrying mosquitoes. Ordinarily, mosquito netting is made of cotton or nylon with 23 to 26 meshes per inch. Netting should not be allowed to lie loosely on the head, because mosquitoes can feed through the net wherever it touches the skin. Generally, people construct a crude frame over the bed or end of the bed to fit the net on (Figure 3). For added protection, the net can be sprayed lightly with permethrin (Permanone®) before getting inside. This permethrin treatment should be effective for several months if not rinsed or washed out. In tropical countries, insecticide-impregnated bed nets are an important component of malaria prevention campaigns. Generally, the nets are treated at the factory with either permethrin or deltamethrin, which may last for the duration of the net (i.e., a few years). Millions of these bed nets are purchased and distributed each year as part of malaria control programs.

II. PROTECTION FROM TICKS

A. Boots, Trousers, and Tape

Personal protection techniques for tick bites include avoiding tick-infested woods, tucking pant legs into boots or socks, and using repellents on pant legs and socks (Figure 4 and Figure 5). Tucking trousers into boots or socks forces ticks to crawl up the outside of one's pants, thus making them easier to spot and remove. Wide masking tape may be used to "tape down" pants legs at the ankles, or they can even be placed around the ankles or thighs with the sticky side exposed to protect against ticks crawling up the legs (Figure 6). One research study showed that significantly fewer nymphal deer

Figure 1
Protective jacket made of netting.

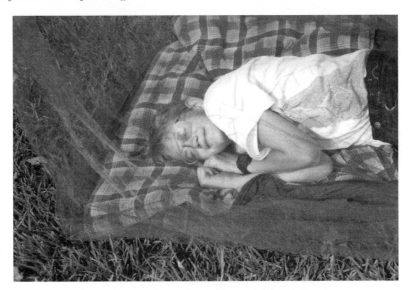

Figure 2
Netting for protection against mosquito bites when camping. (Photo courtesy of Joseph Goddard.)

ticks were picked up on 5-min walks in the woods when boots were worn with ankles taped than when sneakers were worn with socks exposed.[1]

B.   Unorthodox Methods

Many unorthodox (or at least, questionable) methods of tick protection are commonly used by people. Flea collars worn around the ankles, Avon's Skin-So-Soft®, vitamin C,

Figure 3
Bed net used for protection against mosquitoes at night. (Photo courtesy of Rachel Freeman.)

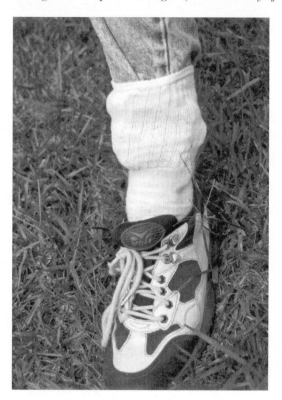

Figure 4
Tucking pant legs into socks for tick protection.

garlic, sulfur powder, panty hose, and others have been reported to me as "great tick repellents." Although these efforts may possibly provide some degree of protection, tucking the trousers with proper use of Permanone is much more effective.[2] Besides, using pet products such as flea collars may be dangerous owing to absorption of pesticide through the skin (Figure 7); they have not been approved by the Food and Drug Administration (FDA) for human wear.

Figure 5
Tucking pants legs into rubber boots for tick protection.

Figure 6
"Taping up" with masking tape for tick protection. The last layer or two should be with the sticky side out.

C. Tick Removal

What is the best way to remove a tick after it gets on you? The answers are varied, depending on whom you ask and what part of the country you are in, because many folklore methods are available. Hard ticks attach themselves firmly to a host for a feeding period of several days and are especially difficult to remove. Methods such as touching attached ticks with a hot match, coating them with mineral oil, petroleum jelly, or some other substance, or "unscrewing" them are but a few of the home remedies that supposedly induce them to "back out."

The theory behind coating a tick with fingernail polish, petroleum jelly, or mineral oil is that covering the spiracles with a substance will interfere with their breathing and make the ticks back out. However, ticks are able to shut off their spiracles (at least temporarily) and certainly do not breathe via the mouthparts. Even if coating a tick with a substance causes it to back out, the time required to accomplish this is unacceptable (usually 1 to 4 h).

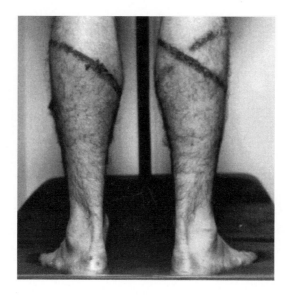

Figure 7
Skin lesions from human use of animal flea collars for tick protection. (Photo courtesy of S. Evans and R. Fitzsimmons, U.S. Army Environmental Hygiene Agency.)

Because the lengthy feeding period is an important factor in disease transmission by ticks, it is crucial that a tick be removed as soon as possible to reduce chances of infection by disease organisms. During several years of field research with ticks, the author often had to remove them from himself or others, and pulling them straight off with blunt forceps (tweezers) seemed to work best (Figure 8). There has been some research in this area. Glen Needham at Ohio State University did a very good study of this problem.[3] He evaluated five methods commonly used for tick removal: (1) petroleum jelly, (2) fingernail polish, (3) 70% isopropyl alcohol, (4) hot kitchen match, and (5) forcible removal with forceps. Needham found that the commonly advocated methods are either ineffective, or worse, actually created greater problems. If petroleum jelly or some other substance causes the tick to back out on its own (and most often it does not), the cement surrounding the mouthparts used for attachment remains in the skin, where it continues to cause irritation. Touching the tick with a hot match may cause it to burst, increasing risk of disease pathogen exposure. Furthermore, hot objects may induce ticks to salivate or regurgitate infected fluids into the wound. Unscrewing a tick is likely to leave broken mouthparts in the host's skin.

Needham recommended the following procedure for tick removal: (1) use blunt forceps or tweezers, (2) grasp the tick as close to the skin surface as possible and pull upward with steady, even pressure, (3) take care not to squeeze, crush, or puncture the tick, (4) do not handle the tick with bare hands because infectious agents may enter via mucous membranes or breaks in the skin, and (5) after removing the tick, disinfect the bite site and wash hands thoroughly with soap and water.

Many tick-borne diseases can be successfully treated with antibiotics in their initial stages; therefore, early diagnosis is imperative. For this reason, marking the day of a tick bite on a calendar is a good idea. If unexplained disease symptoms occur within 2 weeks from this day, persons should be reminded to see their physicians and specifically inform him or her of the tick bite. This method has proved to be very helpful in diagnosis of tick-borne disease. Although there are a number of well-known tick

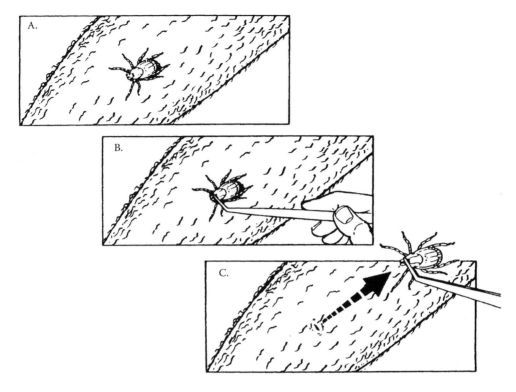

Figure 8
Recommended method for tick removal: grab tick with forceps as close to the skin as possible and pull straight off. (From USAF Publication USAFSAM-SR-89-2.)

removal methods (mostly folklore), the best one seems to be the simplest: pull them straight off with blunt forceps and disinfect the bite site.

III. PROTECTION FROM OTHER ARTHROPODS

A. Biting Midges

Biting midges (no-see-ums), the tiny slender gnats that are extremely common along the Atlantic and Gulf Coasts, are small enough to pass through ordinary screen wire used to cover windows and doors. Accordingly, a finer mesh screen wire must be used to prevent entry of these flies into dwellings. Outdoors, DEET repellents and long-sleeved shirts and long pants can provide relief in infested areas. Avon's Skin-So-Soft is often used as a repellent for biting midges, and controlled studies indicate that the product provides some protection.[4] However, the effectiveness is probably because the oiliness of Skin-So-Soft traps the tiny midges on the skin surface.

B. Sand Flies

Sand flies are small, delicate, mosquitolike insects (usually less than 5 mm long) that inflict painful bites. Although the most serious problems from sand flies occur in tropical countries, there are many species in temperate zones, as well. The author often collects them in mosquito light traps in Mississippi.[5] Because sand flies do not bite through clothing, long sleeves, trousers, and socks should be worn in areas where sand flies are active.[6] In heavily infested areas, head nets, gloves, and repellent-treated net jackets and hoods can provide additional protection. Campsites should be chosen that are high, breezy, open, and dry. Fine-mesh bed nets should be used in areas where sand flies are present.

C. Chiggers (Red Bugs)

Chigger mites (sometimes called *red bugs*) occur in grass, weeds, or leaves and get onto passing vertebrate hosts. Therefore, personal protection measures for chiggers are similar to those for ticks. Tucking pants legs into socks or boots and spraying clothing with DEET-based repellents provide good protection. Treating clothing with Permanone is also very effective against chiggers. After exposure to infested outdoor areas, hot soapy baths, or showers will help remove any chiggers, attached or unattached.

D. Kissing Bugs

Kissing bugs (*Triatoma* species), the vectors of Chagas' disease in Mexico and Central and South America, are nocturnal insects that seek refuge by day in the cracks and crevices of poorly constructed houses or in the loose roof thatching of huts. Personal protection from the bugs involves avoidance (if possible) — not sleeping in thatched roof huts in endemic areas — and exclusion methods such as bed nets. Prevention and control of domestic species of triatomines can be accomplished by proper construction of houses, wise choice of building materials, sealing cracks and crevices, and precision-targeting with insecticides in the home.

E. Fire Ants

Fire ants are extremely aggressive ants that currently inhabit much of the southern U.S. They were accidentally imported into the U.S. between 1918 and 1940 and continue to spread. Outdoors, fire ants can be recognized by their mounds, which are elevated earthen mounds 8 to 90 cm high, surrounded by relatively undisturbed vegetation. When disturbed, the ants characteristically boil out of their mounds in great numbers, stinging their victims. Personal protection from fire ants primarily involves recognition and avoidance of their mounds. Insect repellents are apparently of no value against fire ant attacks.[7] However, there is apparently some benefit in wearing socks, at least in delaying ant attacks.[7]

REFERENCES

1. Carroll, J.F. and Kramer, M., Different activities and footwear influence exposure to host-seeking nymphs of *Ixodes scapularis* and *Amblyomma americanum*, *J. Med. Entomol.*, 38, 596, 2001.

2. Evans, S.R., Korch, G.W., Jr., and Lawson, M.A., Comparative field evaluation of permethrin and DEET-treated military uniforms for personal protection against ticks, *J. Med. Entomol.*, 27, 829, 1990.

3. Needham, G.R., Evaluation of five popular methods for tick removal, *Pediatrics*, 75, 997, 1985.

4. Schreck, C.E. and Kline, D.L., Repellency determinations of four commercial products against six species of ceratopogonid biting midges, *Mosq. News*, 41, 7, 1981.

5. Goddard, J., New records for the phlebotomine sand fly *Lutzomyia shannoni* (Dyar) in Mississippi, *J. Miss. Acad. Sci.*, 50, 195, 2005.

6. Rutledge, L.C. and Gupta, R.K., Moth flies and sand flies, in *Medical and Veterinary Entomology*, Mullen, G.R. and Durden, L.A., Eds., Academic Press, New York, 2002, p. 147.

7. Goddard, J., Personal protection measures against fire ant attacks, *Ann. Allerg. Asthma Immunol.*, 95, 344, 2005.

GLOSSARY

Abdomen: The hindmost of the three body divisions in insects.

Acaracide: A substance poisonous to ticks or mites.

Aculeate: Pertaining to the members of the order Hymenoptera, which sting — the bees, wasps, and ants.

Alate: inged form; in ants alates are the winged reproductive forms.

Anal: The posterior wing veins.

Antennae (sing. Antenna): A pair of sensory segmented appendages located on the head above the mouthparts.

Anterior: Toward the front end.

Anthropophilic: Describes any bloodsucking arthropod that prefers humans to lower animals as its food source.

Apiary: Any place where honey bees are kept.

Apical: At the tip or end.

Apterous: Without wings.

Arbovirus: An arthropod-borne virus.

Arista: A large bristle located on the last antennal segment of some flies.

Atopy: IgE-dependent allergy often arising from an unknown exposure to an antigen.

Babesiosis: Infection with a protozoan organism in the genus *Babesia*; often a malaria-like illness.

Basis capitulum: Basal portion of the capitulum on which the tick mouthparts are attached. Various shapes (hexagonal, rectangular, subtriangular, etc.) are possible in hard ticks.

Beak: The proboscis of a sucking insect.

Bifid: Clearly divided into two parts.

Brood: All the immature members of a colony including eggs, larvae, and pupae.

Brood Cell: A cell made by a worker bee in which to lay an egg.

Bubo: An enlargement of a lymph gland caused by an infection.

Bug: Loosely used to denote any arthropod; technically, meaning only members of the insect order Hemiptera.

Cantharidin: A chemical produced by certain beetles (especially, family Meloidae) that causes blistering on the skin of humans.

Calypter: A basal lobe or lobes on the posterobasal portion of the axillary membrane of the wings of some Diptera.

Carapace: The sclerotized (hardened) plate forming the dorsal surface of an arachnid cephalothorax.

Carnivorous: Feeding on animals.

Caste: A group of morphologically distinct individuals within a colony often having distinctive behavior (e.g., workers, queens, and males).

Caterpillar: The larva of a butterfly or moth having a cylindrical body, a well-developed head, thoracic legs, and abdominal prolegs.

Cell: Terminology used in describing areas of an insect wing. An area of the wing enclosed by veins.

Cephalothorax: Head and thorax combined; characteristic of arachnids.

Cerci (sing. Cercus): Paired appendages at the posterior end of the abdomen, as in a cockroach.

Chelae (sing. Chela): The second pair of appendages (pedipalps) of scorpions and pseudoscorpions; modified into pincers.

Chelicerae: Paired appendages of an arachnid, highly variable in shape and size. In scorpions they are short, chelate, and lacking a poison gland; in spiders they terminate in a sharp tip with a venom duct; and in ticks they lie dorsally to the hypostome, completing the cylindrical mouthparts that are inserted when a tick feeds.

Chitin: A complex nitrogenous carbohydrate forming the main skeletal substance of arthropods.

Class: A grouping used in classification. A division of a phylum.

Classification: The arrangement of species in a hierarchical system of categories and taxa.

Clypeus: That part of the insect head below the front to which the labrum is attached.

Cocoon: A silken enclosure secreted by a larva just before pupation.

Colony: Individuals, other than a single mated pair, that cooperate to build a nest or rear offspring.

Complete metamorphosis: Type of insect development in which there are egg, larva (caterpillar), pupa (resting stage), and adult stages.

Compound eye: Insect eye composed of many individual elements represented externally by hexagonal facets.

Contiguous: Touching one another.

Cornua: Small projections extending from the dorsal, posterolateral angles of basis capituli in ticks.

Costa: The thickened anterior vein of the insect wing.

Coxae (sing. Coxa): Basal segments of the leg. In ticks, small sclerotized plates on the venter representing the first segment of the leg to which the trochanters are movably attached. From anterior to posterior, the coxae are designated by Roman numerals I, II, III, and IV. Bifid coxae are those that are cleft, divided, or forked.

Dentition: Refers to the presence and arrangement of denticles (teeth). In ticks, the numerical arrangement of the rows of denticles on the ventral side of the hypostome is expressed by dentition formulas. Thus, dentition 3/3 means that there are three longitudinal rows of denticles on each side of the median line of the hypostome.

Dermatitis: Inflammation or eruption of the skin.

Desensitization (also called *Immunotherapy*): Elimination or reduction of allergic sensitivity; usually accomplished through a programmed course of antigen treatment.

Deutonymph: The third stage of a mite.

Dichotomous: Divided into two parts. Insect identification keys are often dichotomous, giving the reader two choices after each question.

Diurnal: Active in the daytime.

Dorsal: Pertaining to the back or top of the body.

Ecdysis: The process of shedding the skin or exoskeleton.

Ectoparasite: A parasite that feeds on the surface of the body and is usually bloodsucking.

Ehrlichiosis: Disease caused by one of several rickettsial organisms in the genus *Ehrlichia*.

Elytra: Thick or leathery front wing of beetles.

Envenomization (also called *Envenomation*): The poisonous effects caused by the bites, stings, secretions, stinging hairs of insects, other arthropods, certain invertebrate animals, or the bites of reptiles.

Enzootic: Disease of animals constantly present in an area.

Epizootic: Describes any disease of animals, the number of cases of which exceeds that normally expected.

Erucism: Urtication caused by moth or butterfly larvae.

Exuvium: The cast exoskeleton of an arthropod.

Eyes: In insects, either simple (singular) or compound, variously arranged on the head. In ticks, eyes, when present, are located on the edges of the scutum. They are about even with the site of leg I attachment in hard ticks; soft ticks may have eyes on their lateral margins near coxae I and II.

Facet: The external surface of one of the individual units of a compound eye, as in the fly.

Family: A category in the hierarchy of classification. A division of an order.

Femur: The third leg segment outward from the insect body.

Festoons: Uniform rectangular areas, separated by distinct grooves, located on the posterior margin of most genera of hard ticks.

Filariasis: Disease caused by filarial worms (Nematodes), which invade lymphatic tissues and are transmitted by mosquitoes.

Flagellum: The third and succeeding segments of most antennae.

Flagellomere: A division of the flagellum of the antennae.

Foci (sing. Focus): With reference to a disease, specific areas in which the disease is prevalent.

Foundress: An individual, usually a fertilized female, that founds a new colony. All subsequent offspring are her daughters and sons.

Galea: Portion of some insect mouthparts. Specifically, the outer lobe of the maxilla.

Gaster: The prominent part of wasp or ant abdomen, separated from the other body parts by a thin connecting segment called a petiole or pedicel.

Genera (sing. Genus): Categories in the classification hierarchy to which species are assigned.

Gradual metamorphosis: Type of insect development in which there are egg, nymph, and adult stages; no wormlike larval stage present.

Gravid: Full of ripe eggs; ready to lay eggs.

Grub: Term used for a thick-bodied, sluggish, often white insect larva.

Goblets: Small, round structures located in the spiracular plate of ticks.

Grooves: On ticks, linear depressions or furrows, primarily on the ventral surface.

Halteres (sing. Haltere): Small knoblike structures on each side of the thorax of a fly immediately behind each wing.

Haustellate: Having mouthparts adapted for sucking blood.

Head: The anterior body region of an insect bearing eyes, antennae, and mouthparts. Ticks and mites have no true head.

Hemelytron: The forewing of the true bugs (order Hemiptera).

Hemimetabolous: Simple metamorphosis.

Hemocoel: The major body cavity of insects containing blood.

Hemolymph: Arthropod blood.

Hemolytic anemia: Shortage of red blood cells due to their premature destruction; sometimes a complication as a result of brown recluse spider bite.

Herbivorous: Feeding on plants.

Hibernation: A period of lethargy or suspension of most bodily activities with a greatly reduced respiration rate, occurring mostly during periods of low temperature.

Histamine: Organic substance released from tissues during an allergic reaction to injury or invasion by an antigen, causing dilation of local blood vessels.

Holometabolous: Complete metamorphosis.

Holoptic: In flies, the eyes touching above.

Hood: The anterior projection of the integument in soft ticks above and covering the mouthparts.

Hypognathus: Head and mouthparts situated ventrally (pointed down).

Hypopharynx: Mouthpart structure, located medially, that is anterior to the labium. In many sucking insects this structure contains the salivary channel.

Hypostome: In ticks, the median ventral structure of the mouthparts that lies parallel to and between the palps. It bears the *recurved teeth* or denticles.

Imago: The adult stage of an insect.

Infarction: Formation of a localized area of necrosis produced by occlusion of the arterial blood supply or venous drainage of a part.

Inornate: In ticks, the absence of a color pattern on the scutum.

Instar: An insect between successive molts.

Joint: An articulation. The area of flexion between sections of an appendage.

Labellum: Insect mouthpart; the tip of the labium.

Labial palpi (sing. Labial palpus): Segmented sensory appendages of the labium of insects.

Labium: "Lower lip" of insect mouthparts.

Labrum: "Upper lip" of insect mouthparts.

Laciniae (sing. Lacinia): Insect mouthparts; the inner lobe of the maxilla.

Large local reaction: Reaction to sting or bite that is exaggerated, but still contiguous with the sting site.

Larvae (sing. Larva): An immature stage of an insect having complete metamorphosis but excluding the egg or pupal stage. Also, a six-legged first instar mite or tick.

Larviparous: Method of reproduction by bringing forth larvae that have already hatched in the female's body.

Lepidopterism: Urtication caused by hairs, scales, or spines of adult moths or butterflies.

Lesion: An injury to body tissue. Often a spot or mark.

Maggot: A legless larva (usually limited in usage to Diptera) that has no well developed head region.

Malphigian tubules: Long and slender blind tubes in the hemocoel that open into the beginning of the hind intestine of insets. Excretory in function.

Mammillae (sing. Mammilla): Elevations of various forms found on the integument of *Ornithodoros* tick species.

Mandibles: The most anterior pair of two pairs of insect mouthpart structures.

Maxillae (sing. Maxilla): The pair of mouthpart structures lying behind the mandibles.

Maxillary palpi (sing. Maxillary palpus): Segmented sensory structures located on the maxillae.

Mesosoma: The seven abdominal segments of scorpions.

Metamorphosis: The series of changes through which an insect passes in developing from egg to adult.

Metasoma: The *tail* of scorpions.

Molt: The process of shedding the exoskeleton.

Musciform: Resembling a fly.

Myiasis: The invasion of human tissues by dipterous larvae.

Nearctic: One of the zoogeographical regions of the earth that includes Canada, Alaska, Greenland, the U.S., and the temperate northern part of Mexico.

Neotropical: South America, Central America (including Mexico), and the Antilles.

Nits (sing. Nit): Eggs of lice.

Nocturnal: Being active during night.

Nomenclature: The scientific names of living organisms and the application of these names.

Nymph: An immature insect that does not have a pupal stage (e.g., cockroach, grasshopper, bed bug). Also, an eight-legged immature tick or mite.

Obligate parasite: A parasitic association in which the parasite cannot complete its life cycle without a suitable host.

Ocelli (sing. Ocellus): The simple eyes of arthropods.

Oothecae (sing. Ootheca): Egg case in cockroaches.

Opisthosoma: The entire abdomen in arachnids.

Opisthognathus: Mouthparts situated and directed toward the posterior (backward).

Order: A division of a class in the hierarchy of categories.

Ornate: Definite enamellike color pattern superimposed on the base color of the integument in hard ticks.

Ovipositor: Tubular structure used by many insects to lay eggs; modified into a stinger in the ants, wasps, and bees.

Palpi (sing. Palpus or Palp): In insects, a segmented process on the maxillae or labium. In ticks, paired articulated appendages located on the front and sides of the basis capituli and lying parallel to the hypostome.

Papilloma: Benign tumor derived from epithelium.

Papules (sing. Papule): Small elevations of the skin that are usually inflamed.

Parasite: Any animal or plant that lives in or on, and at the expense of, another animal or plant.

Parthenogenesis: Condition in which egg development can occur without fertilization.

Pectines: Feathery sensory organs of scorpions; believed to sense ground vibrations.

Pedicel: In spiders, the petiole (waist) between the cephalothorax and abdomen; in ants and other Hymenoptera the stalk between thorax and abdomen.

Pedipalps: The second pair of appendages of an arachnid.

Petechiae (sing. Petechia): Pinpoint, flat, round, purplish red spots.

Petiole: The narrowed portion of the abdomen of certain Hymenoptera.

Pharynx: In insects, the anterior region of the foregut, located between the mouth and the esophagus.

Pheromone: A chemical produced by one individual that causes a specific reaction by other members of the same species.

Phylum: A major division of the animal kingdom; a category.

Plumose: As in antennae, meaning featherlike.

Prepupa: The quiescent stage immediately before the pupal stage in insects.

Proboscis: A beaklike projection containing various arrangements of mouthparts and their modifications.

Prognathous: With the head horizontal and the jaws directed forward.

Pronotum: Dorsal shield over the anterior segment of the thorax; in cockroaches the pronotum looks like the head.

Protonymph: In mites, the second instar.

Pubescent: Covered with short, fine hairs; appearing hairy.

Pupae (sing. Pupa): A nonfeeding and inactive stage (except mosquitoes) between the larvae and adults.

Puparium: A shell or case produced by the hardening of the last larval skin.

Quinones: Caustic, highly volatile hydrocarbons secreted by some arthropods as a defensive measure.

Rickettsiae (sing. Rickettsia): Single-celled, very small, intracellular bacteria. Notorious members of this group include the causative agents of Rocky Mountain spotted fever, *Rickettsia rickettsii,* and murine typhus, *Rickettsia typhi.*

Segment: A ring or subdivision of the body, or of an appendage, between areas of flexibility, with muscles attached for movement.

Sclerite: Hardened plate or portion of an insect body.

Scutum: The sclerotized dorsal plate posterior to the capitulum in hard ticks. It covers almost the entire dorsal surface in the male and about one half of the dorsal surface in the unengorged female.

Sensillae (sing. Sensilla): Setae, bristles, or hairs having a sensing function.

Sexual dimorphism: Morphological differences between males and females of a species.

Species complex: A group of closely related species, the taxonomic relationships of which are sometimes unclear, making precise identification difficult.

Spiracle: A breathing pore that marks the external opening of the tracheal system.

Spurs: In ticks, coxal spurs are projections from the posterior surface of the posterior margin of the coxae; they may be rounded or pointed, large or small.

Stylet: A needlelike structure, especially the elongated portion of the piercing–sucking type of insect mouthparts.

Sylvatic: Describes any disease usually acquired in a forest or other uncultivated, unoccupied areas, rather than in an urban environment or other areas developed by humans.

Synonym: Another name used for a species, or other taxon, invalid because it is either of a more recent date or invalidly proposed.

Systemic: Affecting the entire body.

Tarsi (sing. Tarsus): The terminal leg segments.

Taxa (sing. Taxon): Groups of organisms classified together.

Taxonomy: The naming and arranging of species and groups into a system of classification.

Telson: In scorpions, the last segment of the "tail;" bears the sting.

Thorax: A body region in insects, located behind the head, that bears the legs and wings.

Tibia: The fourth segment (from the body outward) of an insect leg.

Trachea: An internal respiratory tube.

Transovarial transmission: The production (by an infected vector) of infected eggs that hatch into individuals likewise infected and capable of transmitting the infecting organism.

Transstadial transmission: The survival of parasites or pathogens through successive stages (larva, nymph, and adult).

Trochanter: The portion of the leg between the coxa and the femur.

Trophallaxis: Exchange of alimentary canal liquid among colony members of social insects.

Urticaria: Intensely itchy wheals; also popularly called *hives*.

Vein: A tube running through the membrane of the wings of insects.

Wheal: A localized area of edema on the skin, often with severe itching. The typical lesion of urticaria.

Wigglers: Mosquito larvae.

Worker: A member of a nonreproductive or sterile caste that contributes to a colony welfare by rearing offspring of reproductives. In Hymenoptera, workers are the ones that sting.

Zoogeographic regions: The six divisions of the earth's surface distinguished by major differences in animal and plant life.

Zoonoses (sing. Zoonosis): Diseases of animals that may be transmitted to humans.

INDEX